FUNDAMENTALS OF CRYPTOLOGY
A Professional Reference and Interactive Tutorial

THE KLUWER INTERNATIONAL SERIES
IN ENGINEERING AND COMPUTER SCIENCE

FUNDAMENTALS OF CRYPTOLOGY
A Professional Reference and Interactive Tutorial

by

Henk C.A. van Tilborg
Eindhoven University of Technology
The Netherlands

KLUWER ACADEMIC PUBLISHERS
Boston / Dordrecht / London

Distributors for North, Central and South America:
Kluwer Academic Publishers
101 Philip Drive; Assinippi Park
Norwell, Massachusetts 02061 USA
Telephone (781) 871-6600
Fax (781) 681-9045
E-Mail: kluwer@wkap.com

Distributors for all other countries:
Kluwer Academic Publishers Group
Post Office Box 322
3300 AH Dordrecht, THE NETHERLANDS
Telephone 31 78 6576 000
Fax 31 78 6576 254
E-Mail: services@wkap.nl

 Electronic Services <http://www.wkap.nl>

Library of Congress Cataloging-in-Publication Data

Tilborg, Henk C. A. van, 1947-
    Fundamentals of cryptology : a professional reference and interactive tutorial / by Henk
C. A. van Tilborg.
        p. cm.-- (The Kluwer international series in engineering and computer science ; SECS 528)
    Updated and improved version of An introduction to cryptology. 1988.
    Accompanied by a full text electronic version on CD-ROM as an interactive
mathematica manuscript.
    Includes bibliographical references.
    Additional material to this book can be downloaded from http://extras.springer.com
    ISBN 0-7923-8675-2 (alk. paper)
    1. Computers--Access control. 2. Cryptography. I. Tilborg, Henk C. A. van, 1947-
Introduction to cryptology. II. Title. III. Series
QA76.9.A25.T52 1999
005.8--dc21   99-048298

Copyright © 2000 by Kluwer Academic Publishers. Third Printing 2003.

The enclosed CD-ROM makes use of Mathematica software. Mathematica is a registered trademark
of Wolfram Research, Inc.

Permission for books published in Europe: permissions@wkap.nl
Permissions for books published in the United States of America: permissions@wkap.com

Printed on acid-free paper.   Printed in the United States of America.

Contents

Preface

The protection of sensitive information against unauthorized access or fraudulent changes has been of prime concern throughout the centuries. Modern communication techniques, using computers connected through networks, make all data even more vulnerable for these threats. Also, new issues have come up that were not relevant before, e.g. how to add a (digital) signature to an electronic document in such a way that the signer can not deny later on that the document was signed by him/her.

Cryptology addresses the above issues. It is at the foundation of all information security. The techniques employed to this end have become increasingly mathematical of nature. This book serves as an introduction to modern cryptographic methods. After a brief survey of classical cryptosystems, it concentrates on three main areas. First of all, stream ciphers and block ciphers are discussed. These systems have extremely fast implementations, but sender and receiver have to share a secret key. Public key cryptosystems (the second main area) make it possible to protect data without a prearranged key. Their security is based on intractable mathematical problems, like the factorization of large numbers. The remaining chapters cover a variety of topics, such as zero-knowledge proofs, secret sharing schemes and authentication codes. Two appendices explain all mathematical prerequisites in great detail. One is on elementary number theory (Euclid's Algorithm, the Chinese Remainder Theorem, quadratic residues, inversion formulas, and continued fractions). The other appendix gives a thorough introduction to finite fields and their algebraic structure.

This book differs from its 1988 version in two ways. That a lot of new material has been added is to be expected in a field that is developing so fast. Apart from a revision of the existing material, there are many new or greatly expanded sections, an entirely new chapter on elliptic curves and also one on authentication codes. The second difference is even more significant. The whole manuscript is electronically available as an interactive Mathematica manuscript. So, there are hyperlinks to other places in the text, but more importantly, it is now possible to work out non-trivial examples. Even a non-expert can easily alter the parameters in the examples and try out new ones. It is our experience, based on teaching at the California Institute of Technology and the Eindhoven University of Technology, that most students truly enjoy the enormous possibilities of a computer algebra notebook. Throughout the book, it has been our intention to make all *Mathematica* statements as transparent as possible, sometimes sacrificing elegant or smart alternatives that are too dependent on this particular computer algebra package.

There are several people that have played a crucial role in the preparation of this manuscript. In alphabetical order of first name, I would like to thank Fred Simons for showing me the full potential of *Mathematica* for educational purposes and for enhancing many the *Mathematica* commands, Gavin Horn for the many typo's that he has found as well as his compilation of solutions, Lilian Porter for her feedback on my use of English, and Wil Kortsmit for his help in getting the manuscript camera-ready and for solving many of my Mathematica questions. I also owe great debt to the following people who helped me with their feedback on various chapters:

Berry Schoenmakers, Bram van Asch, Eric Verheul, Frans Willems, Mariska Sas, and Martin van Dijk.

Henk van Tilborg
Dept. of Mathematics and Computing Science
Eindhoven University of Technology
P.O. Box 513
5600 MB Eindhoven
the Netherlands
email: henkvt@win.tue.nl.

1 Introduction

1.1 Introduction and Terminology

Cryptology, the study of cryptosystems, can be subdivided into two disciplines. *Cryptography* concerns itself with the design of cryptosystems, while *cryptanalysis* studies the breaking of cryptosystems. These two aspects are closely related; when setting up a cryptosystem the analysis of its security plays an important role. At this time we will not give a formal definition of a cryptosystem, as that will come later in this chapter. We assume that the reader has the right intuitive idea of what a cryptosystem is.

Why would anybody use a cryptosystem? There are several possibilities:

Confidentiality: When transmitting data, one does not want an eavesdropper to understand the contents of the transmitted messages. The same is true for stored data that should be protected against unauthorized access, for instance by hackers.

Authentication: This property is the equivalent of a signature. The receiver of a message wants proof that a message comes from a certain party and not from somebody else (even if the original party later wants to deny it).

Integrity: This means that the receiver of certain data has evidence that no changes have been made by a third party.

Throughout the centuries (see [Kahn67]) cryptosystems have been used by the military and by the diplomatic services. The nowadays widespread use of computer controlled communication systems in industry or by civil services, often asks for special protection of the data by means of cryptographic techniques.

Since the storage, and later recovery, of data can be viewed as transmission of this data in the time domain, we shall always use the term transmission when discussing a situation when data is stored and/or transmitted.

1.2   Shannon's Description of a Conventional Cryptosystem

Chapters 2, 3, and 4 discuss several so-called conventional cryptosystems. The formal definition of a conventional cryptosystem as well as the mathematical foundation of the underlying theory is due to C.E. Shannon [Shan49]. In Figure 1.1, the general outline of a conventional cryptosystem is depicted.

In the next section we shall elaborate on concepts like language and text. This will provide a cryptanalist with useful models when describing the output of the sender in the scheme.

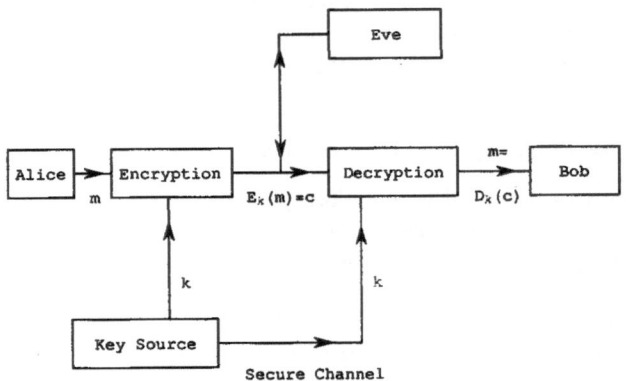

The conventional cryptosystem

Figure 1.1

Let $\mathcal{A}$ be a finite set, which we will call *alphabet*. With $|\mathcal{A}|$ we denote the cardinality of $\mathcal{A}$. We shall often use $\mathbf{Z}_q = \{0, 1, \ldots, q-1\}$ as alphabet, where we work with its elements modulo q (see the beginning of Subsection A.3.1 and Section B.2. The alphabet $\mathbf{Z}_{26}$ can be identified with the set $\{a, b, \ldots, z\}$. In most modern applications q will often be 2 or a power of 2.

A concatenation of n letters from $\mathcal{A}$ will be called an *n-gram* and denoted by $\mathbf{a} = (a_0, a_1, \ldots, a_{n-1})$. Special cases are *bi-grams* ($n = 2$) and *tri-grams* ($n = 3$). The set of all n-grams from $\mathcal{A}$ will be denoted by $\mathcal{A}^n$.

A *text* is an element from $\mathcal{A}^* = \bigcup_{n \geq 0} \mathcal{A}^n$. A *language* is a subset of $\mathcal{A}^*$. In the case of programming languages this subset is precisely defined by means of recursion rules. In the case of spoken languages these rules are very loose.

Let $\mathcal{A}$ and $\mathcal{B}$ be two finite alphabets. Any one-to-one mapping E of $\mathcal{A}^*$ to $\mathcal{B}^*$ is called a *cryptographic transformation*. In most practical situations $|\mathcal{A}|$ will be equal to $|\mathcal{B}|$. Also often the cryptographic transformation E will map n-grams into n-grams (to avoid data expansion during the encryption process).

Let **m** be the message (a text from $\mathcal{A}^*$) that Alice in Figure 1.1 wants to transmit in secrecy to Bob. It is usually called the *plaintext*. Alice will first transform the plaintext into $c = E(\mathbf{m})$, the so-called *ciphertext*. It will be the ciphertext that she will transmit to Bob.

> **Definition 1.1**
> A *symmetric* (or *conventional*) *cryptosystem* $\in$ is a set of cryptographic transformations $\in$ $= \{E_k \mid k \in \mathcal{K}\}$.
> The index set $\mathcal{K}$ is called the *key space*, and its elements k *keys*.

Since E_k is a one-to-one mapping, its inverse must exist. We shall denote it with D_k. Of course, the E stands for *encryption* (or enciphering) and the D for *decryption* (or deciphering). One has

$$D_k(E_k(\mathbf{m})) = \mathbf{m}, \qquad \text{for all plaintexts } m \in \mathcal{A}^* \text{ and keys } k \in \mathcal{K}.$$

If Alice wants to send the plaintext **m** to Bob by means of the cryptographic transformation E_k, both Alice and Bob must know the particular choice of the key k. They will have agreed on the value of k by means of a so-called *secure channel*. This channel could be a courier, but it could also be that Alice and Bob have, beforehand, agreed on the choice of k.

Bob can decipher **c** by computing

$$D_k(\mathbf{c}) = D_k(E_k(\mathbf{m})) = \mathbf{m}.$$

Normally, the same cryptosystem $\in$ will be used for a long time and by many people, so it is reasonable to assume that this set of cryptographic transformations $\in$ is also known to the cryptanalist. It is the frequent changing of the key that has to provide the security of the data. This principle was already clearly stated by the Dutchman Auguste Kerckhoff (see [Kahn67]) in the 19-th century.

The *cryptanalist* (Eve) who is connected to the transmission line can be:

■ *passive* (eavesdropping): The cryptanalist tries to find **m** (or even better k) from **c** (and whatever further knowledge he has). By determining k more ciphertexts may be broken.

■ *active* (tampering): The cryptanalist tries to actively manipulate the data that are being transmitted. For instance, he transmits his own ciphertext, retransmits old ciphertext, substitutes his own texts for transmitted ciphertexts, etc..

In general, one discerns three levels of cryptanalysis:

■ *Ciphertext only attack*: Only a piece of ciphertext is known to the cryptanalist (and often the context of the message).

■ *Known plaintext attack*: A piece of ciphertext with corresponding plaintext is known. If a system is secure against this kind of attack the legitimate receiver does not have to destroy deciphered messages.

- *Chosen plaintext attack*: The cryptanalist can choose any piece of plaintext and generate the corresponding ciphertext. The public-key cryptosystems that we shall discuss in Chapters 7-12 have to be secure against this kind of attack.

This concludes our general description of the conventional cryptosystem as depicted in Figure 1.1.

1.3   Statistical Description of a Plaintext Source

In cryptology, especially when one wants to break a particular cryptosystem, a probabilistic approach to describe a language is often already a powerful tool, as we shall see in Section 2.2.

The person Alice in Figure 1.1 stands for a finite or infinite *plaintext source* $\mathcal{S}$ of text, that was called plaintext, from an alphabet $\mathcal{A}$, e.g. $\mathbf{Z}_q$. It can be described as a finite resp. infinite sequence of random variables M_i, so by sequences

$$M_0, \ M_1, \ \ldots, \ M_{n-1} \quad \text{for some fixed value of } n,$$

resp.

$$M_0, \ M_1, \ M_2, \ \ldots,$$

each described by probabilities that events occur. So, for each letter combination (r-gram) $(m_0, m_1, \ldots, m_{r-1})$ over $\mathcal{A}$ and each starting point j the probability

$$\mathrm{Pr}_{\text{plain}}(M_j = m_0, \ M_{j+1} = m_1, \ \ldots, \ M_{j+r-1} = m_{r-1})$$

is well defined. In the case that $j = 0$, we shall simply write $\mathrm{Pr}_{\text{plain}}(m_0, \ m_1, \ \ldots, \ m_{r-1})$. Of course, the probabilities that describe the plaintext source $\mathcal{S}$ should satisfy the standard statistical properties, that we shall mention below but on which we shall not elaborate.

i) $\mathrm{Pr}_{\text{plain}}(m_0, \ m_1, \ \ldots, \ m_{r-1}) \geq 0$ for all texts $(m_0, \ m_1, \ \ldots, \ m_{r-1})$.

ii) $\sum_{(m_0, m_1, \ldots, m_{r-1})} \mathrm{Pr}_{\text{plain}}(m_0, \ m_1, \ \ldots, \ m_{r-1}) = 1$.

iii) $\sum_{(m_r, m_{r+1}, \ldots, m_{l-1})} \mathrm{Pr}_{\text{plain}}(m_0, \ m_1, \ \ldots, \ m_{l-1}) = \mathrm{Pr}_{\text{plain}}(m_0, \ m_1, \ \ldots, \ m_{r-1})$, for all $l > r$.

The third property is called *Kolmogorov's consistency condition*.

Example 1.1

The plaintext source $\mathcal{S}$ (Alice in Figure 1.1) generates individual letters (1-grams) from $\{a, b, \ldots, z\}$ with an independent but identical distribution, say $p(a)$, $p(b)$, ..., $p(z)$. So,

$$\mathrm{Pr}_{\text{plain}}(m_0, \ m_1, \ldots, m_{n-1}) = p(m_0) \, p(m_1) \cdots p(m_{n-1}), n \geq 1.$$

The distribution of the letters of the alphabet in normal English texts is given in Table 1.1 (see Table 12-1 in [MeyM82]). In this model one has that

$$\mathrm{Pr}_{\text{plain}}(\text{run}) = p(r) \, p(u) \, p(n) = 0.0612 \times 0.0271 \times 0.0709 \approx 1.18 \ 10^{-4}.$$

Note that in this model also $Pr_{plain}(nru) = p(n)\,p(r)\,p(u)$, *etc., so, unlike in a regular English texts, all permutations of the three letters r, u, and n are equally likely in* $\mathcal{S}$.

a	0.0804	h	0.0549	o	0.0760	v	0.0099
b	0.0154	i	0.0726	p	0.0200	w	0.0192
c	0.0306	j	0.0016	q	0.0011	x	0.0019
d	0.0399	k	0.0067	r	0.0612	y	0.0173
e	0.1251	l	0.0414	s	0.0654	z	0.0009
f	0.0230	m	0.0253	t	0.0925		
g	0.0196	n	0.0709	u	0.0271		

Probability distributions of 1-grams in English.

Table 1.1

Example 1.2

$\mathcal{S}$ *generates 2-grams over the alphabet* $\{a, b, , ..., z\}$ *with an independent but identical distribution, say* $p(s, t)$, *with* $s, t \in \{a, b, ..., z\}$. *So, for* $n \geq 1$

$$Pr_{plain}(m_0, m_1, ..., m_{2n-1}) = p(m_0\,m_1)\,p(m_2, m_3) \cdots p(m_{2n-2}\,m_{2n-1}).$$

The distribution of 2-grams in English texts can be found in the literature (see Table 2.3.4 in [Konh81]).

Of course, one can continue like this with tables of the distribution of 3-grams or more. A different and more appealing approach is given in the following example.

```
ed["a"] = 0.0723; ed["j"] = 0.0006; ed["s"] = 0.0715;
ed["b"] = 0.0060; ed["k"] = 0.0064; ed["t"] = 0.0773;
ed["c"] = 0.0282; ed["l"] = 0.0396; ed["u"] = 0.0272;
ed["d"] = 0.0483; ed["m"] = 0.0236; ed["v"] = 0.0117;
ed["e"] = 0.1566; ed["n"] = 0.0814; ed["w"] = 0.0078;
ed["f"] = 0.0167; ed["o"] = 0.0716; ed["x"] = 0.0030;
ed["g"] = 0.0216; ed["p"] = 0.0161; ed["y"] = 0.0168;
ed["h"] = 0.0402; ed["q"] = 0.0007; ed["z"] = 0.0010;
ed["i"] = 0.0787; ed["r"] = 0.0751;
```

Equilibrium distribution in English.

Table 1.2

	a	b	c	d	e	f	g	h	i	j	k	l	m
a	0.0011	0.0193	0.0388	0.0469	0.002	0.01	0.0233	0.002	0.048	0.002	0.0103	0.1052	0.0281
b	0.0931	0.0057	0.0016	0.0008	0.3219	0	0	0	0.0605	0.0057	0	0.1242	0.0049
c	0.1202	0	0.0196	0.0004	0.1707	0	0	0.1277	0.0761	0	0.0324	0.0369	0.0015
d	0.1044	0.002	0.0026	0.0218	0.3778	0.0007	0.0132	0.0007	0.1803	0.0033	0	0.0125	0.0178
e	0.066	0.0036	0.0433	0.1194	0.0438	0.0142	0.0125	0.0023	0.0158	0.0005	0.0036	0.0456	0.034
f	0.0838	0	0	0	0.1283	0.0924	0	0	0.1608	0	0	0.0299	0.0009
g	0.1078	0	0	0.0018	0.2394	0	0.0177	0.1281	0.0839	0	0	0.0203	0.0027
h	0.1769	0.0005	0.0014	0.0008	0.5623	0	0	0.0005	0.1167	0	0	0.0016	0.0016
i	0.038	0.0082	0.0767	0.0459	0.0437	0.0129	0.028	0.0002	0.0016	0	0.005	0.0567	0.0297
j	0.1259	0	0	0	0.1818	0	0	0	0.035	0	0	0	0
k	0.0395	0.0028	0	0.0028	0.5282	0.0028	0	0.0198	0.1582	0	0.0113	0.0198	0.0028
l	0.1342	0.0019	0.0022	0.0736	0.1918	0.0105	0.0108	0	0.1521	0	0.0079	0.1413	0.0082
m	0.1822	0.0337	0.0026	0	0.2975	0.001	0	0	0.1345	0	0	0.001	0.0654
n	0.055	0.0004	0.0621	0.1681	0.1212	0.0102	0.1391	0.0013	0.0665	0.0009	0.0066	0.0073	0.0106
o	0.0085	0.0101	0.0162	0.0231	0.0037	0.1299	0.0082	0.0025	0.0092	0.0014	0.0078	0.0416	0.0766
p	0.1359	0	0.0006	0	0.1747	0	0	0.0237	0.0423	0	0	0.0812	0.0073
q	0	0	0	0	0	0	0	0	0	0	0	0	0
r	0.1026	0.0033	0.0172	0.0282	0.2795	0.0031	0.0175	0.0017	0.1181	0	0.0205	0.0164	0.0303
s	0.0604	0.0012	0.0284	0.0027	0.1795	0.0024	0	0.0561	0.1177	0	0.0091	0.0145	0.0112
t	0.0619	0.0003	0.0036	0.0002	0.1417	0.0007	0.0002	0.3512	0.1406	0	0	0.0101	0.0044
u	0.0344	0.0415	0.0491	0.0243	0.0434	0.0052	0.0382	0.001	0.0258	0	0.0014	0.1097	0.0329
v	0.0749	0	0	0.0023	0.6014	0	0	0	0.2569	0	0	0	0.0012
w	0.2291	0.0008	0	0.0032	0.1942	0	0	0.1422	0.2104	0	0	0.0041	0
x	0.0672	0	0.1119	0	0.1269	0	0	0.0075	0.1119	0	0	0	0.0075
y	0.0586	0.0034	0.0103	0.0069	0.2897	0	0	0	0.069	0	0.0034	0.0172	0.0379
z	0.2278	0	0	0	0.4557	0	0	0	0.2152	0	0	0.0127	0

	n	o	p	q	r	s	t	u	v	w	x	y	z
a	0.1878	0.0008	0.0222	0	0.118	0.1001	0.1574	0.0137	0.0212	0.0057	0.0026	0.0312	0.0023
b	0	0.0964	0	0	0.0662	0.0229	0.0049	0.0727	0.0016	0	0	0.1168	0
c	0.0011	0.2283	0	0.0004	0.0426	0.0087	0.0893	0.0347	0	0	0	0.0094	0
d	0.0053	0.0733	0	0.0007	0.0324	0.0495	0.0013	0.0601	0.0099	0.004	0	0.0264	0
e	0.1381	0.004	0.0192	0.0034	0.1927	0.1231	0.0404	0.0048	0.0215	0.0205	0.0152	0.0121	0.0004
f	0.0009	0.2789	0	0	0.1215	0.0026	0.0496	0.0462	0	0	0	0.0043	0
g	0.0451	0.114	0	0	0.1325	0.0256	0.0247	0.0512	0	0	0	0.0053	0
h	0.0038	0.0786	0	0	0.0153	0.0027	0.0233	0.0085	0	0.0011	0	0.0041	0
i	0.2498	0.0893	0.01	0.0008	0.0342	0.1194	0.1135	0.0011	0.025	0	0.0023	0.0002	0.0079
j	0	0.3147	0	0	0.007	0	0	0.3357	0	0	0	0	0
k	0.0565	0.0198	0	0	0.0085	0.1102	0.0028	0.0028	0	0	0	0.0113	0
l	0.0004	0.0778	0.0041	0	0.0034	0.0389	0.0254	0.0269	0.0056	0.0011	0	0.0819	0
m	0.0042	0.1246	0.0722	0	0.0026	0.0244	0.0005	0.0337	0.0005	0	0	0.0192	0
n	0.0194	0.0528	0.0004	0.0007	0.0011	0.0751	0.1641	0.0124	0.0068	0.0018	0.0002	0.0157	0.0004
o	0.219	0.0222	0.0292	0	0.153	0.0357	0.0396	0.0947	0.0334	0.0345	0.0012	0.0041	0.0004
p	0.0006	0.1511	0.0581	0	0.2306	0.018	0.0287	0.0457	0	0	0	0.0017	0
q	0	0	0	0	0	0	0	1	0	0	0	0	0
r	0.0325	0.1114	0.0055	0	0.0212	0.0655	0.0596	0.0192	0.0142	0.0017	0.0002	0.0306	0
s	0.0021	0.0706	0.0386	0.0009	0.0027	0.0836	0.2483	0.0579	0	0.0039	0	0.0081	0
t	0.0015	0.1229	0.0003	0	0.0479	0.0418	0.0213	0.0195	0.0005	0.0088	0	0.0203	0.0005
u	0.1517	0.0019	0.0386	0	0.146	0.1221	0.1255	0.0029	0.0014	0	0.001	0.0014	0.0005
v	0	0.053	0	0	0	0.0023	0	0.0012	0.0012	0	0	0.0058	0
w	0.0357	0.1292	0	0	0.0106	0.0366	0.0016	0	0	0	0	0.0024	0
x	0	0.0075	0.3507	0	0	0	0.1716	0	0	0	0.0373	0	0
y	0.0172	0.2207	0.031	0	0.031	0.1517	0.0172	0.0138	0	0.0103	0	0.0069	0.0034
z	0	0.0506	0	0	0	0	0	0.0127	0	0	0	0	0.0253

Transition probabilities $p(t \mid s)$, row s, column t, in English.

Table 1.3

Example 1.3

In this model, the plaintext source $\mathcal{S}$ generates 1-grams by means of a Markov process. This process can be described by a transition matrix $P = (p(t \mid s))_{s,t}$ which gives the probability that a letter s in the text is followed by the letter t. It follows from the theory of Markov processes that P has 1 as an eigenvalue. Let $\underline{\mathbf{p}} = (p(a), p(b), \ldots, p(z))$, be the corresponding eigenvector (it is called the equilibrium distribution of the process).

Assuming that the process is already in its equilibrium state at the beginning, one has

$$\text{Pr}_{\text{plain}}(m_0, m_1, ..., m_{n-1}) = p(m_0)\, p(m_1 \mid m_0)\, p(m_2 \mid m_1) \cdots p(m_{n-1} \mid m_{n-2}).$$

Let $\underline{p}$ and P be given by Table 1.2 and Table 1.3 from [Konh81] (here they are denoted by "ed" resp. "TrPr"). Then, one obtains the following, more realistic probabilities of occurrence:

$$\text{Pr}_{\text{plain}}(\text{run}) = p(r)\, p(u \mid r)\, p(n \mid u) = 0.0751 \times 0.0192 \times 0.1517 \approx 2.19\ 10^{-4},$$

$$\text{Pr}_{\text{plain}}(\text{urn}) = p(u)\, p(r \mid u)\, p(n \mid r) = 0.0272 \times 0.1460 \times 0.0325 \approx 1.29\ 10^{-4},$$

$$\text{Pr}_{\text{plain}}(\text{nru}) = p(n)\, p(r \mid n)\, p(u \mid r) = 0.0814 \times 0.0011 \times 0.0192 \approx 1.72\ 10^{-6},$$

By means of the Mathematica functions `StringTake`, `ToCharacterCode`, *and* `StringLength`, *these probabilities can be computed in the following way (first enter the input Table 1.2 and Table 1.3, by executing all initialization cells)*

```
sourcetext = "run";
ed[StringTake[sourcetext, {1}]] *
StringLength[sourcetext]-1
     ∏          TrPr[[
    i=1
             ToCharacterCode[
       StringTake[sourcetext, {i}]] - 96,
     ToCharacterCode[StringTake[sourcetext, {i + 1}]] - 96]]
```

```
{{0.000218448}}
```

Better approximations of a language can be made, by considering transition probabilities that depend on more than one letter in the past.

Note, that in the three examples above, the models are all *stationary*, which means that $\text{Pr}_{\text{plain}}(M_j = m_0,\ M_{j+1} = m_1,\ ...,\ M_{j+n-1} = m_{n-1})$ is independent of the value of j. In the middle of a regular text one may expect this property to hold, but in other situations this is not the case. Think for instance of the date at the beginning of a letter.

1.4 Problems

Problem 1.1
What is the probability that the text "apple" occurs, when the plaintext source generates independent, identically distributed 1-grams, as described in Example 1.1.
Answer the same question when the Markov model of Example 1.3 is used?

Problem 1.2 [M]
Use the *Mathematica* function `Permutations` and the input formula at the end of Section 1.3 to determine for each of the 24 orderings of the four letters e, h, l, p the probability that it occurs in a language generated by the Markov model of Example 1.3.

2   Classical Cryptosystems

2.1   Caesar, Simple Substitution, Vigenère

In this chapter we shall discuss a number of classical cryptosystems. For further reading we refer the interested reader to ([BekP82], [Denn82], [Kahn67], [Konh81], or [MeyM82]).

2.1.1   Caesar Cipher

One of the oldest cryptosystems is due to Julius Caesar. It shifts each letter in the text cyclicly over k places. So, with $k = 7$ one gets the following encryption of the word cleopatra (note that the letter z is mapped to a):

cleopatra $\xrightarrow{+1}$ dmfpqbusb $\xrightarrow{+1}$ engqrcvtc $\xrightarrow{+1}$ fohrsdwud $\xrightarrow{+1}$ gpistexve $\xrightarrow{+1}$ hqjtufywf $\xrightarrow{+1}$ irkuvgzxg $\xrightarrow{+1}$ jslvwhayh

By using the *Mathematica* functions `ToCharacterCode` and `FromCharacterCode`, which convert symbols to their ASCI code and back (letter a has value 97, letter b has value 98, etc.), the Caesar cipher can be executed by the following function:

```
CaesarCipher[plaintext_, key_] :=
    FromCharacterCode[
  Mod[ ToCharacterCode[plaintext] - 97 + key, 26] + 97]
```

An example is given below.

```
plaintext = "typehereyourplaintextinsmallletters";
key = 24;
CaesarCipher[plaintext, key]
```

```
rwncfcpcwmspnjyglrcvrglqkyjjjcrrcpq
```

In the terminology of Section 1.2, the *Caesar cipher* is defined over the alphabet {0, 1, ..., 25} by:

$$E_k(m) = ((m + k) \bmod 26), \quad 0 \le m < 26,$$

and

$$\in = \{E_k \mid 0 \le k < 26\},$$

where $(i \bmod n)$ denotes the unique integer j satisfying $j \equiv i \pmod{n}$ and $0 \le j < n$. In this case, the key space $\mathcal{K}$ is the set $\{0, 1, \ldots, 25\}$ and $D_k = E_{q-1-k}$.

An easy way to break the system is to try out all possible keys. This method is called *exhaustive key search*. In Table 2.1 one can find the cryptanalysis of the ciphertext "xyuysuyifvyxi".

```
x y u y s u y i f v y x i
w x t x r t x h e u x w h
v w s w q s w g d t w v g
u v r v p r v f c s v u f
t u q u o q u e b r u t e
```

Cryptanalysis of the Caesar cipher

Table 2.1

To decrypt the ciphertext yhaklwpnw., one can easily check all keys with the caesar function defined above.

```
ciphertext = "yhaklwpnw";
Table[CaesarCipher[ciphertext, -key], {key, 1, 26}]
```

```
{xgzjkvomv, wfyijunlu, vexhitmkt, udwghsljs, tcvfgrkir, sbuefqjhq,
 ratdepigp, qzscdohfo, pyrbcngen, oxqabmfdm, nwpzalecl,
 mvoyzkdbk, lunxyjcaj, ktmwxibzi, jslvwhayh, irkuvgzxg,
 hqjtufywf, gpistexve, fohrsdwud, engqrcvtc, dmfpqbusb,
 cleopatra, bkdnozsqz, ajcmnyrpy, ziblmxqox, yhaklwpnw}
```

2.1.2   Simple Substitution

□ **The System and its Main Weakness**

With the method of a *simple substitution* one chooses a fixed permutation π of the alphabet $\{a, b, \ldots, z\}$ and applies that to all letters in the plaintext.

Example 2.1

In the following example we only give that part of the substitution π that is relevant for the given plaintext. We use the Mathematica function StringReplace.

```
StringReplace["plaintext",
    {"a" -> "k", "e" -> "z", "i" -> "b", "l" -> "r",
        "n" -> "a", "p" -> "v", "t" -> "q", "x" -> "d"}]
```

```
vrkbaqzdq
```

A more formal description of the simple substitution system is as follows: the key space $\mathcal{K}$ is the set S_q of all permutations of $\{0, 1, ..., q-1\}$ and the cryptosystem $\mathcal{E}$ is given by

$$\mathcal{E} = \{E_\pi \mid \pi \in S_q\},$$

where

$$E_\pi (m) = \pi (m), \quad 0 \le m < q.$$

The decryption function D_π is given by $D_\pi = E_{\pi^{-1}}$, as follows from

$$D_\pi (E_\pi (m)) = D (\pi (m)) = E_{\pi^{-1}} (\pi (m)) = \pi^{-1} (\pi (m)) = m, \quad 0 \le m < q.$$

Unlike Caesar's cipher, this system does not have the drawback of a small key space. Indeed, $|\mathcal{K}| = |S_{26}| = 26! \approx 4.03 \ 10^{26}$. This system however does demonstrate very well that a large key space should not fool one into believing that a system is secure! On the contrary, by simply counting the letter frequencies in the ciphertexts and comparing these with the letter frequencies in Table 1.1, one very quickly finds the images under π of the most frequent letters in the plaintext. Indeed, the most frequent letter in the ciphertext will very likely be the image under π of the letter e. The next one is the image of the letter n, etc. After having found the encryptions of the most frequent letters in the plaintext, it is not difficult to fill in the rest. Of course, the longer the cipher text, the easier the cryptanalysis becomes. In Chapter 5, we come back to the cryptanalysis of the system, in particular how long the same key can be used safely.

□ **Cryptanalysis by The Method of a Probable Word**

In the following example we have knowledge of a very long ciphertext. This is not necessary at all for the cryptanalysis of the ciphertext, but it takes that long to know the full key. Indeed, as long as two letters are missing in the plaintext, one does not know the full key, but the system is of course broken much earlier than that.

Apart from the ciphertext, given in Table 2.2, we shall assume in this example that the plaintext discusses the concept of "bidirectional communication theory". Cryptanalysis will turn out to be very easy.

```
zhjeo  ndize  hicle  osiol  digic  lmhzq  zolyi  zehdp  zhjeo  ndize
hycdh  hlpvs  uczyc  dhzhj  eondi  zehge  moylk  zhjpm  lhylg  gidiz
gizyd  ppsdo  lylzr  losye  nnmhz  ydize  hicle  osceu  lrloq  lgyoz
vlgic  lneol  flhlo  dpydg  lzhuc  zyciu  eeone  olzhj  eondi  zehge
moylg  zhjpm  lhyll  dycei  clogi  dizgi  zydpp  siclq  zolyi  zehej
iczgz  hjpml  hylzg  lkaol  gglqv  sqzol  yilqi  odhgj  eondi  zehxm
dhizi  zlguc  zycyd  hehps  vlqlo  zrlqz  jiclp  duejy  dmgdp  ziszg
evglo  rlqqz  gizhf  mzgcz  hficl  ldopz  loydm  gljoe  niclp  dilol
jjlyi  zhvze  pefsd  hqgey  zepef  syenn  mhzyd  izehi  cleos  gllng
iecdr  luzql  daapz  ydize  hgqml  ieicl  jdyii  cdipz  rzhfv  lzhfg
dolvs  iclzo  dyize  hggem  oylge  jzhje  ondiz  ehucz  yczhj  pmlhy
lldyc  eiclo  zhdpp  aeggz  vplqz  olyiz  ehgic  laolg  lhiad  aloql
gyzvl  gicly  dglej  vzqzo  lyize  hdpye  nnmhz  ydize  hicle  osdaa
pzlqi  eiclg  eyzdp  vlcdr  zemoe  jneht  lsg…
```

Ciphertext obtained with a simple substitution

Table 2.2

Assuming that the word "communication" will occur in the plaintext, we look for strings of 13 consecutive letters, in which letter 1 = letter 8, letter 2 = letter 12, letter 3 = letter 4, letter 6 = letter 13 and letter 7 = letter 11.

Indeed, we find the string "yennmhzydizeh" three times in the ciphertext. This gives the following information about π.

$$
\begin{array}{ccccccc}
c & o & m & u & n & i & a & t \\
\downarrow & \downarrow & \downarrow & \downarrow & \downarrow & \downarrow & \downarrow \\
y & e & n & m & h & z & d & i
\end{array}
$$

Assuming that the word "direction" does also occur in the plaintext, we need to look for strings of the form "*z**yizeh" in the ciphertext, because of the information that we already have on π. It turns out that "qzolyizeh" appears four times, giving:

$$
\begin{array}{ccc}
d & r & e \\
\downarrow & \downarrow & \downarrow \\
q & o & l
\end{array}
$$

If we substitute all this information in the ciphertext one easily obtains π completely. For instance, the text begins like

> in*ormationt*eor*treat*t*eunid…,

which obviously comes from

> information theory treats the unid(irectional) …,

This gives the π-image of the letters f, h, y and s.

Continuing like this, one readily obtains π completely.

a	b	c	d	e	f	g	h	i	j	k	l	m	n	o	p	q	r	s	t	u	v	w	x	y	z
↓	↓	↓	↓	↓	↓	↓	↓	↓	↓	↓	↓	↓	↓	↓	↓	↓	↓	↓	↓	↓	↓	↓	↓	↓	↓
d	v	y	q	l	j	f	c	z	w	t	p	n	h	e	a	x	o	g	i	m	r	u	k	s	b

Example 2.2

Mathematica makes is quite easy to find a substring with a certain pattern.For instance, to test where in a text one can find a substring of length 6 with letters 1 and 4 equal and also letters 2 and 5 (as in the Latin word "quoque"), one can use the Mathematica functions If, StringTake, StringLength, Do, Print *and the following:*

```
ciphertext = "xyuysuyifvyxi";
Do[
  If[StringTake[ciphertext, {i + 1}] == StringTake[ciphertext,
     {i + 4}] ∧ StringTake[ciphertext, {i + 2}] ==
     StringTake[ciphertext, {i + 5}],
   Print[i + 1, "   ", StringTake[ciphertext, {i + 1, i + 6}]]],
  {i, 0, StringLength[ciphertext] - 6}]
```

```
3    uysuyi
```

This example was taken from Table 2.1.

2.1.3 Vigenère Cryptosystem

The *Vigenère cryptosystem* (named after the Frenchman B. de Vigenère who in 1586 wrote his Traicté des Chiffres, describing a more difficult version of this system) consists of r Caesar ciphers applied periodically. In the example below, the key is a word of length $r = 7$. The i-th letter in the key defines the particular Caesar cipher that is used for the encryption of the letters $i, i + r, i + 2r, \ldots$ in the plaintext.

Example 2.3

We identify $\{0, 1, \ldots, 25\}$ with $\{a, b, \ldots, z\}$. The so-called Vigenère Table (see Table 2.3) is a very helpful tool when encrypting or decrypting. With the key "michael" one gets the following encipherment:

```
plaintext   a c r y p t o s y s t e m o f t e n i s a c
key         m i c h a e l m i c h a e l m i c h a e l m
ciphertext  m k t f p x z e g u a e q z r b g u i w l o
```

```
 0  a  b  c  d  e  f  g  h  i  j  k  l  m  n  o  p  q  r  s  t  u  v  w  x  y  z
 1  b  c  d  e  f  g  h  i  j  k  l  m  n  o  p  q  r  s  t  u  v  w  x  y  z  a
 2  c  d  e  f  g  h  i  j  k  l  m  n  o  p  q  r  s  t  u  v  w  x  y  z  a  b
 3  d  e  f  g  h  i  j  k  l  m  n  o  p  q  r  s  t  u  v  w  x  y  z  a  b  c
 4  e  f  g  h  i  j  k  l  m  n  o  p  q  r  s  t  u  v  w  x  y  z  a  b  c  d
 5  f  g  h  i  j  k  l  m  n  o  p  q  r  s  t  u  v  w  x  y  z  a  b  c  d  e
 6  g  h  i  j  k  l  m  n  o  p  q  r  s  t  u  v  w  x  y  z  a  b  c  d  e  f
 7  h  i  j  k  l  m  n  o  p  q  r  s  t  u  v  w  x  y  z  a  b  c  d  e  f  g
 8  i  j  k  l  m  n  o  p  q  r  s  t  u  v  w  x  y  z  a  b  c  d  e  f  g  h
 9  j  k  l  m  n  o  p  q  r  s  t  u  v  w  x  y  z  a  b  c  d  e  f  g  h  i
10  k  l  m  n  o  p  q  r  s  t  u  v  w  x  y  z  a  b  c  d  e  f  g  h  i  j
11  l  m  n  o  p  q  r  s  t  u  v  w  x  y  z  a  b  c  d  e  f  g  h  i  j  k
12  m  n  o  p  q  r  s  t  u  v  w  x  y  z  a  b  c  d  e  f  g  h  i  j  k  l
13  n  o  p  q  r  s  t  u  v  w  x  y  z  a  b  c  d  e  f  g  h  i  j  k  l  m
14  o  p  q  r  s  t  u  v  w  x  y  z  a  b  c  d  e  f  g  h  i  j  k  l  m  n
15  p  q  r  s  t  u  v  w  x  y  z  a  b  c  d  e  f  g  h  i  j  k  l  m  n  o
16  q  r  s  t  u  v  w  x  y  z  a  b  c  d  e  f  g  h  i  j  k  l  m  n  o  p
17  r  s  t  u  v  w  x  y  z  a  b  c  d  e  f  g  h  i  j  k  l  m  n  o  p  q
18  s  t  u  v  w  x  y  z  a  b  c  d  e  f  g  h  i  j  k  l  m  n  o  p  q  r
19  t  u  v  w  x  y  z  a  b  c  d  e  f  g  h  i  j  k  l  m  n  o  p  q  r  s
20  u  v  w  x  y  z  a  b  c  d  e  f  g  h  i  j  k  l  m  n  o  p  q  r  s  t
21  v  w  x  y  z  a  b  c  d  e  f  g  h  i  j  k  l  m  n  o  p  q  r  s  t  u
22  w  x  y  z  a  b  c  d  e  f  g  h  i  j  k  l  m  n  o  p  q  r  s  t  u  v
23  x  y  z  a  b  c  d  e  f  g  h  i  j  k  l  m  n  o  p  q  r  s  t  u  v  w
24  y  z  a  b  c  d  e  f  g  h  i  j  k  l  m  n  o  p  q  r  s  t  u  v  w  x
25  z  a  b  c  d  e  f  g  h  i  j  k  l  m  n  o  p  q  r  s  t  u  v  w  x  y
```

The Vigenère Table.

Table 2.3

Because of the redundancy in the English language one reduces the effective size of the key space tremendously by choosing an existing word as the key. Taking the name of a relative, as we have done above, reduces the security of the encryption more or less to zero.

In Mathematica, addition of two letters as defined by the Vigenère Table can be realized in a similar way, as our earlier implementation of the Caesar cipher:

```
AddTwoLetters[a_, b_] :=
FromCharacterCode[Mod[(ToCharacterCode[a] - 97) +
                 (ToCharacterCode[b] - 97), 26] + 97]
```

By means of the Mathematica functions StringTake *and* StringLength *, and the function AddTwoLetters, defined above, encryption with the Vigenère cryptosystem can be realized as follows:*

```
plaintext = "typehereyourplaintextinsmallletters";
key = "keyword";
ciphertext = "";
Do[ciphertext = ciphertext <>
          AddTwoLetters[StringTake[plaintext, {i}],

              StringTake[
      key, {Mod[i - 1, StringLength[key]] + 1}]],
   {i, 1, StringLength[plaintext]}];
ciphertext
```

```
dcnavvuocmqfgokmlpsowsrqiocovirpsiv
```

A more formal description of the Vigenère cryptosystem is as follows

$$\in \ = \ \{E_{(k_0, k_1, \dots, k_{r-1})} \mid (k_0, \ k_1, \ \dots, \ k_{r-1}) \in \mathcal{K} = \mathbb{Z}_{26}^r\}$$

and

$$E_{(k_0, k_1, \dots, k_{r-1})}(m_0, m_1, m_2, \dots) = (c_0, c_1, c_2, \dots)$$

with

$$c_i = ((m_i + k_{(i \bmod r)}) \bmod 26). \tag{2.1}$$

Instead of using r Caesar ciphers periodically in the Vigenère cryptosystem, one can of course also use r simple substitutions. Such a system is an example of a so-called *polyalphabetic substitution*. For centuries, no one had an effective way of breaking this system, mainly because one did not have a technique of determining the key length r. Once one knows r, one can find the r simple substitutions by grouping together the letters $i, i+r, i+2r, \dots$, for each i, $0 \le i < r$, and break each of these r simple substitutions individually. In 1863, the Prussian army officer, F.W. Kasiski, solved the problem of finding the key length r by statistical means. In the next section, we shall discuss this method.

2.2   The Incidence of Coincidences, Kasiski's Method

2.2.1   The Incidence of Coincidences

Consider a ciphertext $\mathbf{c} = c_0, c_1, ..., c_{n-1}$ which is the result of a Vigenère encryption of an English plaintext $\mathbf{m} = m_0, m_1, ..., m_{n-1}$ under the key $\mathbf{k} = k_0, k_1, ..., k_{r-1}$ (see also (2.1)). As explained at the end of the previous section, the key to breaking the Vigenère system is to determine the key length r.

In our analysis we are going to assume the very simple model of a plaintext source outputting independent, individual letters, each with probability distribution given by Table 1.1 (see Example 1.1). We further assume that the letters k_i in the key are chosen with independent and uniform distribution from $\{a, b, ..., z\}$ (so, with probability $1/26$).

Let $\mathbf{c}_{\text{left}}^{(i)}$ and $\mathbf{c}_{\text{right}}^{(i)}$ the substrings of c consisting of the i left most resp. right most symbols of $\mathbf{c}$, so:

$$\mathbf{c}_{\text{left}}^{(i)} = c_0, c_1, ..., c_{i-1} \qquad \text{and} \qquad \mathbf{c}_{\text{right}}^{(i)} = c_{n-i}, c_{n-i+1}, ..., c_{n-1}.$$

Let us now count the number of agreements between $\mathbf{c}_{\text{left}}^{(i)}$ and $\mathbf{c}_{\text{right}}^{(i)}$, i.e. the number of coordinates j where $(\mathbf{c}_{\text{left}}^{(i)})_j = (\mathbf{c}_{\text{right}}^{(i)})_j$. We shall show in Lemma 2.1 that the expected value of this number divided by the string length i will be 0.06875 or $1/26 \approx 0.03846$, depending on whether the (unknown) key length r divides $n - i$ or does not divide $n - i$.

Let us show by example how this difference in expected values can be used to determine the unknown key length r.

Example 2.4

In this example we consider the ciphertext

> *"glrtnhklttbrxbxwnnhshjwkcjmsmrwnxqmvehuimnfxbzcwixbrnhxqhhclgcipcgimglrtnhklttbrshvil gwcmwyejqbxbmlywimbkhhjwkcjmsmrwnxqmplceiwkcjmehtpslmmlxowmylxbxflxeebrahjwkcjms mrwnxqm".*

By means of the Mathematica functions `StringTake`*,* `StringLength`*,* `Characters`*, and* `Table`*, we can easily compute the number of agreements between* $\mathbf{c}_{\text{left}}^{(i)}$ *and* $\mathbf{c}_{\text{right}}^{(i)}$ *in any range of values of i:*

```
ciphertext =
  "ubsyvkmhvyrrtsbbcrdsndwrtshxmbufrmxgabnvmircewerucamlyz\
    brvfwivvmlyzwapspyogsslechbgcubsvyczqrcwrmhvcxgooyvcy\
    dspomtqfpyqkgbcmerucadlcaflrsuqjrbhceqesfcehuoqmdstor\
    cdoymeqqwaglgovggamdabbigztbbqyfwbxwmgfpowgztyeilosrk\
    gfahuovqfogswruqnvpwfvrnmpqqgsslatgrmqubsvyczqrswcjde\
    owqqroihqdspdibffnxwgztbbqyfwbxus";
L = StringLength[ciphertext];
Table[N[Count[Characters[StringTake[ciphertext, i]] -
    Characters[StringTake[ciphertext, -i]], 0]/i,
  1], {i, L - 20, L - 1}]
```

```
{0.03, 0.04, 0.08, 0.02, 0.05, 0.04, 0.04, 0.03, 0.06, 0.07,
 0.06, 0.04, 0.02, 0.05, 0.08, 0.04, 0.05, 0.02, 0.01, 0.05}
```

The (relative) higher values in this listing at places −6 and −18 indicate that the key length r is 6. Indeed, the key that has been used to generate this example is the word "monkey", which has 6 letters.

This can be checked with the following analogue of the Vigenère encryption of Example 2.3.

```
SubTwoLetters[a_, b_] :=
FromCharacterCode[
  Mod[(ToCharacterCode[a] - 97) - (ToCharacterCode[b] - 97),
    26] + 97]
```

```
ciphertext =
   "ubsyvkmhvyrrtsbbcrdandwrtshxmbufrmxgabnvmircewerucamlyz\
    brvfwivvmlyzwapspyogsslechbgcubsvyczqrcwrmhvcxgooyvcy\
    dspomtqfpyqkgbcmerucadlcaflrsuqjrbhceqesfcehuoqmdstor\
    cdoymeqqwaglgovggsmdabbigztbbqyfwbxwmgfpowgztyeilosrk\
    gfahuovqfogswruqnvpwfvrnmpqqgsslatgrmqubsvyczqrswcjde\
    owqqroihqdspdibffnxwgztbbqyfwbxus";
key = "monkey";
plaintext = "";
Do[plaintext = plaintext <>
           SubTwoLetters[StringTake[ciphertext, {i}],

               StringTake[
     key, {Mod[i - 1, StringLength[key]] + 1}]],
    {i, 1, StringLength[ciphertext]}]
plaintext
```

```
informationtheorytreatstheunidirectionalikformationchannelbywhichaninfo
   rmationsourceinfluencesstatisticallyareceivercommunpcationtheoryhowe
   verdescribesthemoregeneralcaseinwhichtwoormoreinformationsourcesinfl
   uenceeachotherstatisticallythedirectionofthisinfluenceisexpressedbyc
   srectedtransinformationqu
```

Lemma 2.1

Let c be a ciphertext which is the result of a Vigenère encryption of a plaintext m of length n with key k of length r.

Suppose that m is generated by the plaintext source of Example 1.1. So, all the letters in m are generated independently of each other, all with the frequency distribution $p(m)$ given by Table 1.1. Suppose further that the letters k_i in the key are chosen with independent and uniform distribution from $\{a, b, ..., z\}$ (so, with probability 1/26). Then, for each $1 \leq i < j \leq n$,

$$\Pr[c_i = c_j] = \begin{cases} \sum_m p(m)^2 \approx 0.06875, & \text{if } r \text{ divides } j - i, \\ 1/26 \approx 0.03846, & \text{if } r \text{ does not divide } j - i. \end{cases}$$

Proof:

If $j - i$ is divisible by r, then $c_i = c_j$ if and only if $m_i = m_j$. This follows directly from formula (2.1), since $(j \bmod r)$ equals $(i \bmod r)$. So,

$$\Pr[c_i = c_j] = \Pr[m_i = m_j] = \sum_m \Pr[m_i = m_j = m] =$$

$$\sum_m \Pr[m_i = m] \Pr[m_j = m] = \sum_m p(m)^2 \approx 0.06875.$$

If $j - i$ is not divisible by r, then by (2.1) $c_i = c_j$ if and only if $m_i + k_{(i \bmod r)} = m_j + k_{(j \bmod r)}$. Since $(j \bmod r) \neq (i \bmod r)$, it follows that $k_{(j \bmod r)}$ takes on the value $m_i + k_{(i \bmod r)} - m_j$ with probability $1/26$. We conclude that

$$\Pr[c_i = c_j] = 1/26 \approx 0.03846.$$

$\square$

It may be clear that with increasing length of the ciphertext, it is easier to determine the key length from the relative number of agreements between $\mathbf{c}_{\text{left}}^{(i)}$ and $\mathbf{c}_{\text{right}}^{(i)}$.

2.2.2 Kasiski's Method

Kasiski based his cryptanalysis of the Vigenère cryptosystem on the fact that when a certain combination of letters (a frequent plaintext fragment) is encrypted more than once with the same segment of the key (because they occur at a multiple of the key length r), one will see a repetition of the corresponding ciphertext at those places.

We quote an example from [Baue97]:

Example 2.5

Consider the following plaintext and ciphertext pair (where the key "comet" has been used):

```
plaintext  t h e r e i s a n o t h e r f a m o u s p i a n o p l a y .
    key    c o m e t c o m e t c o m e t c o m e t c o m e t c o m e .
ciphertext v v q v x k g m r h v v q v y c a a y l r w m r h z m c .
```

In the ciphertext one can find the substring "vvqv" (of length 4) repeated twice, namely starting at positions 1 and 11. This indicates that r divides 10. The substring "mrh" (of length 3) also occurs twice: at positions 8 and 23. So, it seems likely that r also divides 15. Combining these results, we conclude that r = 5, which is indeed the case.

See [Baue97] for a further analysis of the Vigenère cryptosystem.

2.3   Vernam, Playfair, Transpositions, Hagelin, Enigma

In this section, we shall briefly discuss a few more cryptosystems, without going deep into their structure.

2.3.1   The One-Time Pad

The *one-time pad*, also called the *Vernam cipher* (after the American A.T. & T. employee G.S. Vernam, who introduced the system in 1917), is a Vigenère cipher with key length equal to the length of the plaintext. Also, the key must be chosen in a completely random way and can only be used once. In this way the system is unconditionally secure, as is intuitively clear and will be proved in Chapter 5. The "hot line" between Washington and Moscow uses this system. The major drawback of this system is the length of the key, which makes this system impractical for most applications.

2.3.2   The Playfair Cipher

The *Playfair cipher* (1854, named after the Englishman L. Playfair) was used by the British in World War I. It operates on 2-grams. First of all, one has to identify the letters i and j. The remaining 25 letters of the alphabet are put rowwise in a 5×5 matrix K, as follows. Put the first letter of a keyword in the top-left position. Continue rowwise from left to right. If a letter occurs more than once in the keyword, use it only once. The remaining letters of the alphabet are put into K in their natural order. For instance, the keyword "hieronymus" gives rise to

$$\begin{pmatrix} h & i & e & r & o \\ n & y & m & u & s \\ a & b & c & d & f \\ g & k & l & p & q \\ t & v & w & x & z \end{pmatrix}$$

The 2-gram $(x, y) = (K_{i,j}, K_{m,n})$ with $x \neq y$ will be encrypted into

$$\begin{aligned}
(K_{i,n}, K_{m,j}), &\quad \text{if } i \neq m \text{ and } j \neq n, \\
(K_{i,j+1}, K_{i,n+1}), &\quad \text{if } i = m \text{ and } j \neq n, \\
(K_{i+1,j}, K_{m+1,j}), &\quad \text{if } i \neq m \text{ and } j = n,
\end{aligned}$$

where the indices are taken modulo 5. If the symbols x and y in the 2-gram (x, y) are the same, one first inserts the letter q and enciphers the text $\ldots x\, q\, y \ldots$.

2.3.3 Transposition Ciphers

A completely different way of enciphering is called *transposition*. This system breaks the text up into blocks of fixed length, say n, and applies a fixed permutation σ to the coordinates. For instance, with $n = 5$ and $\sigma = (1, 4, 5, 2, 3)$, one gets the following encryption:

$$\texttt{crypt ograp hical } \dots \xrightarrow{\sigma} \texttt{ytrcp rpgoa cliha } \dots$$

Often the permutation is of a geometrical nature, as is the case with the so-called *column transposition*. The plaintext is written rowwise in a matrix of given size, but will be read out columnwise in a specific order depending on a keyword. For instance, after having identified letters $a, b, \dots, z$ with the numbers $1, 2, \dots, 26$ the keyword "right" will dictate you to read out column 3 first (being the alphabetically first of the 5 letters in "right"), followed by columns 4, 2, 1 and 5. So, the plaintext

```
computing science has had very little influence on computing
                         practice
```

when encrypted with a 5×5 matrix and keyword "right" will first be filled in rowwise as depicted below

```
4 3 1 2 5        4 3 1 2 5        4 3 1 2 5
c o m p u        y l i t t        n g p r a
t i n g s        l e i n f        c t i c e
c i e n c        l u e n c        . . .
e h a s h        e o n c o
a d v e r        m p u t i
```

and then read out (columnwise in the indicated order) to give the ciphertext:

```
mneav pgnse oiihd ctcea uschr iienu tnnct leuop yllem tfcoi ....
```

Since transpositions do not change letter frequencies, but destroy dependencies between consecutive letters in the plaintext, while Vigenère etc. do the opposite, one often combines such systems. Such a combined system is called a *product cipher*. Shannon used the words confusion and diffusion in this context.

Ciphersystems that encrypt the plaintext symbol for symbol in a way that depends on previous input symbols are often called *stream ciphers* (they will discussed in Chapter 3). Cryptosystems that encrypt blocks of symbols (of a fixed length) simultaneously but independent of previous encryptions, they are called *block ciphers* (see Chapter 4).

During World War II both sides used so called rotor machines for their encryption. Several variations of the machines described in the next two subsections were in use at that time. We shall give a rough idea of each one.

2.3.4 Hagelin

The Hagelin

Figure 2.1

The *Hagelin*, invented by the Swede B. Hagelin and used by the U.S. Army, has 6 rotors with 26, resp. 25, 23, 21, 19 and 17 pins. Each of these pins can be put into an active or passive position by letting it stick out to the left or right of the rotor. After encryption of a letter (depending on the setting of these pins and a rotating cylinder), the 6 rotors all turn one position. So, after 26 encryptions the first rotor is back in its original position. For the sixth rotor this takes only 17 encryptions.

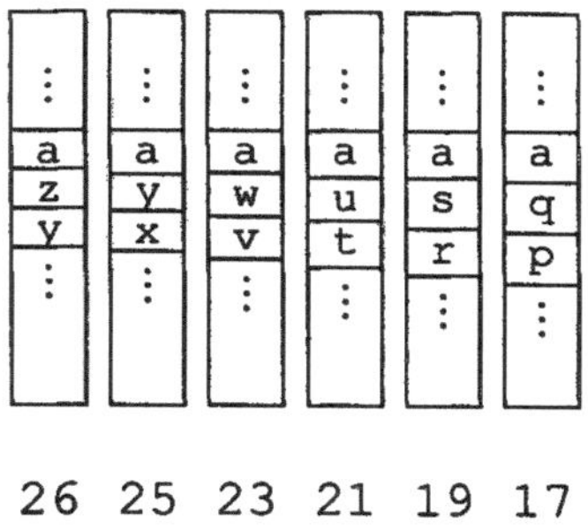

26 25 23 21 19 17

The six rotors in the Hagelin machine,
each with its own number of positions.

Figure 2.2

Since the number of pins on the rotors are coprime, the Hagelin can be viewed as a mechanical
Vigenère cryptosystem with period $26 \times 25 \times 23 \times 21 \times 19 \times 17 = 101,405,850$. We refer the reader
who is interested in the cryptanalysis of the Hagelin to Section 2.3 in [BekP82].

2.3.5   Enigma

The Enigma

Figure 2.3

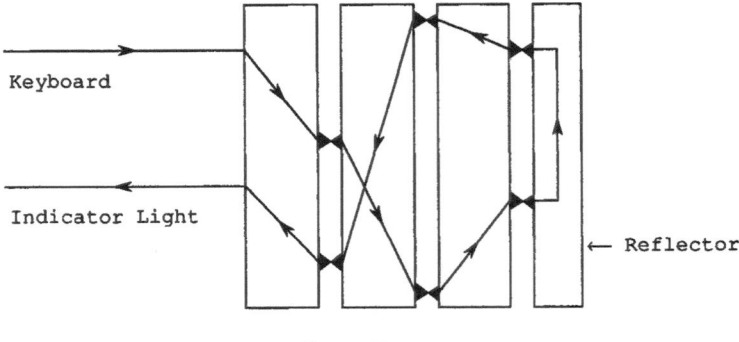

A Schematic Description of the Enigma

Figure 2.4

The electro-mechanical *Enigma*, used by Germany and Japan, was invented by A. Scherbius in 1923. It consists of three rotors and a reflector. See Figure 2.4. When punching in a letter, an electronic current will enter the first rotor at the place corresponding with that letter, but will leave it somewhere else depending on the internal wiring of that rotor. The second and third rotors do the same, but have a different wiring. The reflector returns the current at a different place and the current will go through rotors 1, 2 and 3 again but in reverse order. The current will light up a letter, which gives the encryption of the original letter.

Simultaneously, the first rotor will turn position. After 26 rotations of the first rotor the second will turn one position. When the second rotor has made a full cycle, the third rotor will rotate over one position.

The key of the Enigma consists of

i) the choice and order of the rotors,
ii) their initial position and
iii) a fixed initial permutation of the alphabet.

For an idea about the cryptanalysis of the Enigma the reader is referred to Chapter 5 in [Konh81].

2.4   Problems

Problem 2.1
The following ciphertext about president Kennedy has been made with a simple substitution. What is the corresponding
plaintext?

"rgjjg mvkto tzpgt stbgp catjw pgocm gjs"

Problem 2.2
Decrypt the following ciphertext, which is made with the Playfair cipher and the key "hieronymus" (as in Subsection 2.3.2).

"erohh mfimf ienfa bsesn pdwar gbhah ro"

Problem 2.3
Encrypt the following plaintext using the Vigenère system with the key "vigenere".

"who is afraid of virginia woolf"

Problem 2.4^M
Consider a ciphertext obtained through a Caesar encryption. Write a *Mathematica* program to find all substrings of length 5 in the ciphertext that could have been obtained from the word "Brute".
Test this program on the text "xyuysuyifvyxi" from Table 2.1. (See also the input in Example 2.2)

3   Shift Register Sequences

3.1   Pseudo-Random Sequences

During and after World War II, the introduction of logical circuits made completely electronic cryptosystems possible. These turned out to be very practical in the sense of being easy to implement and very fast. The analysis of their security is not so easy! Working with logical circuits often leads to the alphabet $\{0, 1\}$. There are only two possible permutations (substitutions) of the set $\{0, 1\}$. One action interchanges the two symbols. This can also be described by adding 1 (modulo 2) to the two elements. The other permutation leaves the two symbols invariant, which is the same as adding 0 (modulo 2) to these two elements.

Since the Vernam cipher is unconditionally secure but not very practical, it is only natural that people came up with the following scheme.

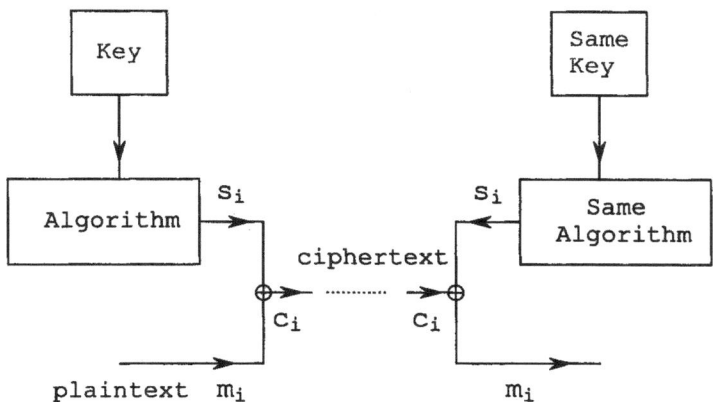

A binary cryptosystem with pseudo-random $\{s_i\}_{i\geq 0}$-sequence.

Figure 3.1

Of course one would like the sequence $\{s_i\}_{i\geq 0}$ to be random, but with a finite state machine and a deterministic algorithm one can not generate a random sequence. Indeed, one will always generate a sequence, which is ultimately periodic. This observation shows that (apart from a beginning segment) the scheme is a special case of the Vigenère cryptosystem. On the other hand, one can try to generate sequences that appear to be random, have long periods and have the right cryptographic properties. Good reference books for this theory are [Bek82], [Gol67], and [Ruep86].

In [Gol67], S.W. Golomb formulated three postulates that a binary, periodic sequence $\{s_i\}_{i \geq 0}$ should satisfy to be called *pseudo-random*. Before we can give these, we have to introduce some terminology.

> **Definition 3.1**
> A sequence $\{s_i\}_{i \geq 0}$ is called *periodic* with *period p*, if p is the smallest positive integer for which
> $$s_{i+p} = s_i \qquad \text{for all } i \geq 0.$$

A *run* of length k is a subsequence of $\{s_i\}_{i \geq 0}$ consisting of k identical symbols, bordered by different symbols. If the run starts at moment t, one has in formula:

$$s_{t-1} \neq s_t = s_{t+1} = \ldots = s_{t+k-1} \neq s_{t+k}.$$

One makes the following distinction:

$$\text{a block of length } k: \qquad 0 \overset{k}{\overleftrightarrow{11 \ldots 1}} 0$$

$$\text{a gap of length } k: \qquad 1 \overset{k}{\overleftrightarrow{00 \ldots 0}} 1$$

The *autocorrelation $AC(k)$* of a periodic sequence $\{s_i\}_{i \geq 0}$ with period p is defined by:

$$AC(k) = \frac{A(k) - D(k)}{p}, \tag{3.1}$$

where $A(k)$ and $D(k)$ denote the number of agreements resp. disagreements over a full period between $\{s_i\}_{i \geq 0}$ and $\{s_{i+k}\}_{i \geq 0}$, which is $\{s_i\}_{i \geq 0}$ shifted over k positions to the left. So

$$A(k) = |\{0 \leq i < p \mid s_i = s_{i+k}\}|,$$

$$D(k) = |\{0 \leq i < p \mid s_i \neq s_{i+k}\}|.$$

Note that one can also write $AC(k) = (2 . A(k) - p)/p$.

Example 3.1

Consider a sequence that is periodic with period p given by its first p elements.

With the Mathematica functions `Count`, `Length`, `Mod`, `RotateLeft`, *and* `Table` *one easily computes all values of the autocorrelation function $AC(k)$, $0 \leq k \leq p-1$.*

```
segment = {1, 1, 0, 1, 0, 0, 0, 0};
p = Length[segment];
Table[
 (2 * Count[Mod[segment - RotateLeft[segment, k], 2], 0] - p) / p,
 {k, 0, p-1}]
```

$$\left\{1, 0, 0, 0, -\frac{1}{2}, 0, 0, 0\right\}$$

If k is a multiple of p one has that $A(k) = p$, $D(k) = 0$, so AC = 1. One speaks of the *in-phase* autocorrelation.

If p does not divide k, one speaks of the *out-of-phase* autocorrelation. The value of AC now lies between -1 and $+1$.

Definition 3.2   *Golombs Randomness Postulates*

G1: The number of zeros and the number of ones are as equal as possible per period, i.e. both are $p/2$ if p is even and they are $(p \pm 1)/2$ if p is odd.

G2: Half of the runs in a cycle have length 1, one quarter of the runs have length 2, one eight of the runs have length 3, and so forth. Moreover half of the runs of a certain length are gaps, the other half are blocks.

G3: The out-of-phase autocorrelation $AC(k)$ has the same value for all values of k.

G1 states that zeros and ones occur with roughly the same probability. One can count these occurrences quite easily with the *Mathematica* function `Count`.

```
segment = {1, 1, 0, 1, 0, 0, 0, 0};
Count[segment, 0]
Count[segment, 1]
```

```
5
```

```
3
```

G2 implies that after 011 the symbol 0 (leading to a block of length 2) has the same probability as the symbol 1 (leading to a block of length ≥ 3), etc. So, **G2** says that certain n-grams occur with the right frequencies. These frequencies can be computed by means of the *Mathematica* functions `Count`, `Length`, `RotateLeft`, `Table`, and `Take`.

```
segment = {0, 1, 1, 0, 1, 0, 0, 0, 1, 1, 0, 0, 0, 1, 0, 1, 1};
p = Length[segment];
ngram = {1, 0, 1}; k = Length[ngram];
Count[
  Table[Take[RotateLeft[segment, i], k] == ngram, {i, p}], True]
```

```
3
```

The interpretation of **G3** is more difficult. It does say that counting the number of agreements between a sequence and a shifted version of that sequence does not give any information about the period of that sequence, unless one shifts over a multiple of the period. A related situation is described in Lemma 2.1, where such a comparison made it possible to determine the length of the key used in the Vigenère cipher. In cryptographic applications p will be too large for such an approach.

> **Lemma 3.1**
> Let $\{s_i\}_{i \geq 0}$ be a binary sequence with period p, $p > 2$, which satisfies Golomb's randomness postulates.
> Then p is odd and $AC(k)$ has the value $-1/p$ when k is not divisible by p.

Proof: Consider a $p \times p$ cyclic matrix with top row $s_0, s_1, \ldots, s_{p-1}$. We shall count in two different ways the sum of all the agreements minus the disagreements between the top row and all the other rows. Counting rowwise we get by G3 for each row i, $2 \leq i \leq p$, the same contribution $p.AC(k)$. This gives a total value of $p(p-1).AC(k)$.

We shall now evaluate the above sum, by counting columnwise, the number of agreements minus the number of disagreements between all lower entries with the top entries.

<u>Case: p even.</u>

By G1, the contribution of each column will be $(p/2 - 1) - p/2 = -1$, since each column counts exactly $p/2 - 1$ agreements of a lower entry with the top entry and exactly $p/2$ disagreements. Summing this value over all columns gives $-p$ for the total sum. Equating the two values yields $(p-1)AC(k) = -1$. However, Equation (3.1) implies that $p.AC(k)$ is an integer. This is not possible when $AC(k) = -1/(p-1)$, unless $p = 2$.

<u>Case: p odd.</u>

One gets for $(p+1)/2$ columns the contribution $(p-1)/2 - (p-1)/2$, which is 0, and for $(p-1)/2$ columns the contribution $(p-3)/2 - (p+1)/2$, which is -2. Hence one obtains the value $-(p-1)$ for the summation. Putting this equal to $p(p-1).AC(k)$ yields the value $AC(k) = -1/p$.

□

The well known χ^2-test and the spectral test, [CovM67], yields ways to test the pseudo-randomness properties of a given sequence. We shall not discuss these methods here. The interested reader is referred to [Golo67], Chapter IV, [Knut81], Chapter 3, or Maurer's universal statistical test [Maur92].

There are also properties of a cryptographic nature which the sequence $\{s_i\}_{i \geq 0}$ in Figure 3.1 should satisfy.

C1: The period p of $\{s_i\}_{i \geq 0}$ has to be taken very large (about the order of magnitude of 10^{50}).

C2: The sequence $\{s_i\}_{i \geq 0}$ should be easy to generate.

C3: Knowledge of part of the plaintext with corresponding ciphertext should not enable a cryptanalist to generate the whole $\{s_i\}_{i \geq 0}$-sequence (known plaintext attack).

3.2 Linear Feedback Shift Registers

3.2.1 (Linear) Feedback Shift Registers

Feedback shift registers are very fast implementations to generate binary sequences. Their general form is depicted in Figure 3.2.

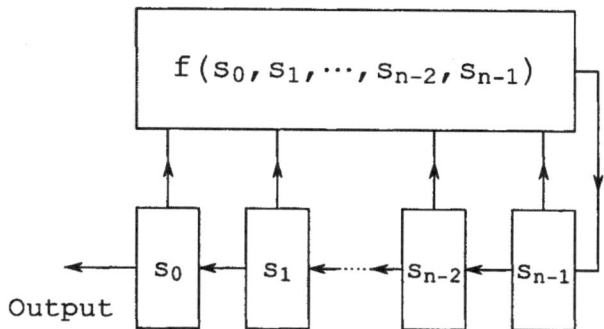

General Form of a Feedback Shift Register

Figure 3.2

A *feedback shift register* (FSR) of *length n* contains n memory cells, which together form the (beginning) *state* $(s_0, s_1, ..., s_{n-1})$ of the shift register. The function f is a mapping of $\{0, 1\}^n$ in $\{0, 1\}$ and is called the *feedback function* of the register. Since f can be represented as a Boolean function, it can easily be made with elementary logical functions.

After the first time unit, the shift register will output s_0 and go to state $(s_1, s_2, ..., s_n)$, where $s_n = f(s_0, s_1, ..., s_{n-1})$.

Continuing in this way, the shift register will generate an infinite sequence $\{s_i\}_{i \geq 0}$.

Example 3.2

Consider the case that $n = 3$ and that f is given by $f(s_0, s_1, s_2) = s_0 s_1 + s_2$. Starting with an initial state (s_0, s_1, s_2), one can quite easily determine the successive states with the Mathematica functions Mod, Do, *and* Print *as follows:*

```
Clear[f];
f[x_, y_, z_] := Mod[x * y + z, 2];
{s0, s1, s2} = {0, 1, 1};
Do[  {s0, s1, s2} = {s1, s2, f[s0, s1, s2]};
         Print[{s0, s1, s2}], {i, 1, 6}]
```

{1, 1, 1}

{1, 1, 0}

{1, 0, 1}

{0, 1, 1}

{1, 1, 1}

{1, 1, 0}

In this section, we shall study the special case that f is a linear function, say:

$$f(s_0, s_1, \ldots, s_{n-1}) = c_0 s_0 + c_1 s_1 + \ldots + c_{n-1} s_{n-1},$$

where all the c_i's are binary and all the additions are taken modulo 2.

The general picture of a *linear feedback shift register*, which we shall shorten to *LFSR*, is depicted in the figure below.

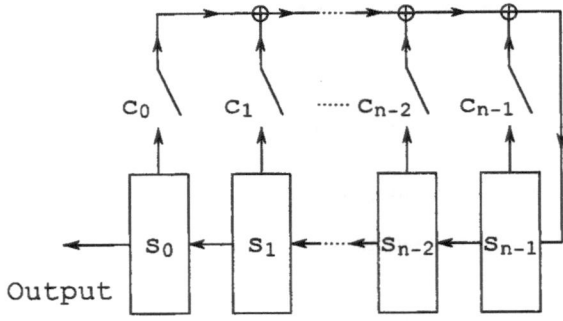

General linear feedback shift register (LFSR)

Figure 3.3

The output sequence $\{s_i\}_{i \geq 0}$ of such a LFSR can be described by the starting state $(s_0, s_1, \ldots, s_{n-1})$ and the linear recurrence relation:

$$s_{k+n} = \sum_{i=0}^{n-1} c_i s_{k+i}, \qquad k \geq 0. \tag{3.2}$$

or, equivalently

$$\sum_{i=0}^{n} c_i s_{k+i} = 0, \qquad k \geq 0. \tag{3.3}$$

where $c_n = 1$ by definition. Let $\underline{s}^{(i)}$ denote the state at time i, i.e. $\underline{s}^{(i)} = (s_i, s_{i+1}, \ldots, s_{i+n-1})$. Then, similarly to (3.2) one has the following recurrence relation for the successive states of the LFSR:

$$\underline{s}^{(k+n)} = \sum_{i=0}^{n-1} c_i \underline{s}^{(k+i)}, \ k \geq 0. \tag{3.4}$$

The coefficients c_i in (3.2) and Figure 3.3 are called the *feedback coefficients* of the LFSR. If $c_i = 0$ then the corresponding switch in Figure 3.3 is open, while if $c_i = 1$ this switch is closed. We shall always assume that $c_0 = 1$, because otherwise the output sequence $\{s_i\}_{i \geq 0}$ is just a delayed version of a sequence, generated by a LFSR with its c_0 equal to 1.

As a consequence, any state of the LFSR not only has a unique successor state, as is natural, but also has a unique predecessor. Indeed, for any $k \geq 0$ the value of s_k is uniquely determined by $s_{k+1}, \ldots, s_{k+n}$ by means of (3.2). Later on (in Thm. 3.22) we shall prove this property in a more general situation.

Example 3.3

With $n = 4$, $c_0 = c_1 = 1$, $c_2 = c_3 = 0$, we get the following LFSR:

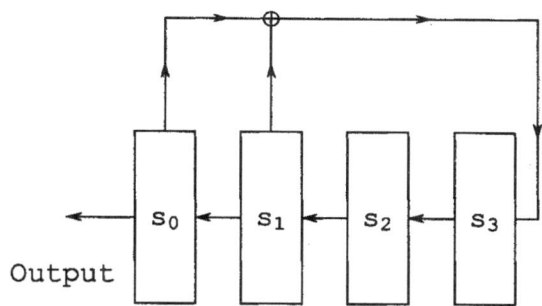

Example of LFSR with $n = 4$.

Figure 3.4

With starting state (1,0,0,0) one gets the subsequent list of successive states:

```
{s0, s1, s2, s3} = {1, 0, 0, 0}
Do[{s0, s1, s2, s3} = {s1, s2, s3, Mod[s0 + s1, 2]};
      Print[i, "      ", {s0, s1, s2, s3}], {i, 15}]
```

```
{1, 0, 0, 0}

 1      {0, 0, 0, 1}

 2      {0, 0, 1, 0}

 3      {0, 1, 0, 0}

 4      {1, 0, 0, 1}
```

5	$\{0, 0, 1, 1\}$
6	$\{0, 1, 1, 0\}$
7	$\{1, 1, 0, 1\}$
8	$\{1, 0, 1, 0\}$
9	$\{0, 1, 0, 1\}$
10	$\{1, 0, 1, 1\}$
11	$\{0, 1, 1, 1\}$
12	$\{1, 1, 1, 1\}$
13	$\{1, 1, 1, 0\}$
14	$\{1, 1, 0, 0\}$
15	$\{1, 0, 0, 0\}$

Note that the state at $t = 15$ is identical to the state at $t = 0$, so the output sequence $\{s_i\}_{i \geq 0}$ has period 15.

One can easily determine the output sequence of a LFSR with the *Mathematica* Functions `Table`, `Mod`, and `Do` as follows:

```
Clear[s]; {s[0], s[1], s[2], s[3]} = {1, 0, 0, 0};
s[j_] := Mod[ s[j - 4] + s[j - 3], 2]
Table[s[j], {j, 0, 15}]
```

```
{1, 0, 0, 0, 1, 0, 0, 1, 1, 0, 1, 0, 1, 1, 1, 1}
```

Since there are precisely $2^n - 1$ different states in a LFSR of length n and the all-zero state always goes over into itself, one can conclude that the period of $\{s_i\}_{i \geq 0}$ will never exceed $2^n - 1$.

3.2.2 PN-Sequences

Definition 3.3
A *PN-sequence* or pseudo-noise sequence is an output sequence of an n-stage LFSR with period $2^n - 1$.

If an n-stage LFSR does not run cyclically through all $2^n - 1$ non-zero states, it certainly does not generate a PN-sequence. As a consequence we have the following theorem.

Lemma 3.2
An n-stage LFSR that generates a PN-sequence $\{s_i\}_{i \geq 0}$ runs cyclically through all $2^n - 1$ non-zero states.
Any non-zero output sequence of this LFSR is a shift of $\{s_i\}_{i \geq 0}$.

We want to classify all LFSR's which generate PN-sequences. To this end, we associate with an LFSR with feedback coefficients $c_0, c_1, \ldots, c_{n-1}$ its *characteristic polynomial* $f(x)$, which is defined as follows:

$$f(x) = c_0 + c_1 x + \ldots + c_{n-1} x^{n-1} + x^n = \sum_{i=0}^{n} c_i x^i, \tag{3.5}$$

where $c_n = 1$ by definition and $c_0 = 1$ by assumption.

Definition 3.4
Let $f = \sum_{i=0}^{n} c_i x^i$. Then

$$\Omega(f) = \{ \{s_i\}_{i \geq 0} \mid \{s_i\}_{i \geq 0} \text{ satisfies (3.2)} \}.$$

In words, $\Omega(f)$ is the set of all output sequences of the LFSR with characteristic polynomial $f(x)$.

Lemma 3.3
Let f be the characteristic polynomial of an n-stage LFSR. Then $\Omega(f)$ is a binary vector space of dimension n.

Proof: Since (3.2) is a linear recurrence relation, $\Omega(f)$ obviously is a linear vectorspace. Also, each $\{s_i\}_{i \geq 0}$ in $\Omega(f)$ is uniquely determined by its first n entries $s_0, s_1, \ldots, s_{n-1}$ (the beginning state), so the dimension of $\Omega(f)$ is at most n. On the other hand, the n different sequences starting with

$$\overset{i}{\overline{00 \ldots 0}} \, 1 \, \overset{n-i-1}{\overline{00 \ldots \ldots 00}},$$

$0 \leq i \leq n - 1$, are clearly independent. So, the dimension of $\Omega(f)$ is at least n.

$\square$

Let f be a polynomial of degree n, say $f(x) = \sum_{i=0}^{n} c_i x^i$ with $c_n \neq 0$. Then, the *reciprocal polynomial* of $f(x)$ is defined by

$$f^*(x) = x^n f(1/x) = c_0 x^n + c_1 x^{n-1} + \ldots + c_{n-1} x + c_n = \sum_{i=0}^{n} c_{n-i} x^i, \tag{3.6}$$

With a sequence $\{s_i\}_{i \geq 0}$ we associate the *power series* (also called *generating function*)

$$S(x) = \sum_{i=0}^{\infty} s_i x^i. \tag{3.7}$$

Instead of writing $\{s_i\}_{i \geq 0} \in \Omega(f)$, we shall also use the notation $S(x) \in \Omega(f)$. We know that $S(x)$ is uniquely determined by the beginning state $(s_0, s_1, \ldots, s_{n-1})$ and the characteristic polynomial $f(x)$. In the following theorem and corollary, we shall now make this dependency more explicit.

Theorem 3.4

Let $\{s_i\}_{i \geq 0} \in \Omega(f)$, with f given by (3.5). Further, let $S(x)$ be the generating function of $\{s_i\}_{i \geq 0}$. Then, $S(x) f^*(x)$ is a polynomial of degree less than n.

Proof:

$$S(x) f^*(x) \stackrel{(3.6) \& (3.7)}{=} \left(\sum_{k=0}^{\infty} s_k x^k\right).\left(\sum_{l=0}^{n} c_{n-l} x^l\right) = \sum_{j=0}^{\infty} \left(\sum_{l=0}^{\min(j,n)} c_{n-l} s_{j-l}\right) x^j =$$

$$\sum_{j=0}^{n-1} \left(\sum_{l=0}^{j} c_{n-l} s_{j-l}\right) x^j + \sum_{j=n}^{\infty} \left(\sum_{l=0}^{n} c_{n-l} s_{j-l}\right) x^j =$$

$$\sum_{j=0}^{n-1} \left(\sum_{l=0}^{j} c_{n-l} s_{j-l}\right) x^j + \sum_{j=n}^{\infty} \left(\sum_{i=0}^{n} c_i s_{(j-n)+i}\right) x^j \stackrel{(3.3)}{=}$$

$$\sum_{j=0}^{n-1} \left(\sum_{l=0}^{j} c_{n-l} s_{j-l}\right) x^j.$$

□

Remark:

Note that the proof above implies that $S(x) = \frac{u(x)}{f^*(x)}$ with $u(x) = \sum_{j=0}^{n-1} \left(\sum_{l=0}^{j} c_{n-l} s_{j-l}\right) x^j$. This polynomial is of degree $< n$ and has coefficients depending on the initial state and the characteristic polynomial.

Note also that the mapping $S(x) \longrightarrow S(x) f^*(x)$ is one-to-one since $f^*(x) \neq 0$.

Example 3.4

Consider the LFSR with $n=5$, $f(x) = 1 + x^2 + x^5$ and take as beginning state $(1,1,0,1,0)$. Then $u(x)$ can be computed with the Mathematica function PolynomialMod *as follows:*

```
{c[0], c[1], c[2], c[3], c[4], c[5]} = {1, 0, 1, 0, 0, 1};
{s[0], s[1], s[2], s[3], s[4]} = {1, 1, 0, 1, 0};

u = PolynomialMod[ sum_{j=0}^{4} sum_{l=0}^{j} c[5 - l] s[j - l] x^j, 2]
```

```
1 + x + x^4
```

To check Theorem 3.4 up to some term x^L, we use (3.2) to compute the s_i's up to L (here we use the Mathematica functions Mod, Print, *and* PolynomialMod*):*

```
{c[0], c[1], c[2], c[3], c[4], c[5]} = {1, 0, 1, 0, 0, 1};
{s[0], s[1], s[2], s[3], s[4]} = {1, 1, 0, 1, 0};
            5
fstar = ∑ c[5 - i] x^i;
           i=0

L = 60;
s[i_] := s[i] = Mod[s[i - 5] + s[i - 3], 2];
        L
S = ∑ s[i] x^i; Print[S];
       i=0

PolynomialMod[S * fstar, {x^2, 2}]
```

$$1 + x + x^3 + x^5 + x^{10} + x^{13} + x^{15} + x^{16} + x^{19} + x^{20} + x^{21} + x^{22} + x^{23} + x^{27} + x^{28} + x^{30} +$$
$$x^{31} + x^{32} + x^{34} + x^{36} + x^{41} + x^{44} + x^{46} + x^{47} + x^{50} + x^{51} + x^{52} + x^{53} + x^{54} + x^{58} + x^{59}$$

$$1 + x + x^4$$

Note that the output is indeed the same as above.

Corollary 3.5

$$\Omega(f) = \left\{ \frac{u(x)}{f^*(x)} \mid \text{degree}(u(x)) < n \right\}.$$

Remark: Writing $S(x) = u(x)/f^*(x)$ means the same as $S(x) f^*(x) = u(x)$.

Proof: From Theorem 3.4 and the remark below it we know that each member of $\Omega(f)$ can be written as $u(x)/f^*(x)$ with degree$(u(x)) < n$ and we know that this $u(x)$ is unique. This proves the $\subset$-inclusion.

On the other hand, $\Omega(f)$ has cardinality 2^n by Lemma 3.3 and there are also exactly 2^n binary polynomials $u(x)$ of degree $< n$.

$\square$

It is now easy to prove the following lemma.

Lemma 3.6
Let f and g be two (characteristic) polynomials and let $\{s_i\}_{i \geq 0} \in \Omega(f)$ and $\{t_i\}_{i \geq 0} \in \Omega(g)$. Let lcm$[f, g]$ denote the least common multiple of f and g. Then

$$\{s_i + t_i\}_{i \geq 0} \in \Omega(\text{lcm}[f, g]).$$

Proof: Write $h = \text{lcm}[f, g]$ and $h = a.f$ and $h = b.g$. Let $S(x)$ and $T(x)$ be the generating functions of $\{s_i\}_{i \geq 0}$, resp. $\{t_i\}_{i \geq 0}$.

Corollary 3.5 implies that $S(x) = u(x)/f^*(x)$ and $T(x) = v(x)/g^*(x)$, where degree$(u(x)) <$ degree$(f(x))$ and degree$(v(x)) <$ degree$(g(x))$. Since

$$S(x) + T(x) = \frac{u(x)}{f^*(x)} + \frac{v(x)}{g^*(x)} = \frac{a^*(x)\,u(x)}{a^*(x)\,f^*(x)} + \frac{b^*(x)\,v(x)}{b^*(x)\,g^*(x)} = \frac{a^*(x)\,u(x)+b^*(x)\,v(x)}{h^*(x)},$$

and both $a^*(x)\,u(x)$ as well as $b^*(x)\,v(x)$ have degree less than degree$(h(x))$, it follows that $S(x) + T(x) \in \Omega(h)$.

□

3.2.3 Which Characteristic Polynomials give PN-Sequences?

The *period* of a polynomial f with $f(0) \neq 0$, is the smallest positive m such that $f(x)$ divides $x^m - 1$, i.e. the smallest positive m such that $x^m \equiv 1 \pmod{f(x)}$. It is well defined, since the sequence of successive powers of x, reduced modulo $f(x)$, has to be periodic. Indeed, if $x^i \equiv x^j \pmod{f(x)}$ and $0 < i < j$ then also $x^{i-1} \equiv x^{j-1} \pmod{f(x)}$, because $\gcd(x, f(x)) = 1$. (The term x has a multiplicative inverse by Corollary B.14, so we can indeed divide by x.) We can repeat this process until we get $1 \equiv x^{j-i} \pmod{f(x)}$.

Example 3.5

Let $f(x) = 1 + x^4 + x^5$. Its period can be computed with the Mathematica functions `While` *and* `PolynomialMod` *in the way described above. So, starting with x (trying m = 1), we compute the successive powers of x by multiplying the previous power by x (this amounts to a cyclic shift), and then reducing the answer modulo f(x), until we arrive at the outcome 1.*

```
f = 1 + x^4 + x^5; m = 1; u = x;
While[u =!= 1, u = PolynomialMod[x*u, {f, 2}]; m = m + 1]
m
```

```
21
```

It follows from Theorem B.35 that a binary, irreducible polynomial of degree n divides $x^{2^n-1} - 1$, so it also follows that the period m of such a polynomial will divide $2^n - 1$.

(This observation can be used to determine the period of a polynomial more efficiently, however we shall not discuss that technique at this moment. See the end of Example 8.2)

Lemma 3.7
Let $\{s_i\}_{i\geq 0} \in \Omega(f)$, where f is a polynomial of degree n and period m. Then $\{s_i\}_{i\geq 0}$ has a period dividing m.

Proof: Write $x^m - 1 = f(x)\,g(x)$. Taking the reciprocal on both sides gives $x^m - 1 = f^*(x)\,g^*(x)$. By Corollary 3.5, there exists a polynomial $u(x)$ of degree $< n$ such that

$$S(x) = \frac{u(x)}{f^*(x)} = \frac{u(x)\,g^*(x)}{f^*(x)\,g^*(x)} = \frac{u(x)\,g^*(x)}{(1-x^m)} = u(x)\,g^*(x)\,(1 + x^m + x^{2m} + \ldots)$$

Since degree$(u(x) g^*(x)) <$ degree$(f^*(x) g^*(x)) =$ degree$(x^m - 1) = m$, we see that $S(x)$ must have period m or a divisor of it.

□

> **Lemma 3.8**
> Let $\{s_i\}_{i \geq 0} \in \Omega(f)$, where f is an irreducible polynomial of degree n and period m. Then $\{s_i\}_{i \geq 0}$ also will have period m.

Proof: Let $\{s_i\}_{i \geq 0}$ have period p. By Lemma 3.7, p divides m. Let $S^{(p)}(x) = \sum_{i=0}^{p-1} s_i x^i$. It follows that

$$S(x) = S^{(p)}(x) (1 + x^p + x^{2p} + \ldots) = \frac{S^{(p)}(x)}{1 - x^p},$$

while on the other hand, $S(x) = u(x) / f^*(x)$ by Corollary 3.5. Equating these two

expressions yields

$$S^{(p)}(x) f^*(x) = u(x) (x^p - 1)$$

and thus

$$(S^{(p)}(x))^* f(x) = u^*(x) (x^p - 1).$$

Since $f(x)$ is irreducible of degree n and degree$(u(x)) < n$, it follows that $f(x)$ divides $(x^p - 1)$. So, m, the period of $f(x)$, must divide p. We conclude that $p = m$.

□

Example 3.6

Consider the irreducible polynomial $f(x) = 1 + x + x^2 + x^3 + x^4$, which has period 5, since $(x^5 - 1) = (x - 1) f(x)$. Output sequences in $\Omega(f)$ also have period 5, by the above lemma, as can easily be checked.

```
{s0, s1, s2, s3} = {1, 1, 0, 0}
Do[{s0, s1, s2, s3} = {s1, s2, s3, Mod[s0 + s1 + s2 + s3, 2]};
    Print[i, "    ", {s0, s1, s2, s3}], {i, 5}]
```

```
{1, 1, 0, 0}
```

```
1      {1, 0, 0, 0}

2      {0, 0, 0, 1}

3      {0, 0, 1, 1}

4      {0, 1, 1, 0}

5      {1, 1, 0, 0}
```

A roundabout way to find an irreducible polynomial of degree n is to factor $x^{2^n-1} - 1$ by means of the *Mathematica* function `Factor`:

```
n = 5;
Factor[x^(2^n-1) - 1, Modulus -> 2]
```

```
(1 + x) (1 + x^2 + x^5) (1 + x^3 + x^5) (1 + x + x^2 + x^3 + x^5)
    (1 + x + x^2 + x^4 + x^5) (1 + x + x^3 + x^4 + x^5) (1 + x^2 + x^3 + x^4 + x^5)
```

In *Mathematica* one can find an irreducible polynomial over $\mathbb{F}_p$, p prime, with the function `IrreduciblePolynomial` for which the package `Algebra`FiniteFields`` needs to be loaded first.

```
<< Algebra`FiniteFields`
```

```
p = 2; deg = 11;
IrreduciblePolynomial[x, p, deg]
```

```
1 + x^9 + x^11
```

Lemma 3.9
Let $\{s_i\}_{i \geq 0}$ be a PN-sequence, generated by a LFSR with characteristic polynomial f. Then f is irreducible.

Proof: Write $f = f_1 f_2$ with f_1 irreducible, say of degree $n_1 > 0$.

By Corollary 3.5, the sequence $1/f_1^*(x) \in \Omega(f_1)$, so the period of $1/f_1^*(x)$ divides $2^{n_1} - 1$ by Lemma 3.7 and Theorem B.35.

On the other hand, $1/f_1^*(x) = f_2^*(x)/f^*(x) \in \Omega(f)$, so by Lemma 3.2 $1/f_1^*(x)$ is a cyclic shift of $\{s_i\}_{i \geq 0}$ and thus its period is $2^n - 1$. This is only possible if $n = n_1$, i.e. if $f(x)$ is equal to the irreducible factor $f_1(x)$.

$\square$

Example 3.7

Consider $f(x) = (1 + x + x^2)(1 + x + x^3) = 1 + x^4 + x^5$. It is easy to check that $1 + x + x^2$ divides $x^3 - 1$ and that $1 + x + x^3$ divides $x^7 - 1$. Since 3 and 7 are relatively prime, it follows that $f(x)$ divides $x^{21} - 1$. We conclude that each output sequence has a period dividing 21.

This can be checked for different beginning states as follows.

```
{s0, s1, s2, s3, s4} = {1, 0, 0, 0, 0}
Do[{s0, s1, s2, s3, s4} = {s1, s2, s3, s4, Mod[s0 + s4, 2]};
    Print[i, "      ", {s0, s1, s2, s3, s4}], {i, 21}]
```

{1, 0, 0, 0, 0}

1	{0, 0, 0, 0, 1}
2	{0, 0, 0, 1, 1}
3	{0, 0, 1, 1, 1}
4	{0, 1, 1, 1, 1}
5	{1, 1, 1, 1, 1}
6	{1, 1, 1, 1, 0}
7	{1, 1, 1, 0, 1}
8	{1, 1, 0, 1, 0}
9	{1, 0, 1, 0, 1}
10	{0, 1, 0, 1, 0}
11	{1, 0, 1, 0, 0}
12	{0, 1, 0, 0, 1}
13	{1, 0, 0, 1, 1}
14	{0, 0, 1, 1, 0}
15	{0, 1, 1, 0, 0}
16	{1, 1, 0, 0, 0}
17	{1, 0, 0, 0, 1}
18	{0, 0, 0, 1, 0}
19	{0, 0, 1, 0, 0}
20	{0, 1, 0, 0, 0}
21	{1, 0, 0, 0, 0}

The reader may want to try the beginning state (1, 1, 1, 0, 0) and see what the period of the output sequence is. This output sequence could also have been generated with the LFSR with characteristic polynomial $1 + x + x^3$ and beginning state (1, 1, 1) (see also Example 3.11).

We are now able to prove the main result of this subsection. We remind the reader of the definitio of a primitive polynomial (of degree n), which is an irreducible polynomial with the property that is a primitive element in $GF(2)[x]/(f(x))$. This translates directly into the equivalent property tha $f(x)$ has (full) period $2^n - 1$.

> **Theorem 3.10**
> A non-zero output sequence of a LFSR with characteristic polynomial $f(x)$ is a PN-sequence if and only if $f(x)$ is a primitive polynomial.

Proof: Let $f(x)$ have degree n.

$\Longrightarrow$ Let $\{s_i\}_{i \geq 0} \in \Omega(f)$ be a PN-sequence. It follows from Lemma 3.9 that $f(x)$ must be irreducible. Lemma 3.8 in turn implies that $f(x)$ must have period $2^n - 1$, which makes it a primitive polynomial.

$\Longleftarrow$ If $f(x)$ is primitive, it certainly is irreducible. By Lemma 3.8, $\{s_i\}_{i \geq 0}$ has the same period as $f(x)$ has, which is $2^n - 1$. It follows that $\{s_i\}_{i \geq 0}$ is a PN-sequence.

$\square$

Mathematica finds a primitive polynomial of degree m over $\mathbb{F}_p$ in the variable x by means of the `FieldIrreducible` function.

```
m = 5; p = 2;
FieldIrreducible[GF[p, m], x]
```

$$1 + x^3 + x^5$$

Let us check that this polynomial indeed defines a PN sequence.

```
{s0, s1, s2, s3, s4} = {1, 0, 0, 0, 0}
Do[{s0, s1, s2, s3, s4} = {s1, s2, s3, s4, Mod[s0 + s3, 2]};
     Print[i, "     ", {s0, s1, s2, s3, s4}], {i, 31}]
```

```
{1, 0, 0, 0, 0}

1     {0, 0, 0, 0, 1}
2     {0, 0, 0, 1, 0}
3     {0, 0, 1, 0, 1}
4     {0, 1, 0, 1, 0}
5     {1, 0, 1, 0, 1}
6     {0, 1, 0, 1, 1}
```

7	{1, 0, 1, 1, 1}
8	{0, 1, 1, 1, 0}
9	{1, 1, 1, 0, 1}
10	{1, 1, 0, 1, 1}
11	{1, 0, 1, 1, 0}
12	{0, 1, 1, 0, 0}
13	{1, 1, 0, 0, 0}
14	{1, 0, 0, 0, 1}
15	{0, 0, 0, 1, 1}
16	{0, 0, 1, 1, 1}
17	{0, 1, 1, 1, 1}
18	{1, 1, 1, 1, 1}
19	{1, 1, 1, 1, 0}
20	{1, 1, 1, 0, 0}
21	{1, 1, 0, 0, 1}
22	{1, 0, 0, 1, 1}
23	{0, 0, 1, 1, 0}
24	{0, 1, 1, 0, 1}
25	{1, 1, 0, 1, 0}
26	{1, 0, 1, 0, 0}
27	{0, 1, 0, 0, 1}
28	{1, 0, 0, 1, 0}
29	{0, 0, 1, 0, 0}
30	{0, 1, 0, 0, 0}
31	{1, 0, 0, 0, 0}

To find all primitive polynomials of degree n one can factor the cyclotomic polynomial $Q^{(2^n-1)}(x$ (see Definition B.19). With the *Mathematica* functions Factor and Cyclotomic this goes ε follows.

```
p = 2; m = 6; n = p^m - 1;
Factor[Cyclotomic[n, x], Modulus -> 2]
```

$$(1 + x + x^6) \ (1 + x + x^3 + x^4 + x^6) \ (1 + x^5 + x^6)$$
$$(1 + x + x^2 + x^5 + x^6) \ (1 + x^2 + x^3 + x^5 + x^6) \ (1 + x + x^4 + x^5 + x^6)$$

The next corollary now follows directly from Theorem 3.10 and Theorem B.40.

> **Corollary 3.11**
> There are $\varphi(2^n - 1)/n$ different n-stage LFSR's generating PN-sequences.
> Here φ stands for Euler's totient function (Definition A.6).

The more or less exponential growth of $\varphi(2^n - 1)/n$ as function of n, makes it for moderate value of n already impossible for a cryptanalist to guess the right primitive polynomial or to check them all exhaustively.

With the *Mathematica* function `EulerPhi` one can easily verify this.

```
n = 100;
EulerPhi[2^n - 1] / n
```

5707676340000000000000000000000

3.2.4   An Alternative Description of $\Omega(f)$ for Irreducible f

We shall now solve recurrence relation (3.2) for the case that the corresponding characteristic polynomial $f = \sum_{i=0}^{n} c_i x^i$ is irreducible. This includes, of course, the case that f is primitive, for which we know that the corresponding LFSR outputs PN-sequences.

We follow the standard mathematical method for solving linear recurrence relations.

Substituting $s_j = A.\alpha^j$, for all $j \geq 0$, in $s_{k+n} = \sum_{i=0}^{n-1} c_i s_{k+i}$ leads to the equation

$$A.\alpha^{k+n} = \sum_{i=0}^{n-1} c_i.A.\alpha^{k+i}.$$

Here A and α are elements from an extension field of GF(2) that will be determined in a moment. Dividing the above relation by $A.\alpha^k$, one arrives at $\alpha^n = \sum_{i=0}^{n-1} c_i \alpha^i$, i.e.

$$f(\alpha) = 0.$$

We shall study the case that f is irreducible in more detail. The Galois Field GF(2^n)= GF(2)$[x]/(f(x))$ (see Theorem B.16) contains a zero of f as an element. Calling this zero α, we note that

$$\mathrm{GF}(2^n) = \{ \textstyle\sum_{i=0}^{n-1} a_i \alpha^i \mid a_i \in \mathrm{GF}(2), \ 0 \leq i < n \},$$

with the normal coefficient-wise addition and with the regular product rule (see (B.3) and (B.4)), but always reducing powers of α with an exponent $\geq n$ by means of the relation $\alpha^n = \sum_{i=0}^{n-1} c_i \alpha^n$ to an expression of degree $< n$ (as shown in the Example B.5, where the letter x is used instead of the symbol α.).

Example 3.8

Consider $f(x) = 1 + x + x^4$ and let α be a zero of $f(x)$, so $\alpha^4 = 1 + \alpha$.

Adding the elements $1 + \alpha + \alpha^3$ and $\alpha + \alpha^2$ in $GF(2)[x]/(f(x))$ gives $1 + \alpha^2 + \alpha^3$. Multiplication gives $\alpha + \alpha^3 + \alpha^4 + \alpha^5$ which is $(\alpha + 1) f(\alpha) + (1 + \alpha + \alpha^2 + \alpha^3)$, so the result is $1 + \alpha + \alpha^2 + \alpha^3$.

This could also have been computed with the Mathematica function `PolynomialMod`, *as follows:*

```
f = 1 + a + a^4;
PolynomialMod[(1 + a + a^3) + (a + a^2), {f, 2}]
PolynomialMod[(1 + a + a^3) * (a + a^2), {f, 2}]
```

```
1 + a^2 + a^3
```

```
1 + a + a^2 + a^3
```

Lemma 3.12

Let f be a binary, irreducible polynomial of degree n and let α be a zero of f in $GF(2^n)$. Further, let L be a non-trivial, linear mapping from $GF(2^n)$ to $GF(2)$. Then

$$\Omega(f) = \{ \{L(A.\alpha^j)\}_{j\geq 0} \mid A \in GF(2^n) \}.$$

Proof: We need to check several things.

i) The sequence $\{s_j\}_{j\geq 0} = \{L(A.\alpha^j)\}_{j\geq 0}$ clearly is a binary sequence, because L maps $GF(2^n)$ to $GF(2)$.

ii) The sequence $\{s_j\}_{j\geq 0} = \{L(A.\alpha^j)\}_{j\geq 0}$ satisfies (3.2). To see this, we check the equivalent condition (3.3). By the linearity of L and the relation $f(\alpha) = \sum_{i=0}^{n} c_i \alpha^i = 0$, it follows that

$$\sum_{i=0}^{n} c_i s_{k+i} = \sum_{i=0}^{n} c_i L(A.\alpha^{k+i}) = L(A.\alpha^k(\sum_{i=0}^{n} c_i \alpha^i)) = L(0) = 0.$$

iii) Each of the 2^n choices of $A \in GF(2^n)$ leads to a different binary solution of (3.3), as we shall now show. By Lemma 3.3, these must constitute all the elements in $\Omega(f)$.

Suppose that the sequences $\{L(A.\alpha^j)\}_{j\geq 0}$ and $\{L(B.\alpha^j)\}_{j\geq 0}$ are identical. It follows from $L(A.\alpha^j) = L(B.\alpha^j)$, $j \geq 0$, and the linearity of L that in particular $L((A - B).\alpha^j) = 0$ for $0 \leq j < n$. However, the elements $1, \alpha, ..., \alpha^{n-1}$ form a basis of $GF(2^n)$, because f is irreducible. It follows from the linearity of L that $L((A - B).\omega) = 0$ for each field element ω in $GF(2^n)$. Since L was a non-trivial mapping, we can conclude that $A = B$.

□

A convenient non-trivial linear mapping L from $GF(2^n)$ to $GF(2)$ to consider is the Trace function Tr, introduced in Problem B.16.

An alternative, is the projection of an element $\sum_{i=0}^{n-1} a_i \alpha^i$ to its constant term a_0.

Example 3.9

Take the irreducible polynomial $f(x) = x^4 + x + 1$ of degree 4 (it even is primitive) and let α a zero of $f(x)$, so $f(\alpha) = 0$. The Trace function is given by $Tr(x) = x + x^2 + x^4 + x^8$.

Any element $A \in GF(2^4) = \{ \sum_{i=0}^{3} a_i \alpha^i \mid a_i \in GF(2), \ 0 \le i \le 3 \}$ defines a unique binary sequence $\{s_j\}_{j \ge 0}$, defined by $s_j = Tr(A.\alpha^j)$. Below, we have taken $A = 1 + \alpha + \alpha^2$.

The output sequence, corresponding with any value of A, can be evaluated with the Mathematica functions `PolynomialMod` *and* `Table`, *as follows:*

```
n = 4; f = 1 + a + a^4; A = 1 + a + a^2;

          n-1
Tr[x_] := ∑ x^(2^i);
          i=0

s[j_] := PolynomialMod[Tr[A * a^j], {f, 2}];
Table[s[j], {j, 0, 2^n - 2}]
```

```
{1, 1, 0, 1, 0, 1, 0, 0, 0, 0, 1, 0, 0, 1, 0}
```

3.2.5 Cryptographic Properties of PN Sequences

We shall now investigate to which extent PN-sequences meet Golomb's randomness postulates G1-G3. After that, we check the cryptographic requirements C1-C3. As always, we let n denote the length of the LFSR.

Ad G1: By Lemma 3.2 each non-zero state occurs exactly once per period. The leftmost bit of each state will be the next output bit. So, the number of ones per period is 2^{n-1} and the number of zeros per period is $2^{n-1} - 1$, as the all-zero state does not occur.

Ad G2: There are $2^{n-(k+2)}$ states whose leftmost $k + 2$ coordinates are of the form $0 \overset{\overleftrightarrow{k}}{1 1 \ldots 1} 0$, resp. $1 \overset{\overleftrightarrow{k}}{0 0 \ldots 0} 1$. Thus, gaps and blocks of the length k, $k \le n - 2$, occur exactly $2^{n-(k+2)}$ times per period.

The state $0 \overset{\overleftrightarrow{n-1}}{1 1 \ldots 1}$ occurs exactly once. Its successor is the all-one state, which in turn is followed

$$\overset{n-1}{\overleftarrow{}}$$

by state $1\,1\ldots1\,0$. Therefore, there is no block of length $n-1$ and one block of length n.

Similarly, there is one gap of length $n-1$ and no gap of length n.

Ad G3: With $\{s_i\}_{i\geq0}\in\Omega(f)$ also $\{s_{i+k}\}_{i\geq0}\in\Omega(f)$ by Lemma 3.2. The linearity of $\Omega(f)$ implies that also $\{s_i+s_{i+k}\}_{i\geq0}\in\Omega(f)$ The number of agreements per period between $\{s_i\}_{i\geq0}$ and $\{s_{i+k}\}_{i\geq0}$ equal the number of zeros in one period of $\{s_i+s_{i+k}\}_{i\geq0}$ which is $2^{n-1}-1$ by Lemma 3.2 and G1 Similarly, the number of disagreements is 2^{n-1}. Thus, the out-of-phase autocorrelation $AC(k)$ i $-1/(2^n-1)$ for all $1\leq k<2^n-1$.

We conclude that PN-sequences meet Golomb's randomness postulates in a most satisfactory way Let us now check C1-C3.

Ad C1: Since the period of a PN-sequence generated by an n-stage LFSR is 2^n-1, one can easil get sufficient large periods. For instance, with $n=166$ the period is already about 10^{50}.

Ad C2: LFSR's are extremely simple to implement.

Ad C3: PN-sequences are very unsafe! Indeed, knowledge of $2n$ consecutive bits, sa $s_k,s_{k+1},\ldots,s_{k+2n-1}$, enables the cryptanalist to determine the feedback coefficients $c_0,c_1,\ldots,c_{n-}$ uniquely and thus the whole $\{s_i\}_{i\geq0}$-sequence. This follows from the matrix equation:

$$\begin{pmatrix} s_k & s_{k+1} & \cdots & \cdots & s_{k+n-1} \\ s_{k+1} & s_{k+2} & \cdots & \cdots & s_{k+n} \\ \cdot & \cdot & \cdots & \cdots & \cdot \\ \cdot & \cdot & \cdots & \cdots & \cdot \\ \cdot & \cdot & \cdots & \cdots & \cdot \\ s_{k+n-1} & s_{k+n} & \cdots & \cdots & s_{k+2n-2} \end{pmatrix} \begin{pmatrix} c_0 \\ c_1 \\ \cdot \\ \cdot \\ \cdot \\ c_{n-1} \end{pmatrix} = \begin{pmatrix} s_{k+n} \\ s_{k+n+1} \\ \cdot \\ \cdot \\ \cdot \\ s_{k+2n-1} \end{pmatrix}. \tag{3.8}$$

The above system has a unique solution as we shall now show. If n consecutive states of the LFSl exist that are linearly dependent, i.e. if n consecutive states span a $\leq(n-1)$ dimensional subspace then this remains so because of (3.4). This, however, contradicts the linear independence of stat $(0,0,\ldots,0,1)$ and its $n-1$ successor states. We conclude that any n consecutive states (and i particular the n rows in the matrix above) are linearly independent. Therefore, the unknow feedback coefficients $c_0,c_1\ldots,c_{n-1}$ can easily be determined.

Example 3.10

Assume that we know the following substring of length 10: 1,1,0,1,1,1,0,1,0,1. Assuming that $n=5$, we can solve (3.9) by means of the Mathematica function `LinearSolve` *as follows:*

```
     ( 1  1  0  1  1 )        ( 1 )
     ( 1  0  1  1  1 )        ( 0 )
m =  ( 0  1  1  1  0 ) ; b =  ( 1 ) ;
     ( 1  1  1  0  1 )        ( 0 )
     ( 1  1  0  1  0 )        ( 1 )

LinearSolve[m, b, Modulus -> 2]
```

```
{{1}, {0}, {1}, {0}, {0}}
```

The feedback coefficients are: $c_0 = 1$, $c_1 = 0$, $c_2 = 1$, $c_3 = 0$, $c_4 = 0$. One can check this quite easily with the Mathematica Functions Table*,* Mod*, and* Do *as follows:*

```
n = 5;
{c[0], c[1], c[2], c[3], c[4]} = {1, 0, 1, 0, 0};
{s[0], s[1], s[2], s[3], s[4]} = {1, 1, 0, 1, 1};
Do[s[k] = Mod[∑_{i=0}^{n-1} c[i] * s[k - n + i], 2], {k, n, 2^n}];
Table[s[k], {k, 0, 2^n - 2}]
```

```
{1, 1, 0, 1, 1, 1, 0, 1, 0, 1, 0, 0, 0, 0,
 1, 0, 0, 1, 0, 1, 1, 0, 0, 1, 1, 1, 1, 1, 0, 0, 0}
```

Of course, one does not know in general what the length n is of the LFSR in use. We shall address that problem in a more general setting in Subsection 3.3.1.

If only a string of $2n - 1$ consecutive bits of a PN-sequence is known, the feedback coefficients are not necessarily unique, as follows from the example $n = 4$ and the subsequence 1101011. This remains true even if we had used the additional information that $c_0 = 1$. Below we have added NullSpace to show the dependency in the linear relations.

```
m = (1 1 0 1)      (0)
    (1 0 1 0) ; b = (1) ;
    (0 1 0 1)      (1)
NullSpace[m, Modulus -> 2]
LinearSolve[m, b, Modulus -> 2]
```

```
{{0, 1, 0, 1}}
```

```
{{1}, {1}, {0}, {0}}
```

We have the solutions $(1, 1, 0, 0) + \lambda(0, 1, 0, 1)$ with $\lambda \in \{0, 1\}$.

Since sequences generated by LFSR's fail to meet requirement C3, the next step will be to study nonlinear shift registers. However, since so much is known about PN-sequences, it is quite natural that one tries to combine LFSR's in a non-linear way in order to get pseudo-random sequences with the right cryptographic properties.

3.3 Non-Linear Algorithms

3.3.1 Minimal Characteristic Polynomial

As already mentioned at the beginning of Section 3.1, any deterministic algorithm in a finite state machine will generate a sequence $\{s_i\}_{i \geq 0}$, which is ultimately periodic, say with period p. This means that, except for a beginning part, $\{s_i\}_{i \geq 0}$ will be generated in a trivial way by the LFSR with characteristic polynomial $1 + x^p$. Therefore, the sequence $\{s_i\}_{i \geq 0}$ which was possibly made in a non-linear way, can also be made by a LFSR (except for a finite beginning part). If this beginning part is non empty, not every state has a unique predecessor and the output sequence certainly will not have maximal period. We shall address this problem in Theorem 3.22. Here, we shall assume that the output sequence is periodic right from the start. The discussion above justifies the following definition.

> **Definition 3.5**
> The *linear complexity* (or *linear equivalence*) of a periodic sequence $\{s_i\}_{i \geq 0}$ is the length of the smallest LFSR that can generate $\{s_i\}_{i \geq 0}$.

The following two lemmas are needed to prove explicit statements about the linear complexity of periodic sequences.

> **Lemma 3.13**
> Let h and f be the characteristic polynomials of an m-stage, resp. n-stage LFSR. Then,
> $$\Omega(h) \subset \Omega(f) \iff h \mid f.$$

Proof:

$\Longrightarrow$ Since $1/h^* \in \Omega(h) \subset \Omega(f)$, it follows from Corollary 3.5 that a polynomial $u(x)$ of degree $< n$ exists, such that one has $1/h^*(x) = u(x)/f^*(x)$. We conclude that $f^*(x) = h^*(x) u(x)$ and thus that $f(x) = h(x) u^*(x)$, which means that $h \mid f$.

$\Longleftarrow$ Writing $f(x) = a(x) h(x)$ with $\mathrm{degree}(a(x)) = n - m$, one has by the same Corollary 3.5 that

$$\Omega(h) \quad = \left\{ \frac{v(x)}{h^*(x)} \;\middle|\; \mathrm{degree}(v(x)) < m \right\} = \left\{ \frac{a^*(x) v(x)}{a^*(x) h^*(x)} \;\middle|\; \mathrm{degree}(v(x)) < m \right\}$$

$$= \left\{ \frac{a^*(x) v(x)}{f^*(x)} \;\middle|\; \mathrm{degree}(a^*(x) v(x)) < n \right\} \subset \Omega(f).$$

$\square$

Example 3.11

The sequence $\{s_i\}_{i \geq 0} = 100101110 \ldots$ is the output sequence of the LFSR with $h(x) = 1 + x + x^3$ and beginning state $(1, 0, 0)$, as can be checked by

```
n = 3;
{s[0], s[1], s[2]} = {1, 0, 0};
{c[0], c[1], c[2]} = {1, 1, 0};
Do[s[k] = Mod[∑_{i=0}^{n-1} c[i] * s[k - n + i], 2], {k, n, 2^n}];
Table[s[k], {k, 0, 2^n}]
```

```
{1, 0, 0, 1, 0, 1, 1, 1, 0}
```

However, since $h(x)(1 + x + x^2) = 1 + x^4 + x^5$, the same output sequence can also be obtained from the LFSR with characteristic polynomial $f(x) = 1 + x^4 + x^5$ (see also Example 3.7). As beginning state one now has to take the first five terms of $\{s_i\}_{i \geq 0}$.

```
n = 5;
{s[0], s[1], s[2], s[3], s[4]} = {1, 0, 0, 1, 0};
{c[0], c[1], c[2], c[3], c[4]} = {1, 0, 0, 0, 1};
Do[s[k] = Mod[∑_{i=0}^{n-1} c[i] * s[k - n + i], 2], {k, n, 2^n}];
Table[s[k], {k, 0, 2^n}]
```

```
{1, 0, 0, 1, 0, 1, 1, 1, 0, 0, 1, 0, 1, 1, 1,
 0, 0, 1, 0, 1, 1, 1, 0, 0, 1, 0, 1, 1, 1, 0, 0, 1, 0}
```

Let $\{s_i\}_{i \geq 0} \in \Omega(f)$ for some f and suppose that one is looking for a polynomial h of smallest degree such that $\{s_i\}_{i \geq 0} \in \Omega(h)$. Then, Lemma 3.13 suggests to check the divisors of f. That this is sufficient will be proved later. The next lemma says when one does not need to check the divisors of f.

Lemma 3.14
Let $\{s_i\}_{i \geq 0} \in \Omega(f)$ and $S(x) = u(x) / f^*(x)$. Then,

$$\exists_{h \mid f, \, h \neq f} \left[\{s_i\}_{i \geq 0} \in \Omega(h) \right] \quad \Longleftrightarrow \quad \gcd(u(x), f^*(x)) \neq 1.$$

Proof: Let $d(x)$ divide $\gcd(u(x), f^*(x))$ with degree($d(x)$) > 1.

Then, $S(x) = \frac{u(x)}{f^*(x)} = \frac{u(x)/d(x)}{f^*(x)/d(x)}$, so $\{s_i\}_{i \geq 0} \in \Omega(f/d^*)$. It follows that there exists a proper divisor h

of f, namely f/d^* with $\{s_i\}_{i\geq 0} \in \Omega(h)$.

The proof in the reverse direction goes exactly the same.

$\square$

> **Theorem 3.15**
> Let $\{s_i\}_{i\geq 0}$ be a binary, periodic sequence, say with period p. Let the first p terms of $\{s_i\}_{i\geq 0}$ be given by $S^{(p)}(x) = s_0 + s_1 x + \ldots + s_{p-1} x^{p-1}$.
> Then there exists a unique polynomial $m(x)$ with the following two properties:
> i) $\{s_i\}_{i\geq 0} \in \Omega(m)$,
> ii) $\forall_h \left[\{s_i\}_{i\geq 0} \in \Omega(h) \Longrightarrow m \mid h \right]$.
> The reciprocal $m^*(x)$ of $m(x)$ is given by
> $$m^*(x) = \frac{1-x^p}{\gcd(S^{(p)}(x), 1-x^p)}.$$
> The polynomial $m(x)$ is called the *minimal characteristic polynomial* of $\{s_i\}_{i\geq 0}$.

Example 3.12

Let $\{s_i\}_{i\geq 0}$ have period 15 and let $S^{(15)}(x) = 1 + x^4 + x^7 + x^8 + x^{10} + x^{12} + x^{13} + x^{14}$. Then

$$\gcd(x^{15} - 1, S^{(15)}(x)) = (1+x)(1+x+x^2)(1+x+x^2+x^3+x^4)(1+x+x^4).$$

So, $m^*(x) = (x^{15} - 1)/\gcd(x^{15} - 1, S^{(15)}(x)) = 1 + x^3 + x^4$ and thus $m(x) = 1 + x + x^4$. Indeed, this $S(x)$ is the output sequence of the LFSR in Figure 3.4.

The above calculations can be executed with the Mathematica functions `PolynomialGCD`, `PolynomialQuotient`, *and* `PolynomialMod`.

```
p = 15;
S = 1 + x^4 + x^7 + x^8 + x^10 + x^12 + x^13 + x^14;
g = PolynomialGCD[S, x^p - 1, Modulus -> 2];
MSTAR = PolynomialMod[PolynomialQuotient[x^p - 1, g, x], 2]
```

```
1 + x^3 + x^4
```

Proof of Theorem 3.15:

Let $\{s_i\}_{i\geq 0} \in \Omega(m)$. If $\{s_i\}_{i\geq 0} \in \Omega(h)$ for some divisor h of m, replace m by h and continue with this procedure until it can be assumed that $\{s_i\}_{i\geq 0} \notin \Omega(h)$ for any divisor of m.

We shall show that such an m is unique and of the form given in Theorem 3.15.

Since the period of $\{s_i\}_{i\geq 0}$ is p, Corollary 3.5 implies that for some $u(x)$ with degree($u(x)$) < degree($m(x)$),

$$\frac{S^{(p)}(x)}{1-x^p} = S^{(p)}(x)(1 + x^p + x^{2p} + \ldots) = S(x) = \frac{u(x)}{m^*(x)}.$$

By our assumption on m and by Lemma 3.14, $\gcd(m^*(x), u(x)) = 1$, so

$$\gcd\left(m^*(x), \frac{m^*(x)\,S^{(p)}(x)}{1-x^p}\right) = 1.$$

It follows that

$$\gcd(m^*(x)\,(1 - x^p),\, m^*(x)\,S^{(p)}(x)) = 1 - x^p.$$

i.e.

$$m^*(x).\gcd(1 - x^p,\, S^{(p)}(x)) = 1 - x^p.$$

Hence

$$m^*(x) = \frac{1-x^p}{\gcd(1-x^p, S^{(p)}(x))}.$$

$\square$

Corollary 3.16
The linear complexity of a binary, periodic sequence $\{s_i\}_{i\geq0}$ with period p and initial segment $S^{(p)}(x) = \sum_{i=0}^{p-1} s_i\, x^i$ is equal to

$$p - \mathrm{degree}(\gcd(x^p - 1,\, S^{(p)}(x))).$$

3.3.2   The Berlekamp-Massey Algorithm

Corollary 3.16 may be of help to the designer of a non-linear system to determine how safe his system is against the kind of attack described in the discussion "Ad C3" in Subsection 3.2.5.

A cryptanalist, on the other hand, who knows a segment of the output sequence, say $s_0, s_1, \ldots, s_{k-1}$, can try the following strategy:

i) find the smallest LFSR that generates $s_0, s_1, \ldots, s_{k-1}$,

ii) determine the next output bit of this LFSR and hope that it correctly "predicts" the next bit s_k of the sequence.

Definition 3.6
$L_k(\{s_i\}_{i\geq0})$ is the length of the shortest LFSR that generates $s_0, s_1, \ldots, s_{k-1}$.
When it is clear from the context which $\{s_i\}_{i\geq0}$ is involved we shall simply write L_k. The polynomial $f^{(k)}(x)$ will denote the characteristic polynomial of any L_k-stage LFSR that generates the sequence $s_0, s_1, \ldots, s_{k-1}$.

Clearly $L_k(\{s_i\}_{i\geq0}) \leq k$ for any sequence $\{s_i\}_{i\geq0}$, since any k-state LFSR will generate $s_0, s_1, \ldots, s_{k-1}$, simply by taking $s_0, s_1, \ldots, s_{k-1}$ as starting state.

Lemma 3.17

Let $\{t_i\}_{i\geq0}$ be an output sequence starting with $\overset{k-1}{\overbrace{00\ldots0}}\,1$. Then,

$$L_k(\{t_i\}_{i\geq0}) = k.$$

Proof: Any LFSR of length n, $n < k$, that is filled with the first n symbols of $\{t_i\}_{i\geq0}$ (which are all zero) will output the all-zero sequence, so t_{k-1} will not be 1.

□

Lemma 3.18

Let $\{s_i\}_{i\geq0}$ and $\{t_i\}_{i\geq0}$ be two output sequences. Then, for all $k \geq 0$

$$L_k(\{s_i + t_i\}_{i\geq0}) \leq L_k(\{s_i\}_{i\geq0}) + L_k(\{t_i\}_{i\geq0}).$$

Proof: This is a direct consequence of Lemma 3.6. Indeed, let the LFSR's with characteristic polynomial $f^{(k)}(x)$ and $g^{(k)}(x)$ generate the first k terms of $\{s_i\}_{i\geq0}$, resp. $\{t_i\}_{i\geq0}$. Then by Lemma 3.6, the first k terms of $\{s_i + t_i\}_{i\geq0}$ will be generated by the LFSR with characteristic polynomial $\text{lcm}[f^{(k)}(x), g^{(k)}(x)]$. This lcm has degree at most the sum of the degrees of $f^{(k)}(x)$ and $g^{(k)}(x)$.

□

It follows from Definition 3.6 that $L_{k+1} \geq L_k$ for any sequence $\{s_i\}_{i\geq0}$. More can be said.

Lemma 3.19

Let $\{s_i\}_{i\geq0}$ be an output sequence. Suppose that the LFSR with characteristic polynomial $f^{(k)}(x)$ does not output s_k correctly. Then

$$L_{k+1} \geq \max\{L_k, k+1 - L_k\}.$$

Proof: We already know that $L_{k+1} \geq L_k$.

Let $\{t_i\}_{i\geq0}$ be a sequence starting with $\overset{k}{\overleftarrow{00\ldots01}}$ as beginning sequence. Since the LFSR with characteristic polynomial $f^{(k)}(x)$ does generate $s_0, s_1, \ldots, s_{k-1}$, but not $s_0, s_1, \ldots, s_k$, it follows that this LFSR will generate $\{s_i + t_i\}_{i=0}^{k}$. Since $L_{k+1} \geq L_k$, we can conclude that $L_{k+1}(\{s_i + t_i\}_{i\geq0}) = L_k(\{s_i + t_i\}_{i\geq0}) = L_k(\{s_i\}_{i\geq0})(= L_k)$.

The statement now follows with Lemma 3.17 and Lemma 3.18 from

$$k + 1 = L_{k+1}(\{t_i\}_{i\geq0}) \leq L_{k+1}(\{s_i\}_{i\geq0}) + L_{k+1}(\{s_i + t_i\}_{i\geq0}) = L_{k+1} + L_k.$$

□

The following theorem shows that in fact equality holds in the above lemma. The proof follows from the Berlekamp-Massey algorithm, that constructs $f^{(k)}(x)$ recursively, cf. [Mass69]. This algorithm is well known in algebraic coding theory for the decoding of BCH codes and Reed-Solomon codes (see [Berl68], Chapter 7).

> **Theorem 3.20**
> Let $\{s_i\}_{i \geq 0}$ be an output sequence. Suppose that the LFSR with characteristic polynomial $f^{(k)}(x)$ does not output s_k correctly. Then
>
> $$L_{k+1} = \max\{L_k, k+1-L_k\}.$$

Proof: In view of Lemma 3.19, it suffices to find a polynomial $f(x)$ of degree equal to $\max\{L_k, k+1-L_k\}$ that does output the first $k+1$ terms of $\{s_i\}_{i \geq 0}$ correctly. This is exactly what the Berlekamp-Massey algorithm does in a very efficient way.

We shall prove the theorem by induction.

Getting the induction argument started.

Define $L_0 = 0$ and $f^{(0)}(x) = 1$.

The sequence $\overset{k}{\overleftrightarrow{00\ldots0}}$ of length k can be generated by the (degenerate) LFSR with characteristic polynomial $f^{(k)}(x) = 1$ of degree $L_k = 0$.

The sequence $\overset{k}{\overleftrightarrow{00\ldots0}}1$ of length $k+1$ can be generated by any $(k+1)$-stage LFSR, but not by a shorter LFSR, as we already saw in Lemma 3.17. In this case,

$$L_{k+1} = k+1 = k+1-L_k = \max\{L_k, k+1-L_k\}.$$

This proves the first induction step.

The induction step: $k \longrightarrow k+1$.

By putting $k+n = j$, $c_i = f_i^{(k)}$, and $n = L_k$ in (3.2), the induction hypothesis for k can be formulated as:

$$\sum_{i=0}^{L_k-1} f_i^{(k)} s_{j-L_k+i} = s_j, \qquad L_k \leq j \leq k-1. \tag{3.9}$$

If (3.9) also holds for $j = k$, then $L_{k+1} = L_k$, $f^{(k+1)}(x) = f^{(k)}(x)$ and there remains nothing to prove.

If (3.9) does not hold, then

$$\sum_{i=0}^{L_k-1} f_i^{(k)} s_{j-L_k+i} = s_j + 1, \quad j = k. \tag{3.10}$$

Let m be the unique integer smaller than k defined by

    i) $L_m < L_k$,
    ii) $L_{m+1} = L_k$,

so m is the index of the last increase of L.

Because we have already proved the start of the induction argument, this number is well defined. It follows from the induction hypothesis and the above definition of m that:

$$\sum_{i=0}^{L_m-1} f_i^{(m)} s_{j-L_m+i} = \begin{cases} s_j, & \text{if } L_m \le j \le m-1, \\ s_m + 1, & j = m. \end{cases} \tag{3.11}$$

Notice that $L_k = L_{m+1} = \max\{L_m, m+1-L_m\} = m+1-L_m$.

Define $L = \max\{L_k, k+1-L_k\}$. We claim that

$$\begin{aligned} f(x) &= & x^{L-L_k} f^{(k)}(x) + x^{L-(k+1-L_k)} f^{(m)}(x) \\ &= & x^{L-L_k} f^{(k)}(x) + x^{L-k+m-L_m} f^{(m)}(x) \end{aligned} \tag{3.12}$$

will be a suitable choice for $f^{(k+1)}(x)$.

Clearly, the first term in (3.12) has degree $(L - L_k) + L_k = L$ and the second term has degree $(L - k + m - L_m) + L_m < L$. So, $f(x)$ has the right degree. But also, by (3.9), (3.10), (3.11),

$$\sum_{i=0}^{L-1} f_i s_{j-L+i}$$

$$\stackrel{(3.12)}{=} \sum_{i=L-L_k}^{L-1} f_{i-(L-L_k)}^{(k)} s_{j-L+i} + \sum_{i=L-(L-k+m-L_m)}^{L-k+m} f_{i-(L-k+m-L_m)}^{(m)} s_{j-L+i}$$

$$\stackrel{\text{subst. } i}{=} \sum_{i=0}^{L_k-1} f_i^{(k)} s_{j-L_k+i} + \sum_{i=0}^{L_m-1} f_i^{(m)} s_{j-L_m-k+m+i} + s_{j-k+m}$$

$$= \begin{cases} s_j + 0 = s_j, & L \le j \le k-1, \\ (s_k + 1) + 1 = s_k, & j = k. \end{cases}$$

This proves that the LFSR with characteristic polynomial $f(x)$ indeed can generate $s_0, s_1, \ldots, s_k$.

□

Theorem 3.20 only proves that the degree L_k of $f^{(k)}(x)$ is unique. In general, the polynomial $f^{(k)}(x)$ itself will not be unique.

The algorithm, described in the proof above, can be executed and summarized as follows:

Algorithm 3.21   *Berlekamp-Massey*

    input   a binary sequence $\{s_i\}_{i \geq 0}$, an index u

    initialization $f = 1, L = 0, j = 0$

    parameters used

           f_{ne}, L_{ne} : stand for the characteristic polynomial and length of the LFSR
                 as desired by the present iteration;

           f_{ol}, L_{ol} : stand for the polynomial and length just before the last change
                 in length;

           diff   : the difference between the present iteration number and
                 the iteration number after the last change in length.

  while $(s_j = 0) \wedge (j \leq u)$ **do** $j = j + 1$

  if $j = u + 1$ **then** STOP

  put $f_{ol} = 1; L_{ol} = 0$
          $f = x^{j+1}; L = \text{degree}(f)$
          $k = j + 1; \text{diff} = 0$

  while $k < u$ **do**
         begin
              if $\sum_{i=0}^{L-1} f_i s_{k-L+i} \neq s_k$ **then**
                  begin
                    $L_{ne} = \max \{L, k + 1 - L\}$
                    $f_{ne} = x^{L_{ne}-L} f + x^{L_{ne}-(\text{diff}+1+L_{ol})} f_{ol}$
                        if $L_{ne} \neq L$ **then**
                              begin
                                $f_{ol} = f; L_{ol} = L;$
                           $L = L_{ne}; \text{diff} = 0;$
                              end
                      else
                          begin
                            $\text{diff} = \text{diff} + 1;$
                          end
                     $f = f_{ne}$
                  end
              else
                  begin
                    $\text{diff} = \text{diff} + 1;$
                  end
          $k = k + 1;$
         end

    output f the characteristic function of the shortest LFSR that can output
$(s_0, s_1, ..., s_u)$.

Example 3.13

Consider the sequence

$$\{s_i\}_{i=0}^{30} = \{0, 0, 0, 0, 0, 1, 1, 1, 1, 1, 1, 1, 1, 1, 1, 1, 1, 0, 0, 0, 0, 0, 0, 0, 0, 0, 1, 1, 1, 1, 0\}.$$

The Mathematica version of the Berlekamp-Massey algorithm that we give below makes use of the functions <u>Do</u>, CoefficientList, <u>Mod</u>, Max, PolynomialMod, <u>Length</u>, *and* <u>Print</u>.

Note that we have combined the two while statements in the algorithm above into a single Do statement. All intermediate functions are also printed.

```
s = {0, 0, 0, 0, 0, 1, 1, 1, 1, 1, 1, 1, 1, 1,
     1, 1, 1, 0, 0, 0, 0, 0, 0, 0, 0, 0, 1, 1, 1, 1, 0};
Lol = 0; fol = 1;
diff = 0; Clear[x];
f = 1; L = 0; g = CoefficientList[f, {x}];
Do[If[Mod[Sum_{i=1}^{L} g[[i]] s[[j - 1 - L + i]], 2] == s[[j]], diff = diff + 1,
        Lne = Max[j - L, L];
        fne = PolynomialMod[x^{Lne-L} f + x^{Lne-Lol-diff-1} fol, 2];
    If[Lne != L, fol = f; Lol = L; L = Lne; diff = 0, diff = diff + 1];
    f = fne; g = CoefficientList[f, {x}]];
  Print["j=", j, ", L=", L, ", f=", f], {j, Length[s]}]
```

j=1, L=0, f=1

j=2, L=0, f=1

j=3, L=0, f=1

j=4, L=0, f=1

j=5, L=0, f=1

j=6, L=6, f=$1 + x^6$

j=7, L=6, f=$1 + x^5 + x^6$

j=8, L=6, f=$1 + x^5 + x^6$

j=9, L=6, f=$1 + x^5 + x^6$

j=10, L=6, f=$1 + x^5 + x^6$

j=11, L=6, f=$1 + x^5 + x^6$

j=12, L=6, f=$x^5 + x^6$

j=13, L=6, f=$x^5 + x^6$

j=14, L=6, f=$x^5 + x^6$

j=15, L=6, f=$x^5 + x^6$

j=16, L=6, f=$x^5 + x^6$

j=17, L=6, f=$x^5 + x^6$

j=18, L=12, f=$1 + x^{11} + x^{12}$

j=19, L=12, f=$1 + x^{10} + x^{12}$

$j=20, \quad L=12, \quad f=1 + x^9 + x^{12}$

$j=21, \quad L=12, \quad f=1 + x^8 + x^{12}$

$j=22, \quad L=12, \quad f=1 + x^7 + x^{12}$

$j=23, \quad L=12, \quad f=1 + x^6 + x^{12}$

$j=24, \quad L=12, \quad f=1 + x^5 + x^{12}$

$j=25, \quad L=13, \quad f=x + x^5 + x^{13}$

$j=26, \quad L=13, \quad f=1 + x + x^{12} + x^{13}$

$j=27, \quad L=14, \quad f=1 + x + x^2 + x^5 + x^{12} + x^{13} + x^{14}$

$j=28, \quad L=14, \quad f=x^2 + x^5 + x^{14}$

$j=29, \quad L=14, \quad f=x^2 + x^5 + x^{14}$

$j=30, \quad L=16, \quad f=1 + x + x^4 + x^7 + x^{12} + x^{13} + x^{16}$

$j=31, \quad L=16, \quad f=1 + x + x^4 + x^7 + x^{12} + x^{13} + x^{16}$

3.3.3   A Few Observations about Non-Linear Algorithms

The problem with non-linear feedback shift registers, in general, is the difficulty of their analysis. One has to answer questions like: how many different cycles of output sequences are there, what is their length, what is their linear complexity, etc. The following theorem will make it clear that it is possible to say at least a little bit about general non-linear feedback shift registers.

Clearly, the output sequence of a non-linear FSR does not have maximal period if there are two different states with the same successor state. A state with more than one predecessor is called a *branch point*.

Theorem 3.22
An n-stage feedback shift register with (non-linear) feedback fuction $f(s_0, s_1, \ldots, s_{n-1})$ has no branch points if and only if a Boolean function $g(s_1, s_2, \ldots, s_{n-1})$ exists such that $f(s_0, s_1, \ldots, s_{n-1}) = s_0 + g(s_1, s_2, \ldots, s_{n-1})$.

Proof: Since f is a Boolean function, one can write

$$f(s_0, s_1, \ldots, s_{n-1}) = g(s_1, s_2, \ldots, s_{n-1}) + s_0\, h(s_1, s_2, \ldots, s_{n-1}).$$

$\implies$ If $h(s_1, s_2, \ldots, s_{n-1}) = 0$ for some $(s_1, s_2, \ldots, s_{n-1})$, then both states $(0, s_1, s_2, \ldots, s_{n-1})$ and $(1, s_1, s_2, \ldots, s_{n-1})$ will have the same successor state. Thus a branch point would exist, contradicting our assumption. We conclude that $h \equiv 1$.

$\Longleftarrow$ The state $(0, s_1, s_2, \ldots, s_{n-1})$ has successor $(s_1, s_2, \ldots, s_{n-1}, s_n)$ with $s_n = g(s_1, s_2, \ldots, s_{n-1})$, while state $(1, s_1, s_2, \ldots, s_{n-1})$ has successor $(s_1, s_2, \ldots, s_{n-1}, s_n + 1)$. Therefore, there are no branch points.

$\square$

There are many ways to use LFSR's in a non-linear way. Below we depict two proposals that are extensively discussed in [Ruep86]. Others ideas can be found in [MeOoV97], Chapter 6.

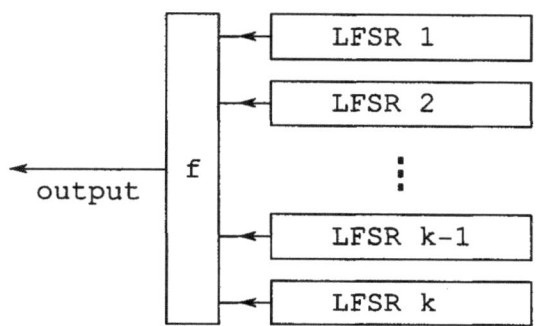

Combining several PN's with one non-linear function f.

Figure 3.5

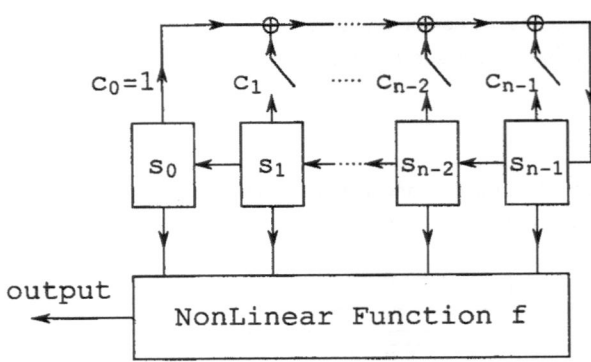

One LFSR with a non-linear output.

Figure 3.6

3.4   Problems

Problem 3.1
Let $\{s_i\}_{i\geq0}$ be binary, periodic sequence of period 17, starting with the sequence 01101000110001011. To which extent does $\{s_i\}_{i\geq0}$ satisfy Golomb's Randomness Postulates?
(Note for the interested reader. The sequence above has its ones at the positions corresponding to the quadratic residues modulo 17 (see also input line above Theorem A.21). The parameters that arise when checking G3 can be predicted by Theorem A.22 and Corollary A.24)

Problem 3.2
Express the polynomial $\gcd(x^m - 1, x^n - 1)$ in terms of x and $\gcd(m, n)$. (See also Problem A.3.)

Problem 3.3
Let $\{u_i\}_{i\geq0}$ and $\{v_i\}_{i\geq0}$ be the output sequences of binary LFSR's of length m resp. n, where $m, n \geq 2$. Assume that $\{u_i\}_{i\geq0}$ and $\{v_i\}_{i\geq0}$ are both PN sequences and that $\gcd(m, n) = 1$. Hence, also $\gcd(2^m - 1, 2^n - 1) = 1$ (see Problem A.3). Let the sequence $\{w_i\}_{i\geq0}$ be defined by $w_i = u_i v_i$, $i \geq 0$, and let p be the period of $\{w_i\}_{i\geq0}$.

a)   Prove that p is a divisor of $(2^m - 1)(2^n - 1)$.
b)   How many zeros and how many ones appear in a subsequence of length $(2^m - 1)(2^n - 1)$ in the sequence $\{w_i\}_{i\geq0}$?
c)   Prove that $(2^m - 1)(2^n - 1)/p$ must divide the two numbers determined in ii).
d)   Prove that $p = (2^m - 1)(2^n - 1)$.
e)   How many gaps of length 1 does the $\{w_i\}_{i\geq0}$-sequence have per period when $m, n \geq 4$?

Problem 3.4
Let $\{s_i\}_{i\geq0}$ be the binary sequence defined by

$$s_i = \begin{cases} 1, & \text{if } i = 2^l - 1, l \in \mathbb{N}, \\ 0, & \text{otherwise.} \end{cases}$$

So, the $\{s_i\}_{i \geq 0}$ starts like 11010001000000010. Let L_k be the linear complexity of $s_0, s_1, \ldots, s_{k-1}$. Prove that

$$L_{2^l} = 2^{l-1}, \ l \geq 1.$$

Problem 3.5^M

Let a binary sequence $\{s_i\}_{i \geq 0}$ have period 15 and start with 010110000101010.
What is the minimal characteristic polynomial of $\{s_i\}_{i \geq 0}$ and what is the linear complexity of this sequence?

Problem 3.6

Consider the binary, periodic sequence $\{s_i\}_{i \geq 0}$ determined by the period $2^{12} - 1$ and the values $s_0 = s_{2^9 - 1} = 1$ and $s_i = 0$ for 0 for $0 \leq i < 2^{12} - 1$, $i \neq 0$, $2^9 - 1$.
What is the minimal characteristic polynomial of $\{s_i\}_{i \geq 0}$? What is the linear complexity of this sequence?

Problem 3.7^M

Consider the binary polynomials $f(x) = 1 + x + x^3$ and $g(x) = 1 + x^2 + x^5$. The corresponding LFSR's are denoted by LFSR(f) resp. LFSR(g). Let $\{s_i\}_{i \geq 0}$ and $\{t_i\}_{i \geq 0}$ denote the output sequences of LFSR(f) resp. LFSR(g).
The sequence $\{u_i\}_{i \geq 0}$ is defined by $u_i = s_i + t_i$, $i \geq 0$.
The 2^8 different initial states $(s_0, s_1, s_2, t_0, t_1, t_2, t_3, t_4)$ generate different periodic sequences $\{u_i\}_{i \geq 0}$.
What are the cycle lengths (=periods) of these periodic sequences? Give an initial state of each cycle.

Problem 3.8

Consider the binary shift register depicted in the figure below.

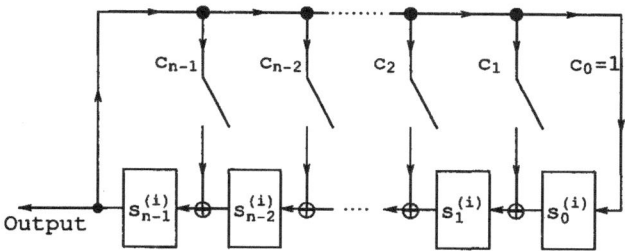

Let $\underline{s}^{(i)} = (s_{n-1}^{(i)}, s_{n-2}^{(i)}, \ldots, s_1^{(i)}, s_0^{(i)})$ be the state of the shift register at time i, $i \geq 0$.
a)   Give the $n \times n$ matrix T satisfying $\underline{s}^{(i+1)} = T\underline{s}^{(i)}$ for all $i \geq 0$.
b)   Prove that the characteristic equation of T over $\mathbb{R}$ is given by

$$\lambda^n = c_{n-1} \lambda^{n-1} + c_{n-2} \lambda^{n-2} + \ldots + c_1 \lambda + 1.$$

c)   From matrix theory we may conclude that over

$$T^n = c_{n-1} T^{n-1} + c_{n-2} T^{n-2} + \ldots + c_1 T + I, \tag{3.13}$$

where I is the $n \times n$ identity matrix.

Since all elements in (3.13) are integer, equation (3.13) also holds modulo 2.

Derive a recurrence relation between $\underline{s}^{(i+n)}$, $\underline{s}^{(i+n-1)}$, ..., $\underline{s}^{(i+1)}$, and $\underline{s}^{(i)}$.

d)   Which LFSR of length n gives the same output sequence as the above shift register? What does the initial state have to be in this LFSR to generate the same output sequence?

Problem 3.9

Let $\alpha \in \mathrm{GF}(2^3)$ be a zero of $f(x) = x^3 + x + 1$. So, by Theorem B.30,

$$f(x) = (x - \alpha)(x - \alpha^2)(x - \alpha^4),$$
$$f^*(x) = (x - \alpha^3)(x - \alpha^5)(x - \alpha^6) = (1 - \alpha x)(1 - \alpha^2 x)(1 - \alpha^4 x).$$

Prove that $\Omega(f)$ consists of all sequences

$$\sum_{i=0}^{\infty}(a.\alpha^i + a^2.\alpha^{2i} + a^4.\alpha^{4i})x^i, \qquad a \in \mathrm{GF}(2^3),$$

(Hint: use Corollary 3.5 and use the partial fraction expansion over $\mathrm{GF}(2^3)$.)

Note that the expression above can be written as $\sum_{i=0}^{\infty} \mathrm{Tr}(a.\alpha^i) x^i$, where Tr stands for the Trace function, as introduced in Problem B.16.

4   Block Ciphers

4.1   Some General Principles

4.1.1   Some Block Cipher Modes

☐ **Codebook Mode**

Block ciphers are conventional cryptosystems that typically handle a fixed number of symbols at a time (under a given key) and do this encryption/decryption independent of past input blocks (see Figure 4.1). For the encryption process, the data (plaintext) enters the block cipher from the left and leaves it on the right as ciphertext. For the decryption, it is exactly the other way around.

In the next section we shall describe a few widely used block ciphers. At this moment, the particular layout of such a cipher is not so important. One should view it as an electronic device that can convert n-tuples of bits to other n-tuples at very high speeds (under a key) in such a way that the reverse process is only feasible if one knows the key.

Assuming that the plaintext is a long binary file, one breaks it up in segments M_i, $i \geq 0$, each n bits long. The result of the encryption of M_i is denoted by C_i and we write

$$C_i = \mathrm{BC}_k(M_i), \ i \geq 0,$$

where k is the key. The decryption process will be denoted by $\mathrm{BC}^{\leftarrow}$, so we have $M_i = \mathrm{BC}_k^{\leftarrow}(C_i)$.

Since an n-tuple of symbols from an alphabet $\mathcal{A}$ can be viewed as one symbol from $\mathcal{A}^n$, the difference between an n-tuple from one alphabet or a single symbol from another alphabet is theoretically of little importance but may be of great practical value.

Therefore, the key property of a block cipher is the lack of memory in the encryption device.

It is clear that as long as the key remains the same, the same plaintext will be encrypted to the same ciphertext. For this reason, encryption in the mode shown in Figure 4.1 is called *codebook mode*. It is as if one uses a codebook or dictionary for the encryption. It may be clear that encrypting the same message twice under the same key is cryptographically insecure, hence, block ciphers are normally not used in codebook mode.

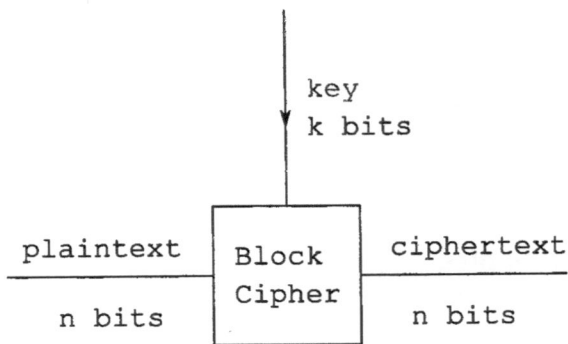

Block Cipher in Codebook Mode

Figure 4.1

□ **Cipher Block Chaining**

There are several standard methods to circumvent the problems mentioned above. One technique is called *cipher block chaining*. We assume again that one is encrypting a long file. Each ciphertext, say C_i at time i, is not only transmitted to the receiver, but it is also added coordinate-wise to the next block of plaintext M_{i+1}.

To this end, the encryption algorithm has to make use of some kind of memory device, commonly called a buffer. See Figure 4.2 below. Of course, the buffer has to be initialized before the encryption process can be started.

Note that by introducing memory to this system it technically has become a stream cipher.

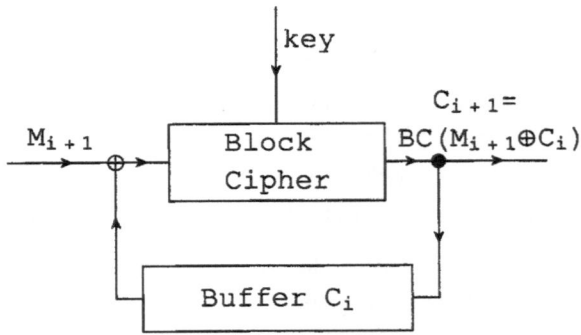

Cipher block chaining - Encryption

Figure 4.2

The decryption process reverses the above process. The buffer has to be initialized with the same initial value as was used to start the encryption. It can be part of the secret key or a just a fixed constant.

The notation $BC^{\leftarrow}$ in Figure 4.3 stands for the inverse of the block cipher used for encryption.

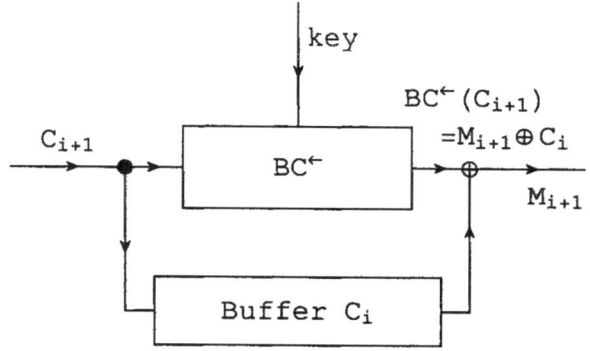

Cipher block chaining - Decryption

Figure 4.3

Remark:

Note, that when $C_i = C_j$, for some $i < j$, in Figure 4.2, one has that $M_i \oplus C_{i-1} = M_j \oplus C_{j-1}$, i.e. $C_{i-1} \oplus C_{j-1} = M_i \oplus M_j$. This means that the modulo sum of the two previous ciphertexts is equal to the sum of the ciphertexts M_i and M_j. In many situations this means that some information about the plaintext leaks away. For instance, as we can deduce from Example 5.2 , the modulo 26 addition of two English texts (with a Vigenère Table (Table 2.3) will still have sufficient structure to enable a unique reversal of the addition process.

The above observation is reason to go to longer block lengths than the ones most commonly in use today (being 64 bits).

□ **Cipher Feedback Mode**

Another way to make sure that a block cipher under the same key encrypts the same plaintext at different moments into different ciphertexts is called the *cipher feedback mode*.

This method is depicted in Figure 4.4 below, but in a more general setting. In many practical situations, for instance in many internet protocols, one wants to transmit only a few bits at a time, say r bits, where r is less than the block length of the block cipher.

Instead of padding the r bits with $n - r$ zeros in order to get an n-tuple that can serve as input for a block cipher, one adds the r-tuple coordinatewise modulo 2 to the r leftmost output bits of the block cipher. The input of the block cipher is given by the contents of a shift register (without feedback) that at each clock pulse shifts r positions to the left to accommodate the r bits of the previous ciphertext.

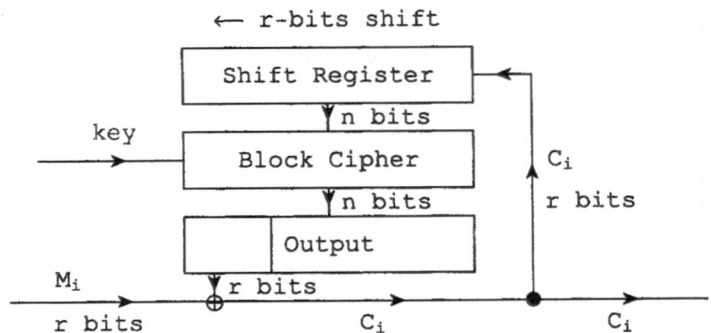

Cipher Feedback Mode

Figure 4.4

4.1.2   An Identity Verification Protocol

In this subsection, we want to give an idea how a block cipher can be used in an *identity verification protocol*. Such a protocol is a discussion between two parties in which one of them wants to convince the other that he is authentic. An application is, for instance, a smart card of a person, say Alice, who wants to withdraw money from her account through a card reader of a bank.

While issuing the card to Alice, the bank stores two numbers on it:

- the identity number Id_A of Alice,

- the secret key k_A of Alice.

The key k_A can not be accessed from the outside world; it does not even have to be known to Alice. The identity number can be accessed by any card reader (it may even be printed or written on the outside). They are related by

$$k_A = BC_{MK}(Id_A), \tag{4.1}$$

where BC stands for a block cipher and MK for the bank's master key. MK is stored in every card reader of the bank. It would be impractical to store the secret keys of all customers in each card reader.

The block cipher BC is also implemented on the card.

When the card is inserted into the card reader, it will be asked to present its identity number (Id_A in our case). A genuine card reader can now compute Alice's secret key k_A from (4.1).

The card reader generates a random string r of n bits and presents it as a *challenge* to the card. The card returns $BC_{k_A}(r)$ as its *response* to the card reader. The card reader simply verifies this calculation. If the card's answer to the challenge r is correct, the card reader "knows" that k_A is stored on the card and it will conclude that the card is authentic. Otherwise, it will not accept the card.

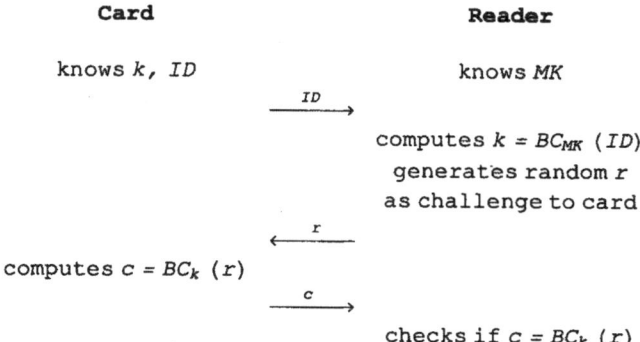

An identity verification protocol.

Figure 4.5

The card can use the same protocol to check that the card reader is genuine. It sends its challenge to the card reader. The reply by the card reader can only be correct if the card reader is able to compute the secret key k_A, i.e. if the card reader knows the bank's master key MK.

Normally, a Personal Identification Code (PIN) is used to link the card to the card holder.

4.2   DES

□ **DES**

In 1974 the National Bureau of Standards (NBS) solicited the American industry to develop a cryptosystem that could be used as a standard in unclassified U.S. Government applications. IBM developed a system called LUCIFER. After being modified and simplified, this system became the *Data Encryption Standard* (*DES* for short) in 1977.

Right away, DES was made available on a fast chip. This made it very suitable for use in large communication systems. The complete design of DES has been made public at the time of its introduction. This has never been done before, although in each textbook one can find the remark that the security of a cryptosystem should not depend on the secrecy of the system.

We shall not give a complete description of DES. The reader is referred to [Konh81], [MeyM82], [MeOoV97], or [Schn96].

DES is a block cipher operating on 64 bits simultaneously (see Figure 4.6).

The key consists of eight groups of 8 bits. One bit in each of these groups is a parity check bit that makes the overall parity in each block odd. So, although the keysize appears to be 64, the effective keysize is 56 bits.

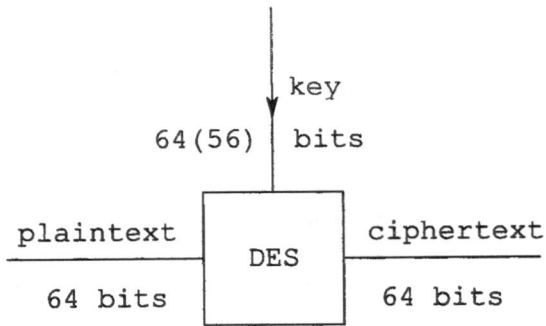

The Data Encryption Standard

Figure 4.6

DES consists of 16 identical rounds. The 64 input bits are divided into two halves: the 32 leftmost bits form L_0 and the 32 rightmost bits form R_0.

In each round, a new L and R are defined by

$$L_i = R_{i-1}, \ 1 \leq i \leq 16,$$
$$R_i = L_{i-1} \oplus f(R_{i-1}, K_i), \ 1 \leq i \leq 16. \tag{4.2}$$

Here, K_i stands for a well-defined subsequence of bits from the key K.

Further, f is function of the previous right-half and this subkey K_i. This function is defined by means of a collection of fixed tables, called substitution tables. The outcome is added coordinatewise modulo 2 to L_{i-1}. Note that L_i is simply the previous right-half. (See Figure 4.7 below.).

The final output of DES is formed from L_{16} and R_{16}.

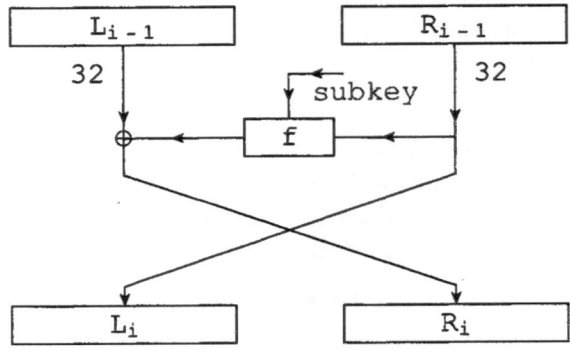

A Typical Round of DES

Figure 4.7

In Figure 4.7 one can see that the inverse algorithm of DES can be computed from the same scheme by simply going from the bottom to the top. Indeed, it follows from (4.2) that for all $1 \leq i \leq 16$

$$R_{i-1} = L_i$$
$$L_{i-1} = R_i \oplus f(R_{i-1}, K_i) = R_i \oplus f(L_i, K_i).$$

Many people have criticized the decision to make DES a standard. The two main objections were:

i) The effective keysize (56 bits) is too small for an organization with sufficient resources. An exhaustive keysearch is, at least in principle, possible.

ii) The design criteria of the tables used in the f-function are not known. Statistical tests however show that these tables are not completely random. Maybe there is a hidden trapdoor in their structure.

During the first twenty years after the publication of the DES-algorithm no effective way of breaking it was published. However, in 1998, for the first time, a DES challenge has been broken by a more or less brute-force attack.

□ **Triple DES**

When it became clear that DES could no longer be used to protect sensitive data, a modification was introduced, called *Triple DES*. It consists of three DES implementations in a row, except that the middle one is orientated the other way around. Thus, one has DES, DES$^{\leftarrow}$, and then again DES. See Figure 4.8 below.

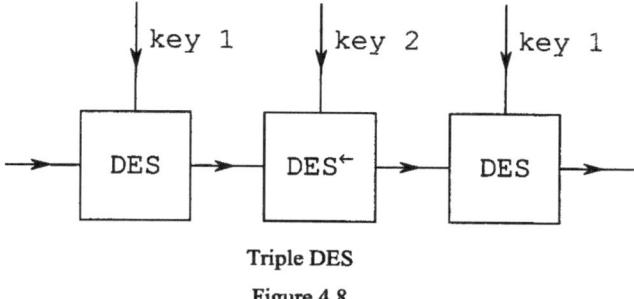

Triple DES

Figure 4.8

There are two interesting things to note about this design. First of all, the third key is the same as the first key. The effective key search is $2 \times 56 = 112$ in this way. This is considered to remain secure for many years to come.

The second observation is that the cipher in the middle is DES$^{\leftarrow}$ instead of DES.

These two features make it possible to keep systems in which Triple DES is implemented compatible with single DES systems. Indeed, by taking the keys 1 and 2 the same, the above system reduces to a single DES scheme.

4.3 IDEA

There are quite a few alternatives to DES. One reason for looking for them may have been the export restrictions by the American government, another, the costs and patent rights. Contrary to DES, which uses well chosen tables in each round, some of the alternatives make use of several mathematical primitives that are algebraically uncorrelated.

IDEA [Lai92] is such a system. The name stands for International Data Encryption Algorithm. IDEA also handles 64 bits at a time (see the remark in Subsection 4.1.1 about this size), but has a key of 128 bits. It consists of 8 identical rounds, which are depicted in Figure 4.9. The 64 bits are equally divided over four blocks of 16 bits each. These blocks are called X_i, $1 \leq i \leq 4$, at the input side of a typical round and Y_i, $1 \leq i \leq 4$, on the output side. The entries K_i, $1 \leq i \leq 6$, denote substrings of the key. Their composition depends on the particular round that has taken place.

The mathematical primitives in IDEA operate on these 16 bits. They are the following operations.

- **Coordinatewise XOR** (addition modulo 2).

In Figure 4.9, this is depicted by $\oplus$.

In *Mathematica* the XOR can be performed with the <u>Mod</u> function (here shown on 4-tuples).

```
Mod[{1, 1, 0, 0} + {1, 0, 1, 0}, 2]
```

```
{0, 1, 1, 0}
```

- **Addition modulo 2^{16}**.

In Figure 4.9, this is depicted by a square with a plus sign in it $\boxplus$.

Interpret the two inputs as the binary representation of two integers. Add these integers modulo 2^{16} and output the binary representation of the sum.

In *Mathematica* this can be performed with the <u>FromDigits</u> and <u>IntegerDigits</u> functions (here shown on 4-tuples).

```
a = FromDigits[{1, 0, 1, 1}, 2]
b = FromDigits[{1, 1, 1, 0}, 2]
su = Mod[a + b, 16]
IntegerDigits[su, 2]
```

```
11
```

```
14
```

```
9
```

```
{1, 0, 0, 1}
```

- **Multiplication modulo $2^{16} + 1$.**

In Figure 4.9, this is depicted by $\otimes$.

Interpret the two inputs (binary 16-tuples) as the binary representation of two integers modulo the prime number $2^{16} + 1 = 65537$. Make an exception for the all-zero word which will be identified with the integer 2^{16}. In this way we have a 1-1 correspondence between binary 16-tuples and the elements of $\mathbf{Z}^*_{65537}$ (see Example B.3).

Multiply these two integers modulo $2^{16} + 1$, and output the binary representation of the product (but map $1 \overset{16}{\overline{0 \ldots 0}}$ to $\overset{16}{\overline{0 \ldots 0}}$).

Since, $2^{16} + 1$ is prime, the multiplication $a \times b$ (as defined above) is a one-to-one mapping for fixed a or b. Below we demonstrate this again for 4-tuples. Note that $2^4 + 1$ is also a prime number.

```
a = FromDigits[{1, 0, 1, 0}, 2];
b = FromDigits[{0, 1, 1, 0}, 2];
a = If[a == 0, 16, a];
b = If[b == 0, 16, b];
pr = Mod[a*b, 17]
pr = If[pr == 16, 0, pr];
IntegerDigits[pr, 2, 4]
```

```
9
```

```
{1, 0, 0, 1}
```

The reader is invited to multiply the sequences {1, 0, 0, 0} and {0, 0, 1, 0}.

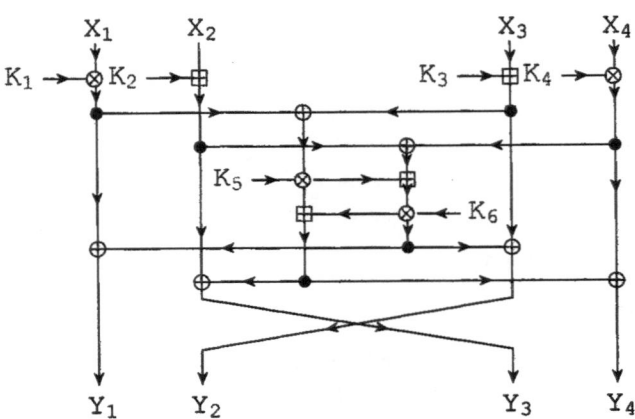

One Round in the International Data Encryption Algorithm
(IDEA)

Figure 4.9

As with DES, IDEA can be inverted by simply going through it from the bottom to the top.

4.4   Further Remarks

RC5 is a scheme that is a little bit similar to IDEA. Its algebraic primitives are again the exclusive or and addition modulo 2^w, where w is the word length, but instead of the multiplication modulo $2^w + 1$, which only works if $2^w + 1$ is prime, RC5 makes use of cyclic shifts.

The word length of RC5 is $2w$, where the user can select w from 16, 32, or 64. An additional advantage of RC5 is the freedom to choose the number of rounds in the scheme. Depending on the required speed and security, the user may opt for many or just a few rounds.

In 1993 two attacks on block ciphers were published, that turned out to be surprisingly strong. These methods are called *linear* and *differential cryptanalysis* (see [MatsY93], resp. [BihS93]) and are in fact known plaintext attacks. Several proposed block ciphers were not strong enough against these attacks, however the DES algorithm could withstand it. Later it became clear that the inventors of DES were already aware of these attacks. For further reading we like to mention [Knud94].

At the time of this writing, a collection of proposals are being studied by the (American) National Institute of Standards and Technology (NIST for short) for a new industrial standard. The names of these proposals are CAST-256, CRYPTON, DFC, DEAL, E2, FROG, HPC, LOKI97, MAGENTA, MARS, RC6, RIJNDAAEL, SAFER+, SERPENT and TWOFISH (see the web page 'Advanced Encryption Standard' http://csrc.nist.gov/encryption/aes/aes_home.htm). The outcome of this study is not yet clear.

4.5   Problems

Problem 4.1
Describe the decryption process for a block cipher used in of cipher feedback mode.

Problem 4.2
Consider a block cipher that is used in cipher block chaining mode. Suppose that during transmission, C_i, the i-th ciphertext block, is corrupted. How many plaintext blocks will be affected?
Answer the same question for the case of cipher feedback mode.

Problem 4.3 [M]
What is the next sensible block length of IDEA, if the same scheme and the same primitives are being used, but only the length of the registers is increased? (This length is 16 in IDEA.)
What is wrong with the intermediate values?

5   Shannon Theory

5.1   Entropy, Redundancy, and Unicity Distance

In Chapter 2, we have seen that the cryptanalysis of a cryptosystem often depends on the structure that is present in most texts. For instance in Table 2.1 we could find the key 22 (or −4), because "tu quoque Brute" was the only possible plaintext that made sense.

This structure in the plaintext remains present in the ciphertext (although in hidden form). If the extra information arising from this structure exceeds our uncertainty about the key, one may be able to determine the plaintext from the ciphertext!

We shall first need to quantify the concept of information. Let X be a random variable defined on a set $X = \{x_1, x_2, ..., x_n\}$ by the probabilities

$$\mathrm{Pr}_X(X = x_i) = p_i, \qquad 1 \le i \le n.$$

So, $\sum_{i=1}^{n} p_i = 1$ and $p_i \ge 0$ for all $1 \le i \le n$.

We shall show that

$$J(p_i) = -\log_2 p_i \tag{5.1}$$

is a good measure for the amount of *information* given by the occurrence of the event x_i, $1 \le i \le n$. The base 2 in (5.1) can be replaced by other choices, but reflects our intuitive notions about information, as we shall see. With 2 as choice for the base in the logarithm the unit of information is a called a *bit*.

Let $X = \{x\}$ above (so $n = 1$). Then $p_1 = 1$. Now the occurrence of an event x that occurs with probability 1 (like the sun will rise again tomorrow) gives no information whatsoever. This corresponds nicely with $J(1) = 0$ in (5.1).

Now consider an event that occurs with probability $1/2$, like the specific sex of a newborn baby. So, now $X = \{b, g\}$. Assuming that both sexes have the same probability $1/2$ of occurring, such an outcome gives precisely one bit of information. For instance, a 1 can denote a boy and a 0 can denote a girl. This one bit of information is again in agreement with $J(1/2) = 1$ in (5.1).

If an event occurs with probability $1/4$, then its occurrence gives two bits of information. This is clear in the case that there are four possible outcomes, each with probability $1/4$. Each outcome can be represented by a different sequence of two bits.

On the other hand, the amount of information that an event gives, when it has a probability of $1/4$

to occur, should be independent of the probabilities of the other possible outcomes. Thus, the value $J(1/4) = 2$ (see (5.1)) agrees again with our intuition. Continuing in this way one gets

$$J(1/2^k) = k, \quad k \geq 0. \tag{5.2}$$

The *expected value* of stochastic variable $J(\Pr_X(X))$, defined over X, is called the *entropy* of X and will be denoted by either $H(X)$ or by $H(\underline{p})$, where $\underline{p} = (p_1, p_2, \ldots, p_n)$. Hence, $H(X) = \text{Exp}(J(\Pr_X(X))) = \sum_{i=1}^{n} p_i J(p_i) = -\sum_{i=1}^{n} p_i \log_2 p_i$:

$$H(\underline{p}) = -\sum_{i=1}^{n} p_i \log_2 p_i. \tag{5.3}$$

When $n = 2$, one often writes $p_1 = p$, $p_2 = 1 - p$, and $h(p)$ instead of $H(\underline{p})$:

$$h(p) = -p.\log_2 p - (1 - p).\log_2(1 - p), \qquad 0 \leq p \leq 1. \tag{5.4}$$

Since $x.\log_2 x$ tends to 0 for $x \to 0$, there are no real problems with the definition and the continuity of the entropy function $H(\underline{p})$ when some of the probabilities are 0 (or 1).

The function $h(p)$ is depicted below (with the *Mathematica* function Plot).

```
p =.;
Entropy[p_] = -p*Log[2, p] - (1-p) Log[2, 1-p];
```

```
Plot[Entropy[x], {x, 0, 1}];
```

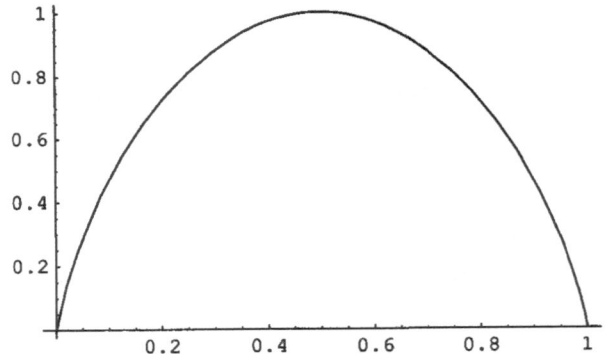

The entropy function $H(\underline{p})$ can be evaluated as follows.

```
MultiEntropy[p_List] := - Sum[ p[[i]] * Log[2, p[[i]]], {i, 1, Length[p]}]
```

```
p = {1/4, 1/4, 1/4, 1/4};
MultiEntropy[p]
```

2

One can give the following interpretations to the entropy $H(X)$ of a random variable X:

- the expected amount of information that a realization of X gives,
- our uncertainty about X,
- the expected number of bits needed to describe an outcome of X.

With these interpretations in mind one expects the entropy function $H(X)$ to have the following properties:

P1: $\quad H(p_1, p_2, ..., p_n) = H(p_1, p_2, ..., p_n, 0)$

P2: $\quad H(p_1, p_2, ..., p_n) = H(p_{\sigma(1)}, p_{\sigma(2)}, ..., p_{\sigma(n)})$,
$\quad\quad$ for any permutation σ of the index set $\{1, 2, ..., n\}$.

P3: $\quad 0 \le H(p_1, p_2, ..., p_n) \le H(1/n, 1/n, ..., 1/n)$.

P4: $\quad H(p_1, p_2, ..., p_n) = H(p_1, p_2, ..., p_{n-2}, p_{n-1} + p_n) + (p_{n-1} + p_n) H\left(\frac{p_{n-1}}{p_{n-1}+p_n}, \frac{p_n}{p_{n-1}+p_n}\right)$.

The interpretations of these properties are straightforward.

P1 says that adding another event to X but one with probability 0 of occurring does not affect the uncertainty about X.

P2 states that renumbering the different events in X leaves the entropy the same.

P3 says that the uncertainty about X is maximal if all events have the same probability of occurring.

Finally, P4 states that the expected number of bits necessary to describe an outcome from X is equal to the number of bits necessary when combining events x_{n-1} and x_n into a single event, say $\hat{x}_{n-1}$, plus the number bits to necessary to distinguish between events x_{n-1} and x_n conditional to the fact that event $\hat{x}_{n-1}$ did occur.

For instance, if $n = 4$, then $H(\frac{1}{4}, \frac{1}{4}, \frac{1}{4}, \frac{1}{4}) = 2$ and also

$$H\left(\frac{1}{4}, \frac{1}{4}, \frac{1}{2}\right) + \frac{1}{2}.H\left(\frac{1}{2}, \frac{1}{2}\right) = \left(\frac{1}{4}.2 + \frac{1}{4}.2 + \frac{1}{2}.1\right) + \frac{1}{2}.1 = 2.$$

Although we shall not prove it here, it can be shown [Khin57] that (5.1) is the only continuous function satisfying (5.2) yielding an entropy function $\sum_{i=1}^{n} p_i J(p_i)$ satisfying the above mentioned properties P1-P4.

Example 5.1

Consider the flipping of a coin. Let Pr(head) = p and Pr(tail) = 1 − p, 0 ≤ p ≤ 1. The entropy is given by (5.4).

That h(1/2) = 1 is of course confirmed by the fact that one needs one bit to represent the outcome of the tossing of a fair coin. For instance, 0 ↔ heads and 1 ↔ tails.

Since h(1/4) ≈ 0.8113 one expects that on the average only 0.8113 bits are needed to represent the outcome of the tossing of an unfair coin with Pr(head) = 1/4. This statement is true in the sense that one can approach the number 0.8113 arbitrarily close. In Chapter 6 we shall show how this is done. The trick will be to represent the outcome of many tossings together by one single string of bits. For instance with two tossings one can represent the outcomes as follows:

```
two tossings  probablity  representation

     hh          1 / 16        111
     ht          3 / 16        110
     th          3 / 16         10
     tt          9 / 16          0
```

The expected length of this representation is

$$\tfrac{1}{16}.3 + \tfrac{3}{16}.3 + \tfrac{3}{16}.2 + \tfrac{9}{16}.1 = \tfrac{27}{16}.$$

But each representation describes two outcomes, so this scheme needs $27/32 \approx 0.843$ bits per tossing. Taking three, four, ... tossings at a time leads to increasingly better approximations of $h(1/4)$.

There is however a problem to address, namely that the receiver of a long string of zeros and ones should be able to determine the outcomes of the tossings in a unique way. One can easily verify that any sequence made up from the subsequences 111, 110, 10 and 0 can only be broken up into these subsequences in just one way . We shall address this problem extensively in Chapter 6.

Example 5.2 (Part 1)

The 26 letters in the English alphabet can be represented with $\log_2 26 \approx 4.70$ bits per letter, by coding sufficiently long strings of letters into binary strings. Indeed, for k letters one needs $\lceil \log_2 26^k \rceil$ bits and thus one needs $\lceil \log_2 26^k \rceil / k$ bits per letter, which converges to $\log_2 26$.

On the other hand, the entropy of 1-grams can easily be computed with the probabilities given in Table 1.1. One obtains 4.15 bits per letter.

Also for bi-grams and tri-grams these computations have been made (see [MeyM82], App.F. One gets the following values:

H(1-grams)	≈ 4.15 bits/letter,
H(2-grams)/2	≈ 3.62 bits/letter,
H(3-grams)/3	≈ 3.22 bits/letter.

According to some tests the asymptotic value for $n \to \infty$ is less than 1.5 bits/letter!

> **Definition 5.1**
>
> Let $(X_0, X_1, ..., X_{n-1})$, $n \geq 1$, denote the plaintext generated by a plaintext source $\mathfrak{S}$ over the alphabet $\mathbb{Z}_2$.
>
> Then the *redundancy D_n* of $(X_0, X_1, ..., X_{n-1})$ is defined by
>
> $$D_n = n - H(X_0, X_1, ..., X_{n-1}).$$
>
> The quantity $\delta = D_n / n$ stands for the average redundancy per letter.

If the alphabet size is q and each symbol is represented by $\log_2 q$ bits, the redundancy is given by $D_n = n.\log_2 q - H(X_0, X_1, ..., X_{n-1})$. If a different representation of the alphabet symbols is used, say with an expected representation length of l bits per symbol, we have $D_n = n.l - H(X_0, X_1, ..., X_{n-1})$.

The redundancy measures to which extent the length of the plaintext exceeds the length that is strictly necessary to carry the information of the text (all measured in bits).

Let us now turn our attention to a cryptosystem $\mathfrak{E}$ consisting of cryptographic transformation E_k indexed by keys k from a key space $\mathcal{K}$. Assume that the unknown plaintext is a regular English text. In the context of this chapter we assume that the cryptanalist has unlimited computing power. So, given a ciphertext a cryptanalist can try out all keys to check for possible plaintexts. As soon as the ciphertext is just a few letters long, some keys can be ruled out because they lead to impossible or improbable letter combinations in the plaintext. The longer the ciphertext, the more keys can be ruled out. They violate the structure or interpretation of English texts. More formally, they violate the redundancy in the plaintext. Sooner or later, only the key that was used for the encryption remains as only candidate.

Let us return to the general setting. Let n be the length of the plaintext (in bits). There are 2^n possible binary sequences, but only $2^{H(X_0, X_1, ..., X_{n-1})}$ represent meaningful messages. The probability that a decryption with the wrong key hits a legitimate message is $2^{H(X_0, X_1, ..., X_{n-1})} / 2^n$. If all keys are tried out and all are equally likely, one expects to find $|\mathcal{K}| 2^{H(X_0, X_1, ..., X_{n-1})} / 2^n$ meaningful plaintexts. Let K denote the uniform distribution over the key space $\mathcal{K}$. Then $|\mathcal{K}| = 2^{H(K)}$ and one can write that $2^{H(K)} 2^{H(X_0, X_1, ..., X_{n-1})} / 2^n$ meaningful messages are expected. If this number is less than 1, very likely it will be just the key used for the encryption that will survive this analysis. The above happens if

$$H(K) + H(X_0, X_1, ..., X_{n-1}) - n \leq 0,$$

i.e. if the redundancy satisfies

$$D_n \geq H(K).$$

If K does not have a uniform distribution, we can still use the interpretation that $H(K)$ denotes the uncertainty about the key to repeat the above reasoning.

Definition 5.2
Consider a ciphertext-only attack on a cryptosystem $\in$ with key-space $\mathcal{K}$ and plaintext source $\ominus$. Then the *unicity distance* of this cryptosystem is defined by

$$\inf \{n \in \mathbb{N}^+ \mid D_n \geq H(K)\},$$

where $H(K)$ is the entropy of the key and D_n the redundancy in the plaintext.

As soon as the redundancy in the plaintext exceeds the uncertainty about the key, the cryptanalist with sufficient resources may be able to determine that plaintext from the ciphertext. Thus, the unicity distance indicates the user of a cryptosystem when to change the key in order to keep the system sufficiently secure.

Example 5.2 (Part 2)

We continue with Example 5.2. Assume that a simple substitution has been applied to an English text (see Subsection 2.1.2). Assuming that all 26! possible substitutions are equally likely, one has

$$H(K) = -\sum_{i=1}^{26!} \frac{1}{26!} \log_2 \frac{1}{26!} = \log_2 26! \approx 88.382 \text{ bits.}$$

If one approximates the redundancy D_n in a text of n letters by $(4.70 - 1.50)\,n = 3.20\,n$ bits, one obtains a unicity distance of $88.4 / 3.2 \approx 28$ characters.

According to Friedman [Frie73]: "practically every example of 25 or more characters representing the mono-alphabetic substitution of a "sensible" message in English can be readily solved." These two numbers are in remarkable agreement.

5.2   Mutual Information and Unconditionally Secure Systems

Quite often random variables contain information about each other. In cryptosystems, the plaintext and the ciphertext are related through the key. In this section we shall give a formal definition (in the information theoretic sense of the word) of an unconditionally secure cryptosystem

Let X and Y be two random variables, defined on $\mathcal{X}$ resp. $\mathcal{Y}$. The *joint distribution* $\Pr_{X,Y}(X = x, Y = y)$ of X and Y is often shortened to just

$$p_{X,Y}(x, y).$$

Similarly, the *conditional probability* $\Pr_{X|Y}(X = x \mid Y = y)$ that $X = x$, given that $Y = y$, is denoted by

$$p_{X|Y}(x \mid y).$$

It satisfies the relation

$$p_{X,Y}(x, y) = p_{X|Y}(x \mid y) \cdot p_Y(y) \tag{5.5}$$

The *uncertainty* about X given $Y = y$ is defined analogous to the entropy function by

$$H(X \mid Y = y) = -\sum_{i=1}^{n} p_{X|y}(x \mid y).\log_2 p_{X|y}(x \mid y). \tag{5.6}$$

It can be interpreted as the expected amount of information that a realization of X gives, when the occurrence of $Y = y$ is already known.

The *equivocation* $H(X \mid Y)$ or *conditional entropy* of X given Y is the expected value of $H(X \mid Y = y)$ over all y. In formula,

$$
\begin{aligned}
H(X \mid Y) &= \sum_{y \in \mathcal{Y}} p_y(y).H(X \mid Y = y) \\[4pt]
&\stackrel{(5.6)}{=} -\sum_{y \in \mathcal{Y}} p_y(y).\sum_{x \in X} p_{X|y}(x \mid y).\log_2 p_{X|y}(x \mid y) \\[4pt]
&\stackrel{(5.5)}{=} -\sum_{x \in X} \sum_{y \in \mathcal{Y}} p_y(y).p_{X|y}(x \mid y).\log_2 p_{X|y}(x \mid y) \\[4pt]
&= -\sum_{x \in X} \sum_{y \in \mathcal{Y}} p_{X,y}(x, y).\log_2 p_{X|y}(x \mid y).
\end{aligned}
\tag{5.7}
$$

Let $H(X, Y)$ be defined analogously to the entropy function H for one variable.

> **Theorem 5.1** *Chain Rule*
>
> $$H(X, Y) = H(X) + H(Y \mid X) = H(Y) + H(X \mid Y)$$

Proof: We use (5.5) and (5.7).

$$
\begin{aligned}
H(X, Y) &= \\
&= -\sum_{x \in X} \sum_{y \in \mathcal{Y}} p_{X,y}(x, y).\log_2 p_{X,y}(x, y) \\[4pt]
&= -\sum_{x \in X} \sum_{y \in \mathcal{Y}} p_{X,y}(x, y).\log_2 p_y(y) - \sum_{x \in X} \sum_{y \in \mathcal{Y}} p_{X,y}(x, y).\log_2 p_{X|y}(x \mid y) \\[4pt]
&= -\sum_{y \in \mathcal{Y}} p_y(y).\log_2 p_y(y) + H(X \mid Y) = H(Y) + H(X \mid Y).
\end{aligned}
$$

The second equality follows by a symmetry argument.

□

In words, the above theorem states that the uncertainty about a joint realization of X and Y equals the uncertainty about X plus the uncertainty about Y given X.

> **Corollary 5.2**
> Let X and Y are independent random variables. Then
>
> i) $H(X, Y) = H(X) + H(Y)$,
> ii) $H(X \mid Y) = H(X)$,
> iii) $H(Y \mid X) = H(Y)$.

Proof: To prove i) we repeat the proof of Theorem 5.1 with $p_{X,y}(x, y) = p_X(x).p_y(y)$.

$$H(X, Y) = -\sum_{x \in X} \sum_{y \in \mathcal{Y}} p_{X,y}(x, y).\log_2 p_{X,y}(x, y)$$

$$= -\sum_{x \in X} \sum_{y \in Y} p_{X,Y}(x).\log_2 p_X(x) - \sum_{x \in X} \sum_{y \in Y} p_{X,Y}(x, y).\log_2 p_Y(y)$$

$$= -\sum_{x \in X} p_X(x).\log_2 p_X(x) - \sum_{y \in Y} p_Y(y).\log_2 p_Y(y)$$

$$= H(X) + H(Y).$$

Statements ii) and iii) follow directly from i) and the chain rule.

□

The amount of information (see (5.1) that a realization $Y = y$ gives about a possible realization $X = x$ can be quantified as the amount of information that the occurrence of $X = x$ gives minus the amount of information that $X = x$ will give when $Y = y$ is already know. We denote this by $I_{X;Y}(x, y)$. It follows that

$$I_{X,Y}(x; y) = (-\log_2 p_X(x)) - (-\log_2 p_{X|Y}(x \mid y))$$

$$= -\log_2 \frac{p_X(x)}{p_{X|Y}(x|y)} \overset{(5.5)}{=} -\log_2 \frac{p_X(x).p_Y(y)}{p_{X,Y}(x,y)} = I_{Y;X}(y; x).$$

Note the symmetry in $I_{X,Y}(x; y) = I_{Y;X}(y; x)$.

The *mutual information* $I(X; Y)$ of X and Y is defined as the expected value of $I_{X,Y}(x; y)$, i.e.

$$I(x; y) = -\sum_{x \in X} \sum_{y \in Y} p_{X,Y}(x, y).I_{X;Y}(x; y)$$

$$= -\sum_{x \in X} \sum_{y \in Y} p_{X,Y}(x, y).\log_2 \frac{p_X(x).p_Y(y)}{p_{X,Y}(x,y)} \tag{5.8}$$

$$= -\sum_{x \in X} \sum_{y \in Y} p_{X,Y}(x, y).\log_2 \frac{p_X(x)}{p_{X|Y}(x|y)} = I(Y; X).$$

Theorem 5.3
$$I(X; Y) = H(X) + H(Y) - H(X, Y) = H(X) - H(X \mid Y) = H(Y) - H(Y \mid X).$$

Proof: From (5.8) it follows that

$$I(X; Y) =$$

$$= -\sum_{x \in X} \sum_{y \in Y} p_{X,Y}(x, y).\log_2 \frac{p_X(x)}{p_{X|Y}(x|y)}$$

$$= -\sum_{x \in X} \sum_{y \in Y} p_{X,Y}(x, y).\log_2 p_X(x) + \sum_{x \in X} \sum_{y \in Y} p_{X,Y}(x, y).\log_2 p_{X|Y}(x \mid y)$$

$$= -\sum_{x \in X} p_X(x).\log_2 p_X(x) - H(X \mid Y) = H(X) - H(X \mid Y).$$

The other statements follow from Theorem 5.1.

□

$I(X; Y)$ can be interpreted as the expected amount of information that Y gives about X (or X about Y).

Example 5.3

The binary symmetric channel can be described as follows. A source sends $X = 0$ or $X = 1$, each with probability $1/2$. The receiver gets $Y = X$ with probability $1 - p$ and $Y = 1 - X$ with probability p. It follows that $\mathcal{X} = \mathcal{Y} = \{0, 1\}$ and that

$$p_Y(0) = p_{Y|X}(0 \mid 0)\, p_X(0) + p_{Y|X}(0 \mid 1)\, p_X(1) = (1-p).\tfrac{1}{2} + p.\tfrac{1}{2} = \tfrac{1}{2}.$$

Similarly, $p_Y(1) = 1/2$. Also $p_{X,Y}(0, 0) = p_{X,Y}(1, 1) = (1 - p)/2$ and $p_{X,Y}(0, 1) = p_{X,Y}(1, 0) = p/2$. So, for the binary symmetric channel we have by (5.8)

$$I(X; Y) = -2\left\{ \tfrac{1-p}{2} \log_2 \tfrac{1/2}{1-p} + \tfrac{p}{2} \log_2 \tfrac{1/2}{p} \right\} =$$

$$= 1 + p.\log_2 p + (1 - p).\log_2(1 - p) = 1 - H(p).$$

We conclude that the receiver gets $1 - H(p)$ bits of information about X per received symbol Y. How to approach this quantity $1 - H(p)$ is the fundamental problem in algebraic coding theory [MacWS77], Section 1.6.

For $p = 1/2$ the receiver gets no information (since $H(1/2) = 1$) about the transmitted symbols, as is to be expected.

Let us now return to the conventional cryptosystem as explained in Chapter 1. Assume that a probability distribution $\Pr_K(K = k)$ is defined on the keyspace $\mathcal{K}$ and let the sequence of random variables

$$M^{(u)} = (M_0, M_1, \ldots, M_{u-1})$$

denote the plaintext, and let

$$C^{(v)} = (C_0, C_1, \ldots, C_{v-1})$$

denote the ciphertext. So, $C^{(v)} = E_k(M^{(u)})$. In most applications v will be equal to u. Since E_k is a one-to-one mapping, the plaintext is uniquely determined by the key and the ciphertext, therefore, one has

$$H(M^{(u)} \mid K, C^{(v)}) = 0. \tag{5.9}$$

Of course the user of the cryptosystem is interested to know how much information $C^{(v)}$ leaks about $M^{(u)}$.

Theorem 5.4

$$I(M^{(u)}; C^{(v)}) \geq H(M^{(u)}) - H(K)$$

In words: the uncertainty about the key together with the information that the ciphertext gives about the plaintext is greater than or equal to the uncertainty about the plaintext. Again, this reflects our intuition.

Proof of Theorem 5.4:

By (5.9) and the chain rule (Thm. 5.1, which also applies to conditional entropies) one has that

$$H(K \mid C^{(v)}) = H(K \mid C^{(v)}) + H(M^{(u)} \mid K, C^{(v)}) = H(M^{(u)}, K \mid C^{(v)})$$

$$= H(M^{(u)} \mid C^{(v)}) + H(K \mid M^{(u)}, C^{(v)}) \geq H(M^{(u)} \mid C^{(v)}).$$

In words: given the ciphertext the uncertainty about the key is at least as great as the uncertainty about the plaintext. This reflects the property that knowing the ciphertext, one can reconstruct the plaintext from the key, but not necessarily the other way around.

It follows that

$$H(M^{(u)} \mid C^{(v)}) \leq H(K \mid C^{(v)}) \leq H(K)$$

and by Theorem 5.3 that

$$I(M^{(u)}; C^{(v)}) = H(M^{(u)}) - H(M^{(u)} \mid C^{(v)}) \geq H(M^{(u)}) - H(K).$$

□

> **Definition 5.3**
> A cryptosystem is called *unconditionally secure* or is said to have *perfect secrecy* if
> $$I(M^{(u)}, C^{(v)}) = 0.$$

> **Corollary 5.5**
> A necessary condition for a cryptosystem to be unconditionally secure is given by
> $$H(M^{(u)}) \leq H(K).$$

In cryptosystem where all keys and all plaintexts are equally likely, Corollary 5.5 states that you need to have at least as many keys as plaintexts.

Example 5.4

Suppose that we have 2^k keys, all with probability $1/2^k$. Then

$$H(K) = -\sum_{i=1}^{2^k} \frac{1}{2^k}.log_2 \frac{1}{2^k} = k \text{ bits.}$$

If the messages are the outcome of u tossings with a fair coin, one has in a similar way that $H(M^{(u)})$, so, for perfect secrecy one needs $k \geq n$.

This can be realized the encryption $c^{(u)} = m^{(u)} \oplus k^{(u)}$, where $k^{(u)}$ stands for the first u bits of the key k and where $\oplus$ stands for a coordinatewise modulo 2 addition. With this encryption, with each ciphertext $c^{(u)}$ each possible plaintext is still equally likely.

5.3 Problems

Problem 5.1
Show that function $-\sum_{i=1}^{n} p_i.\log_2 p_i$ satisfies properties P1-P4 in Section 5.1.

Problem 5.2
Let $\alpha \leq 1/2$.

a) Prove that

$$\frac{1}{n+1} \frac{n^n}{k^k(n-k)^{n-k}} \leq \binom{n}{k} \leq \frac{n^n}{k^k(n-k)^{n-k}}.$$

b) Show that these inequalities imply that

$$\lim_{x\to\infty} \frac{1}{n} \log \sum_{i=0}^{\lfloor \alpha n \rfloor} \binom{n}{i} = h(\alpha),$$

where $h(x)$ is the entropy function defined in (5.4).

Problem 5.3
Assume that the English language has an information rate of 1.5 bits per letter. What is the unicity distance of the Caesar cipher, when applied to an English text?
Answer the same question for the Vigenère cryptosystem with key length r.

Problem 5.4
Consider a memoryless message source that generates an output letter X that is uniformly distributed over the alphabet $\{0, 1, 2\}$.
After transmission over a channel the symbol Y, that is received, will be equal to X with probability $1 - p$, $0 \leq p \leq 1$, and it will be equal to any of the other two letters in the alphabet with probability $p/2$.
Compute the mutual information $I(X, Y)$ between X and Y.

Problem 5.5
Let $\mathfrak{S}$ be a plaintext source that generates independent, identical distributed letters X from $\{a, b, c, d\}$. The probability distribution is given by $\Pr(X = a) = 1/2$, $\Pr(X = b) = 1/4$, and $\Pr(X = c) = \Pr(x = d) = 1/8$. Consider the two coding schemes:

	scheme A		scheme B
a	$\longrightarrow$ 00	a	$\longrightarrow$ 0
b	$\longrightarrow$ 01	b	$\longrightarrow$ 10
c	$\longrightarrow$ 10	c	$\longrightarrow$ 110
d	$\longrightarrow$ 11	d	$\longrightarrow$ 111

The output sequence of the plaintext X is first converted into a $\{0, 1\}$-sequence by means of one of the above coding schemes and subsequently encrypted with the DES algorithm.
What is the unicity distance for both coding schemes?

Problem 5.6
Prove that the one-time pad is an unconditionally secure cryptosystem.

6 Data Compression Techniques

It is clear from Chapter 5 (see Definitions 5.1 and 5.2) that the security of a cryptosystem can be significantly increased by reducing the redundancy in the plaintext. In Example 5.1 such a reduction has been demonstrated.

In this chapter we shall describe two general methods to reduce the redundancy. The process of removing redundancy from plaintexts is called *data compression* or *source coding*.

6.1 Basic Concepts of Source Coding for Stationary Sources

Let a plaintext source $\mathfrak{S}$ output independently chosen symbols from the alphabet $\{m_1, m_2, \ldots, m_n\}$ with respective probabilities $p_1, p_2 \ldots, p_n$. Symbol m_i will be encoded into a binary string c_i of length l_i, $1 \le i \le n$.

The set $\{c_1, c_2, \ldots, c_n\}$ is called a *code C* for source $\mathfrak{S}$. The idea of data compression is to use such a code that the expected value of the length of the encoded plaintext is minimal. Since the symbols generated by the plaintext source are independent of each other, it suffices to minimize the expected length of an encoded symbol

$$L = \sum_{i=1}^{n} p_i \, l_i. \tag{6.1}$$

The minimization has to take place over all possible codes C for source $\mathfrak{S}$. There is however an additional constraint. A receiver (decoder) has to be able to retrieve the individual messages from the concatenation of the successive codewords. Not every code has this property. Indeed let $C = \{0, 01, 10\}$. The sequence 010 can be made in two ways: 0 followed by 10 and 01 followed by 0. This ambiguity has to be avoided.

> **Definition 6.1**
> A code C is called *uniquely decodable* (shortened to *U.D.*) if every concatenation of codewords from C can only in one way be split up into individual codewords.

Example 6.1

Let $n = 4$ and $C = \{0, 01, 011, 111\}$ (this is the code of Example 5.1 in reversed order). This code C is U.D., as we shall now demonstrate.

Consider a concatenation of codewords. If the left most bit is a 1, the left most codeword is 111. If on the other hand the left most bit is a 0, the concatenation either looks like $0\overset{k}{\overline{11\ldots1}}$, for some $k \ge 0$, or it starts with the subsequence $0\overset{k}{\overline{11\ldots1}}0$ for some positive integer k.

Depending on whether $k = 3l$, $3l + 1$, or $k = 3l + 2$, the left most codeword is 0, 01 resp. 011. One *can now remove this codeword and apply the same decoding rule to the remaining, shorter concatenation of codewords.*

Theorem 6.1   *McMillan Inequality* [McMi56]
A necessary and sufficient condition for the existence of a uniquely decodable code C of cardinality n with codewords of length l_i, $1 \leq i \leq n$, is

$$\sum_{i=1}^{n} \frac{1}{2^{l_i}} \leq 1.$$ (6.2)

Proof: We shall only prove that the inequality above is a necessary condition for the existence of a U.D. code with codeword c_i of length l_i, $1 \leq i \leq n$. That it also is a sufficient condition will be proved later in this chapter.

Let $L = \sum_{i=1}^{n} \frac{1}{2^{l_i}}$ and let us assume (without loss of generality) that $l_1 \leq l_2 \leq \ldots \leq l_n$. Then

$$L^N = \left(\sum_{i=1}^{n} \frac{1}{2^{l_i}} \right)^N = \sum_{j=N.l_1}^{N.l_n} \frac{A_j}{2^j},$$

where A_j is the number of ways to write j as $l_{i_1} + l_{i_2} + \ldots + l_{i_N}$, or, alternatively, A_j is the number of ways to make a concatenation of N codewords of total length j.

Because C is U.D., no two different choices of N-tuples of codewords will give rise (when concatenated) to the same string of length j. So, $A_j \leq 2^j$.

Substitution of this inequality in (6.2) implies that for all $N \geq 1$

$$L^N \leq \sum_{j=N.l_1}^{N.l_n} 1 = N(l_n - l_1) + 1.$$

Since the left-hand side grows exponentially in N, while the right hand side is a linear function of N, we conclude that $L \leq 1$.

□

As can be seen in Example 6.1, one may have to look for a much longer prefix of the received sequence than the length of the longest codeword to be able to decode it. This is not very practical.

Definition 6.2
A code C is called a *prefix code* or *instantaneous* if no codeword is a prefix of another codeword.

The code in Example 6.1 is not a prefix code, since the codeword 0 is a prefix of the codeword 01. The code in Example 5.1 clearly is prefix code. For the decoding of a prefix code one simply looks for a prefix of the received sequence that is a codeword. Because the code is a prefix code this codeword is unique. Remove it and proceed in the same way.

Note that when a prefix code is used, one only needs to examine at most l_n bits of the received sequence to determine the first codeword in the received sequence.

The above observation proves the next theorem.

> **Lemma 6.2**
> A prefix code is uniquely decodable.

> **Theorem 6.3**   *Kraft Inequality* [Kraf49]
> A necessary and sufficient condition for the existence of a prefix code with codeword lengths l_i, $1 \le i \le n$, is

$$\sum_{i=1}^{n} \frac{1}{2^{l_i}} \le 1 \tag{6.3}$$

Proof: A prefix code is U.D. by Lemma 6.2. So, it follows from the McMillan inequality (Thm. 6.1) that (6.3) is a necessary condition for a code to be a prefix code.

We shall now prove that (6.3) implies the existence of a prefix code with codewords $\underline{c}_i$ of lengths l_i, $1 \le i \le n$, and a fortiori of a U.D. code with these lengths.

Without loss of generality $l_1 \le l_2 \le \ldots \le l_n$. Because of this ordering and since $\sum_{i=1}^{n-1} \frac{1}{2^{l_i}} < 1$ we can define vectors $\underline{c}_i = (c_{i,1}, c_{i,2}, \ldots, c_{i,l_i})$, $1 \le i \le n$, by the binary expansion of $\sum_{j=1}^{i-1} 1/2^{l_j}$:

$$\sum_{j=1}^{i-1} \frac{1}{2^{l_j}} = \frac{c_{i,1}}{2} + \frac{c_{i,2}}{2^2} + \ldots + \frac{c_{i,l_i}}{2^{l_i}}.$$

For instance, $\underline{c}_1 = (0, 0, \ldots, 0)$ of length l_1, $\underline{c}_2 = (0, \ldots, 0, 1, 0, \ldots, 0)$ of length l_2 with a one on coordinate l_1 etc. By definition, $\underline{c}_i$ has length l_i.

It remains to show that no $\underline{c}_u$ can be the prefix of a codeword $\underline{c}_v$, $u \ne v$. Suppose the contrary. Clearly $l_u \ne l_v$, otherwise the two words would be identical. So, $l_u < l_v$ and thus $u < v$. It also follows that

$$\sum_{j=1}^{v-1} \frac{1}{2^{l_j}} - \sum_{j=1}^{u-1} \frac{1}{2^{l_j}} \overset{\text{def.}}{=} \sum_{j=1}^{l} \frac{c_{v,j}}{2^{l_j}} - \sum_{j=1}^{l} \frac{c_{u,j}}{2^{l_j}} \overset{\text{prefix}}{=}$$

$$\sum_{j=l_u+1_v}^{l} \frac{c_{v,j}}{2^{l_j}} \le \sum_{j=l_u+1_v}^{l} \frac{1}{2^{l_j}} < \sum_{j=l_u+1}^{\infty} \frac{1}{2^{j}} = \frac{1}{2^{l_u}},$$

while on the other hand

$$\sum_{j=1}^{v-1} \frac{1}{2^{l_j}} - \sum_{j=1}^{u-1} \frac{1}{2^{l_j}} = \sum_{j=u}^{v-1} \frac{1}{2^{l_j}} \ge \frac{1}{2^{l_u}}.$$

These two inequalities contradict each other.

$\square$

Example 6.2

Consider $l_1 = 1$, $l_2 = 2$, $l_3 = 3$, and $l_4 = l_5 = 4$.

Since $\frac{1}{2^1} + \frac{1}{2^2} + \frac{1}{2^3} + \frac{1}{2^4} + \frac{1}{2^4} = 1$, the Kraft inequality is satisfied.

The proof above gives the following codewords (we have used the Mathematica functions Length*,* Do*,* Table*,* IntegerDigits*, and* Print*):*

```
l = {1, 2, 3, 4, 4};
L = Length[l]; c =.;
c[1] = Table[0, {l[[1]]}];
Do[c[i] = IntegerDigits[(∑_{j=1}^{i-1} 1/2^{l[[j]]}) 2^{l[[i]]}, 2], {i, 2, L}];
Do[Print[c[i]], {i, 1, L}]
```

{0}

{1, 0}

{1, 1, 0}

{1, 1, 1, 0}

{1, 1, 1, 1}

This code is a prefix code, as one can easily verify.

It is quite remarkable that the McMillan and the Kraft conditions ((6.2) and (6.3) are the same. It follows that the smallest average value of the length of a U.D. code is equal to the smallest average value of the length of a prefix code!

The next two theorems give bounds on the average value of the length of a prefix code (or a U.D. code).

Theorem 6.4
Consider a plaintext source $\mathfrak{S}$ that outputs messages m_i with probability p_i, $1 \leq i \leq n$. Let C be a U.D. code which maps message m_i into codeword c_i of length l_i, $1 \leq i \leq n$. Then the expected value $L = \sum_{i=1}^{n} p_i l_i$ of the length of an encoding satisfies

$$L \geq H(\underline{p}).$$

Proof: It follows from the well-known inequality $\ln x \leq 1 - x$, $x > 0$, and from (6.2) that

$$H(\underline{p}) - L = -\sum_{i=1}^{n} p_i . \log_2 p_i - \sum_{i=1}^{n} p_i l_i = \frac{1}{\ln 2} \sum_{i=1}^{n} p_i . \ln \frac{1}{p_i . 2^{l_i}} \leq$$

$$\frac{1}{\ln 2} \sum_{i=1}^{n} p_i \left(\frac{1}{p_i . 2^{l_i}} - 1 \right) = \frac{1}{\ln 2} \left(\left(\sum_{i=1}^{n} \frac{1}{2^{l_i}} \right) - 1 \right) \leq 0.$$

$\square$

> **Theorem 6.5**
> Consider a plaintext $\mathcal{S}$ that outputs messages m_i with probability p_i, $1 \le i \le n$.
> Then a prefix code C exists for this source with an expected word length L, satisfying
>
> $$L < H(\underline{p}) + 1.$$

Proof: Define l_i by $l_i = \lceil \log_2 1/p_i \rceil$, $1 \le i \le n$. Then $2^{l_i} \ge 1/p_i$ and thus

$$\sum_{i=1}^{n} 1/2^{l_i} \le \sum_{i=1}^{n} p_i = 1.$$

For these values of l_i, $1 \le i \le n$, construct the code C as described in the proof of Theorem 6.3. It is a prefix code and the expected value L of its length satisfies

$$L = \sum_{i=1}^{n} p_i . l_i = \sum_{i=1}^{n} p_i . \lceil \log_2 1/p_i \rceil < \sum_{i=1}^{n} p_i . (\log_2 1/p_i + 1) = H(\underline{p}) + 1.$$

$\square$

> **Corollary 6.6**
> The minimal expected length of all prefix (or U.D.) codes for a plaintext source $\mathcal{S}$ with probability distribution $\underline{p}$ has a value L satisfying
>
> $$H(\underline{p}) \le L < H(\underline{p}) + 1.$$

We shall now apply the above corollary to N-tuples of source symbols. Since the entropy of N independent symbols equals N times the entropy of one symbol, one gets an expected length $L^{(N)}$ for an N-gram that satisfies

$$N . H(\underline{p}) \le L^{(N)} < N . H(\underline{p}) + 1.$$

It follows that

$$H(\underline{p}) \le \frac{L^{(N)}}{N} < H(\underline{p}) + \frac{1}{N}. \tag{6.4}$$

So, $\lim_{N \to \infty} \frac{L^{(N)}}{N} = H(p)$. This confirms the last of the three interpretation of the entropy function H, that were given at the beginning of Chapter 5.

We shall now derive some properties that a prefix code with minimal expected L will satisfy.

Theorem 6.7

Consider the source $\mathfrak{S}$ which outputs independent symbols m_i, $1 \le i \le n$, with probabilities $p_1 \ge p_2 \ge \ldots \ge p_n$.

Among all U.D. codes for this source, let C be one which minimizes the expected value L of the length of an encoding. Let this code C have codewords c_i of length l_i, $1 \le i \le n$. Then, after a suitable reindexing of codewords associated with the messages of the same probability,

P1) $l_1 \le l_2 \le \ldots \le l_n$.
P2) C can be assumed to be a prefix code.
P3) $\sum_{i=1}^{n} \frac{1}{2^{l_i}} = 1$.
P4) $l_{n-1} = l_n$.
P5)   Two of the codewords of length l_n differ only in their last coordinate.

Proof:

P1)   Suppose that $p_u > p_v$ and $l_u > l_v$. Make a new code C^* from C by interchanging c_u and c_v. Then C^* is also an U.D. code. The expected length L^* of C^* satisfies

$$L^* = L + p_u(l_v - l_u) + p_v(l_u - l_v) = L + (p_u - p_v)(l_v - l_u) < L.$$

This contradicts our assumption on the minimality of L.

If $p_u = p_v$, $u < v$, one can obtain $l_u \le l_v$ by a simple renumbering of the indices.

P2)   If a U.D. code exists with expected length L, then a prefix code with the same expected length L also exists because the necessary and sufficient conditions in Theorems 6.1 and 6.2. are the same.

P3)   If $\sum_{i=1}^{n} \frac{1}{2^{l_i}} < 1$ one can decrease l_n by 1 and still satisfy the Kraft inequality (6.3). By Theorem 6.2 a prefix code with smaller expected length would exist. This contradicts our assumption on C.

P4)   If $l_n > l_{n-1}$ then P1 implies that l_n is strictly greater than any of the other codeword lengths. It follows that the left hand side in P3) will be a rational number with denominator 2^{l_n}. For this reason it can not be equal to 1.

P5)   Delete the last coordinate of c_n and call the resulting vector c_n^*. Let C^* be the code $\{c_1, c_2, \ldots, c_{n-1}, c_n^*\}$. It follows from P3) that C^* does not satisfy the Kraft inequality (6.3). So C^* is not a prefix code, while C was. This is only possible if c_n^* is a proper prefix of some codeword c_i, $1 \le i \le n - 1$. This means that this c_i must have length l_n too and also that c_i and c_n^* differ in just their last coordinate.

$\square$

Property P5 gives a clue how to construct a U.D. code with minimal expected codeword length. The method will be described in the next section.

6.2 Huffman Codes

The *Huffman algorithm* [Huff52] constructs for every stationary plaintext source a prefix code that has an average codeword length that is minimal among all U.D. codes for this source. The algorithm has a recursive character.

If the plaintext source has only two possible output symbols, both with a non-zero probability of occurring, the best one can do is to assign the symbols 0 and 1 to them. Clearly, $L = 1 < H(\underline{p}) + 1$ in this case.

Each recursion step consists of two parts: a reduction process and a splitting process.

The reduction process.

Let $\mathfrak{S}$ be a plaintext source which outputs independent symbols m_i, $1 \le i \le n$, with probabilities $p_1 \ge p_2 \ge \ldots \ge p_n$. Replace the two symbols m_{n-1} and m_n by one new symbol m_{n-1}^* with probability $p_{n-1}^* = p_{n-1} + p_n$. In this way, a new source $\mathfrak{S}^*$ is obtained with one output symbol less than $\mathfrak{S}$.

The splitting process.

Let $C^* = \{\underline{c}_1, \underline{c}_2, \ldots, \underline{c}_{n-2}, \underline{c}_{n-1}^*\}$ be a prefix code of minimal expected length L^* for the output symbols $\{m_1, m_2, \ldots, m_{n-2}, m_{n-1}^*\}$ of $\mathfrak{S}^*$ (to find this code in the recursion process, one may want to reindex these symbols in order of non-increasing probabilities).

The code C is given by

$$\begin{aligned}
\underline{c}_i &= \underline{c}_i^* && \text{for } 1 \le i \le n-2, \\
\underline{c}_{n-1} &= (\underline{c}_{n-1}^*, 0), \\
\underline{c}_n &= (\underline{c}_{n-1}^*, 1)
\end{aligned}$$

In words, when the symbol m_{n-1}^* is split up in the two symbols m_{n-1} and m_n, the codeword $\underline{c}_{n-1}^*$ will be extended with a 0 resp. 1 (or the other way around) to distinguish them.

Example 6.3

Let $n = 6$ and let the plaintext source $\mathfrak{S}$ output independent symbols described by the table:

m_1	m_2	m_3	m_4	m_5	m_6
0.3	0.2	0.2	0.1	0.1	0.1

To keep track of the reduction process, we use the notation $(m_{n-1} + m_n)$ for m_{n-1}^*. After applying one reduction and a reordering of the probabilities in non-increasing order we get

m_1	m_2	m_3	$(m_5 + m_6)$	m_4
0.3	0.2	0.2	0.2	0.1

Repeating this process, one gets

m_1	$(m_4 + (m_5 + m_6))$	m_2	m_3
0.3	0.3	0.2	0.2

and

$(m_2 + m_3)$	m_1	$(m_4 + (m_5 + m_6))$
0.4	0.3	0.3

and finally

$(m_1 + (m_4 + (m_5 + m_6)))$	$(m_2 + m_3)$
0.6	0.4

For the splitting process we traverse the above process in opposite direction. We start with the code {0, 1} and at each splitting of a message into two messages, we append a zero resp. a one.

Note, how m_i is replaced by $\underline{c}_i$ at each step. We get

$(\underline{c}_1 + (\underline{c}_4 + (\underline{c}_5 + \underline{c}_6)))$	$(\underline{c}_2 + \underline{c}_3)$
(0)	(1)

and

$(\underline{c}_2 + \underline{c}_3)$	$\underline{c}_1$	$(\underline{c}_4 + (\underline{c}_5 + \underline{c}_6))$
(1)	(0, 0)	(0, 1)

and

$\underline{c}_1$	$(\underline{c}_4 + (\underline{c}_5 + \underline{c}_6))$	$\underline{c}_2$	$\underline{c}_3$
(0, 0)	(0, 1)	(1, 0)	(1, 1)

and

$\underline{c}_1$	$\underline{c}_2$	$\underline{c}_3$	$(\underline{c}_5 + \underline{c}_6)$	$\underline{c}_4$
(0, 0)	(1, 0)	(1, 1)	(0, 1, 0)	(0, 1, 1)

and as code for the source $\mathcal{S}$:

$\underline{c}_1$	$\underline{c}_2$	$\underline{c}_3$	$\underline{c}_4$	$\underline{c}_5$	$\underline{c}_6$
(0, 0)	(1, 0)	(1, 1)	(0, 1, 1)	(0, 1, 0, 0)	(0, 1, 0, 1)

We see that $l_1 = l_2 = l_3 = 2$, $l_4 = 3$, and $l_5 = l_6 = 4$. One can easily check that $\sum_{i=1}^{6} 1/2^{l_i} = 1$ and that $H(\underline{p}) \leq L < H(\underline{p}) + 1$. We use the MultiEntropy function defined in Section 5.1 and further the Mathematica function Length.

```
                                    Length[p]
MultiEntropy[p_List] := -    ∑       p[[i]] * Log[2, p[[i]]]
                                    i=1
```

```
p = {0.3, 0.2, 0.2, 0.1, 0.1, 0.1};
MultiEntropy[p]
l = {2, 2, 2, 3, 4, 4}; len = Length[l];
```

$$\sum_{i=1}^{len} \frac{1}{2^{l[[i]]}} == 1$$

$$\sum_{i=1}^{len} p[[i]] * l[[i]]$$

2.44644

True

2.5

To demonstrate this Huffman code, we apply it to a text made up by the first 6 letters of the alphabet. We first simulate the source with the Mathematica functions Which, Random and Do (note that <> joins two strings).

```
SeedRandom[12321]; randomchar[x_] :=
  Which[x < 0.3, "a", x < 0.5, "b", x < 0.7, "c",
        x < 0.8, "d", x < 0.9, "e", x < 1,    "f"];
sourcetext = ""; n = 10;
Do[sourcetext =
    sourcetext <> randomchar[Random[Real, {0, 1}]], {j, 1, n}];
sourcetext
```

eedcbccaec

To encode we use the Huffman coding determined above and the function StringReplace.

```
code = StringReplace[sourcetext, {"a" → "00", "b" → "10",
    "c" → "11", "d" → "011", "e" → "0100", "f" → "0101"}]
```

0100010011110111100010011

To compare the length of this particular coding with the entropy we use the function MultiEntropy defined above and the Mathematica function StringLength.

```
StringLength[code] / n - MultiEntropy[p]
```

```
0.253561
```

In Mathematica, the decoding can be implemented with the function StringReplace, because this function works from left to right, as follows.

```
st = StringReplace[code, {"0101" -> "f",
    "0100" -> "e", "011" -> "d", "11" -> "c", "10" -> "b",
        "00" -> "a"}]
sourcetext == st
```

```
eedcbccaec
```

```
True
```

In fact, the following figure gives a better way to describe the decoding process. Read the received string bitwise from left to right. Depending on the input symbol follow the tree from its root to the right: a 1 lets you go up and a 0 down. As soon as a leaf (end point) of the tree has been reached, write down the corresponding alphabet symbol and start again at the root with the next.

For instance, the first two symbols in "00010000010000101000010011" are "00" and lead to symbol "a". The next four symbols are "0100" and lead to "e", etc.

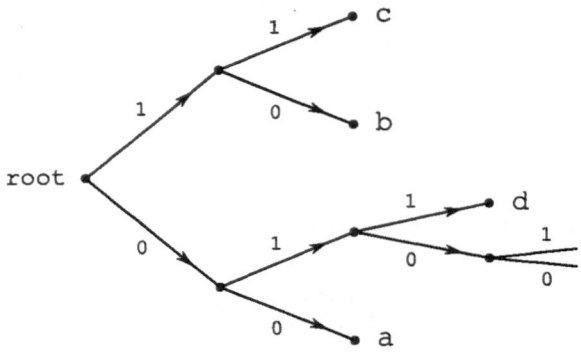

Decoding Tree for Huffman Code

Figure 6.1

> **Lemma 6.8**
> Let $\mathfrak{S}$ be a plaintext source with independent output symbols m_i, $1 \leq i \leq n$, with probabilities $p_1 \geq p_2 \geq \ldots \geq p_n$. Let $\mathfrak{S}^*$ be the reduced plaintext source with independent output symbols m_i^*, $1 \leq i \leq n-1$, with probabilities $p_i^* = p_i$, $1 \leq i \leq n-2$, and $p_{n-1}^* = p_{n-1} + p_n$.
> Assume that C^* is a prefix code for source $\mathfrak{S}^*$ that minimizes the expected value of the length of any prefix encoding for $\mathfrak{S}^*$. Let the words in C^* be denoted by c_i^*, $1 \leq i \leq n-1$. Define code C for $\mathfrak{S}$ by $c_i = c_i^*$ for $1 \leq i \leq n-2$, $c_{n-1} = ((c_{n-1}^*)_1, \ldots, (c_{n-1}^*)_{n-1}, 0)$, and $c_n = ((c_{n-1}^*)_1, \ldots, (c_{n-1}^*)_{n-1}, 1)$.
> Then C is a prefix code for source $\mathfrak{S}$ that minimizes the expected value of the length of any prefix encoding for $\mathfrak{S}$.

Proof: That C is a prefix code is straightforward. Let l_i and l_i^* denote the length of c_i resp. c_i^*. These numbers are related by $l_i = l_i^*$, $1 \leq i \leq n-2$, and $l_{n-1} = l_n = l_{n-1}^* + 1$. The expected lengths L and L^* of C resp. C^* are related by:

$$L = \sum_{i=1}^{n} p_i l_i = \sum_{i=1}^{n-2} p_i l_i + p_{n-1} l_{n-1} + p_n l_n = \sum_{i=1}^{n-2} p_i^* l_i^* + p_{n-1}(l_{n-1}^* + 1) + p_n(l_{n-1}^* + 1) =$$

$$\sum_{i=1}^{n-2} p_i^* l_i^* + (p_{n-1} + p_n) l_{n-1}^* + (p_{n-1} + p_n) = \sum_{i=1}^{n-2} p_i^* l_i^* + p_{n-1}^* l_{n-1}^* + (p_{n-1} + p_n) = L^* + (p_{n-1} + p_n).$$

From Theorem 6.7 and a reasoning like the above, we know that any prefix code $\hat{C}$ for source $\mathfrak{S}$ that minimizes the expected value of the length of an encoding for $\mathfrak{S}$ can be reduced to a code for source $\mathfrak{S}^*$ that has an expected encoding length equal to $\hat{L} - (p_{n-1} + p_n)$. Since L^* was minimal for $\mathfrak{S}^*$, we have $\hat{L} - (p_{n-1} + p_n) \geq L^* = L - (p_{n-1} + p_n)$, i.e. $\hat{L} \geq L$. Since $\hat{L}$ was minimal for $\mathfrak{S}$, we conclude that $\hat{L} = L$, i.e. C realizes the minimal expected length for an encoding of $\mathfrak{S}$.

$\square$

> **Theorem 6.9**
> Let $\mathfrak{S}$ be a plaintext source $\mathfrak{S}$ with independent output symbols m_i, $1 \leq i \leq n$, with probabilities $p_1 \geq p_2 \geq \ldots \geq p_n$.
> Then the Huffman code for this source will have an expected encoding length L that is minimal among all U.D. codes for this source.

Proof: For $n = 2$ the statement is obvious because the Huffman code will be equal to $\{(0), (1)\}$ with $L = 1$. The induction argument is a direct consequence of Lemma 6.8.

$\square$

6.3 Universal Data Compression - The Lempel-Ziv Algorithms

If one wants to compress data from a source with unknown statistics, the Huffman algorithm can not be applied. For such a situation, one needs so-called *universal data compression* techniques. Examples are the Lempel-Ziv algorithms (there are two of them) and a technique called arithmetic coding (see [ZivL77], [ZivL78], resp. [RisL79]).

In [ZivL77], the authors introduce a window of a fixed length that slides over the sequence of source symbols, say from left to right. The *sliding window* consists of two parts: a larger part on the left, called the *search buffer*, and a smaller part on the right, called the *look-ahead buffer*. The source symbols in the search buffer have already been encoded. The encoder encodes as many new source symbols in the look-ahead buffer as possible by looking in the search buffer for the largest match of already encoded symbols. Suppose that the first j unencoded source symbols match with the j symbols in the search buffer that start at position i, but that these j symbols followed by the next source symbol, say a, could not be matched. Then the encoder outputs the triple (i, j, a) and the sliding window will move $j + 1$ characters to the right.

For example, suppose that the search buffer has length 10 and the look-ahead buffer has length 5. Let the sliding window be given by

```
         1  2  3  4  5  6  7  8  9 10    1  2  3  4  5
.. b  b  c[ a| b| a| c| b| c| a| a| c| a][ a| c| b| a| c]b  a  ..
              search buffer                look ahead buffer
```

The largest match that can be found, are the first three letters in the look-ahead buffer with the three letters starting at position 3 in the search buffer. The encoder will send the triple $(2, 3, a)$, where a is the first symbol that could not be matched. The sliding window will move four positions to the right. At the beginning, when the search buffer is empty, the first encoding will start with $(0, 0, x)$, where x is the first symbol of the source.

We shall now discuss a particular variant of the *Lempel-Ziv* codes. We follow [Well99], where also an analysis of the performance can be found. The basic idea is that both sides (sender and receiver) make a dictionary that represents in a smart way substrings that have been transmitted before. If the new string of characters that is to be compressed is already in the dictionary, one encode this string by the index of the corresponding entry in the dictionary. In general, this index will be a lot shorter than the string. If the new string is not in the dictionary, more work has to be done.

The dictionary that sender and receiver are making simultaneously will be (a lot) larger than the alphabet $\mathcal{A}$ of the source $\mathcal{S}$. However, this dictionary will be stored in a very efficient way by means of a so-called linked list.

The reader has to realize that the use of the Lempel-Ziv algorithm involves some overhead. However, for files of moderate length (say, one page of text) it already makes sense to use them.

□ **Initialization**

As already remarked before, the *dictionary* will be stored by means of a *linked list*. Each entry in the list has its own *address* u. The corresponding entry consists of an ordered pair (v, a), where v should be interpreted as a pointer to another entry in the dictionary (so v is again an address) and where a is a letter in the alphabet $\mathcal{A}$. Let A denote the size of $\mathcal{A}$.

To initialize the algorithm we start with a dictionary consisting of the following $A + 1$ entries:

```
address  pointer  letter

   0        0        ∅
   1        0        a₁
   2        0        a₂
   ⋮        ⋮        ⋮
   A        0        aₐ
```

Note that all these entries point to the list element with address 0. The symbol $\emptyset$ is not an element of $\mathcal{A}$. It is an additional symbol, serving as a punctuation mark.

To be ready for the encoding, we set the pointer value v to 0 and the address pointer u to $A+1$ (u is the address of the next empty location in the linked list) .

□ **Encoding**

> **Algorithm 6.10   Encoding for Lempel-Ziv**
> **do   begin** read the next source symbol a
>   **if** (v, a) is already an entry in the dictionary **then** give v the value of the address
> of (v, a)
>   **else   begin**
>   1) transmit v,
>   2) make a new dictionary entry (v, a) with address u,
>   3) $u = u + 1$ (raise pointer u by 1),
>   4) give v the value of the address of $(0, a)$
>   **end**
> **until**   source stops.

The interpretation of the above is the following. If (v, a) is already an entry in the dictionary then the encoder is processing a string of symbols that has occurred at least once before. By assigning to v the value of the address of (v, a), one will be able later on to reconstruct this list.

If (v, a) is not an entry in the dictionary, the encoder is faced with a new string that has not been processed before. It will transmit v to let the receiver know the address of the last source symbol in the preceding string. Further, the encoder makes a new dictionary entry (v, a) with address u. The symbol a will serve as root of a new string. Pointer v is given the value of the address of entry $(0, a)$. The 0 in this entry points at dictionary entry $(0, \emptyset)$ which indicates the beginning of a new string.

Note that the output symbols of the coding process are dictionary indices, more precisely, addresses of the linked list. Their length grows logarithmically in the length of the dictionary. Note also, that each new source symbol will increasingly often not give rise to a new output symbol, because the current string will already have been encoded before.

Example 6.4 (Part 1)

Consider a binary string $\{s_i\}_{i=1}^n$ that we want to compress. So, $\mathcal{A} = \{0, 1\}$ and $A = 2$.

We initialize the coding process by putting

```
Dict = {{0, -1}, {0, 0}, {0, 1}}
u = 3; v = 0; output = {};
```

```
{{0, -1}, {0, 0}, {0, 1}}
```

Note that we have used the negative number -1 instead of the null symbol $\emptyset$.

To demonstrate the coding process, we output for each new source symbol s_i the new dictionary (represented as linked list), the new values of u and v and the complete output sequence.

We use the Mathematica function Position *that finds the place of an element in a list. Because our list contains lists as elements we add [[1]] twice. Note that we subtract 1 from the address, because our numbering starts with 0 instead of 1.*

```
Pos[s_List, el_List] := Position[s, el][[1]][[1]] - 1
```

For instance

```
l = {{3}, {5}, {7}, {2}, {1}};
el = {7};
pos[l, el]
```

```
2
```

Now we are ready for the coding process. We use the Mathematica functions Do, If, MemberQ, Append, *and* Print.

```
s = {1, 1, 0, 0, 0, 1, 0, 1, 1, 0, 0, 1};
Do[If[MemberQ[Dict, {v, s[[i]]}],
                        v = Pos[Dict, {v, s[[i]]}],
   output = Append[output, v];
   Dict = Append[Dict, {v, s[[i]]}];
   v = Pos[Dict, {0, s[[i]]}]];
   Print[Dict, ", v=", v, ", total output is ", output],
   {i, 1, Length[s]}]
```

```
{{0, -1}, {0, 0}, {0, 1}}, v=2, total output is {}
```

```
{{0, -1}, {0, 0}, {0, 1}, {2, 1}}, v=2, total output is {2}

{{0, -1}, {0, 0}, {0, 1}, {2, 1}, {2, 0}}
, v=1, total output is {2, 2}

{{0, -1}, {0, 0}, {0, 1}, {2, 1}, {2, 0}, {1, 0}}
, v=1, total output is {2, 2, 1}

{{0, -1}, {0, 0}, {0, 1}, {2, 1}, {2, 0}, {1, 0}}
, v=5, total output is {2, 2, 1}

{{0, -1}, {0, 0}, {0, 1}, {2, 1}, {2, 0}, {1, 0}, {5, 1}}
, v=2, total output is {2, 2, 1, 5}

{{0, -1}, {0, 0}, {0, 1}, {2, 1}, {2, 0}, {1, 0}, {5, 1}}
, v=4, total output is {2, 2, 1, 5}

{{0, -1}, {0, 0}, {0, 1}, {2, 1}, {2, 0}, {1, 0}, {5, 1}, {4, 1}}
, v=2, total output is {2, 2, 1, 5, 4}

{{0, -1}, {0, 0}, {0, 1}, {2, 1}, {2, 0}, {1, 0}, {5, 1}, {4, 1}}
, v=3, total output is {2, 2, 1, 5, 4}

{{0, -1}, {0, 0}, {0, 1}, {2, 1}, {2, 0}, {1, 0}, {5, 1},
{4, 1}, {3, 0}}, v=1, total output is {2, 2, 1, 5, 4, 3}

{{0, -1}, {0, 0}, {0, 1}, {2, 1}, {2, 0}, {1, 0}, {5, 1},
{4, 1}, {3, 0}}, v=5, total output is {2, 2, 1, 5, 4, 3}

{{0, -1}, {0, 0}, {0, 1}, {2, 1}, {2, 0}, {1, 0}, {5, 1},
{4, 1}, {3, 0}}, v=6, total output is {2, 2, 1, 5, 4, 3}
```

□ Decoding

For a proper decoding, the receiver must be able to reconstruct the same dictionary as was made by the transmitter. He can only act whenever a new output symbol arrives. Let v be this new symbol.

By the encoding algorithm (Alg. 6.10) the arrival of v implies that a new element (say the u-th) has to be added to the dictionary. The pointer of this new entry is given by v.

The source symbol for this entry is not known since it is the root symbol of the next string (which has not been encoded yet by the transmitter). So, only the pair $(v, ?)$ can be added to the dictionary.

The receiver is however able to fill in the missing symbol in the previous dictionary entry (at address $u - 1$).

Further, the receiver can decode the complete source symbol string associated with the received symbol.

We shall demonstrate the above process for the received sequence of Example 6.4.

Example 6.4 (Part 2)

The receiver initializes just as the receiver did. So, $u = 3$, $v = 0$, and the dictionary is given by $\{\{0, \emptyset\}, \{0, 0\}, \{0, 1\}\}$.

He receives the following list of symbols: $\{2, 2, 1, 5, 4, 3\}$.

The first received symbol is $v = 2$.

So, the new dictionary entry will be $\{2, ?\}$ and will have address $u = 3$. The question mark can not be filled in yet.

Pointer 2 in $\{2, ?\}$ points at the entry with address 2 in the dictionary, which is $\{0, 1\}$. This entry tells us that the last symbol of the previous string was a 1 and that for the preceding part we need to go to the dictionary entry with address 0. This entry is $\{0, \emptyset\}$, so we are done.

The new dictionary is given by $\{\{0, \emptyset\}, \{0, 0\}, \{0, 1\}, \{2, ?\}\}$.

The second received symbol is $v = 2$.

To fill in the question mark in the current dictionary, we look at the entry in the dictionary with address $v = 2$. This entry is $\{0, 1\}$. Its source symbol gives the value of the question mark. Therefore, we get the following dictionary $\{\{0, \emptyset\}, \{0, 0\}, \{0, 1\}, \{2, 1\}\}$.

Also, a new dictionary entry has to be added, namely $\{v, ?\}=\{2, ?\}$ at address $u = 4$.

Pointer 2 in this new entry $\{2, ?\}$ points at the entry with address 2 in the dictionary, which is $\{0, 1\}$. This entry tells us that the last symbol of the previous string was a 1 and that for the preceding part we need to go to the dictionary entry with address 0. This entry is $\{0, \emptyset\}$, so we are done. The decoded string is just "1".

The new dictionary is given by $\{\{0, \emptyset\}, \{0, 0\}, \{0, 1\}, \{2, 1\}, \{2, ?\}\}$.

The third received symbol is $v = 1$.

To fill in the question mark in the current dictionary, we look at the entry in the dictionary with address $v = 1$. This entry is $\{0, 0\}$. Its source symbol gives the value of the question mark. So, we get the following dictionary $\{\{0, \emptyset\}, \{0, 0\}, \{0, 1\}, \{2, 1\}, \{2, 0\}\}$.

Also, a new dictionary entry has to be added, namely $\{v, ?\}=\{1, ?\}$ at address $u = 5$.

Pointer 2 in this new entry $\{1, ?\}$ points at the entry with address 1 in the dictionary, which is $\{0, 0\}$. This entry tells us that the last symbol of the previous string was a 0 and that for the preceding part we need to go to the dictionary entry with address 0. This entry is $\{0, \emptyset\}$, so we are done. The decoded string is just "1".

The new dictionary is given by $\{\{0, \emptyset\}, \{0, 0\}, \{0, 1\}, \{2, 1\}, \{2, 0\}, \{1, ?\}\}$.

The fourth received symbol is $v = 5$.

To fill in the question mark in the current dictionary, we look at the entry in the dictionary with address $v = 5$. This entry is $\{1, ?\}$. The pointer 1 in this entry refers to another entry in the dictionary, namely with address 1, so to entry $\{0, 0\}$. Pointer 0 in this entry means that we are at *the root of a string. The source symbol of entry {0, 0} tells us that ? =0. So, we get the following dictionary {{0, Ø}, {0, 0}, {0, 1}, {2, 1}, {2, 0}, {1, 0}}.*

Also, a new dictionary entry has to be added, namely {v, ? }={5,?} at address u = 6.

Pointer 5 in this new entry {5, ?} points at the entry with address 5 in the dictionary, which is {1, 0}. This entry tells us that the last symbol of the previous string was a 0 and that for the preceding part we need to go to the dictionary entry with address 1. This entry is {0, 0}, so the preceding source symbol is 0 and we are pointed to {0, Ø}. This means that we are done and that the decoded string is just "00".

The new dictionary is given by {{0, Ø}, {0, 0}, {0, 1}, {2, 1}, {2, 0}, {1, 0}, {5, ? }}.

The reader is invited to continue this process.

6.4   Problems

Problem 6.1
Decode the string 01100111111111100011, which has been made with the code in Example 6.1.

Problem 6.2
Apply the Huffman algorithm to the plaintext source $\mathfrak{S}$ that generates the symbols a, b, c, d, e, f, g, and h independently with probabilities 1/2, resp. 1/4, 1/8, 1/16 1/32, 1/64, 1/128 and 1/128.
What is the expected number of bits needed for the encoding of one letter? Compare this with the entropy of the source.

Problem 6.3 [M]
Duplicate Example 6.3 for the plaintext source $\mathfrak{S}$ that generates the symbols a, b, c, d, e, f, g, and h independently with probabilities $1/3$, resp. $1/4$, $1/6$, $1/12$, $1/15$, $1/20$, $1/30$, and $1/60$.

Problem 6.4
Apply the Welch variant of the Lempel-Ziv encoding procedure to the binary sequence 0000000000000000.
Demonstrate the first 5 steps of the decoding process.

7 Public-Key Cryptography

7.1 The Theoretical Model

7.1.1 Motivation and Set-up

In modern day communication systems, conventional cryptosystems turned out to have two essential disadvantages.

i) The problem of key management and distribution.

A communication system with n users, who all use a conventional cryptosystem to communicate with each other, implies the need of $\binom{n}{2}$ keys and $\binom{n}{2}$ secure channels.

Whenever a user wants to change his keys or a new user wants to participate in the system $n - 1$ (resp. n) new keys have to be generated and distributed over as many secure channels.

ii) The authentication problem.

In computer controlled communication systems the electronic equivalent of a signature is needed. Conventional cryptosystems do no provide this feature in a natural way, especially when there is a conflict between sender and receiver, it is impossible to decide who is right. Any message made by one of them could also have been made by the other.

These disadvantages prompted researchers to look for a different kind of cryptosystem.

In [DifH76], W. Diffie and M.E. Hellman published their pioneering work on *public-key cryptosystems*. See Figure 7.1, where their system is depicted.

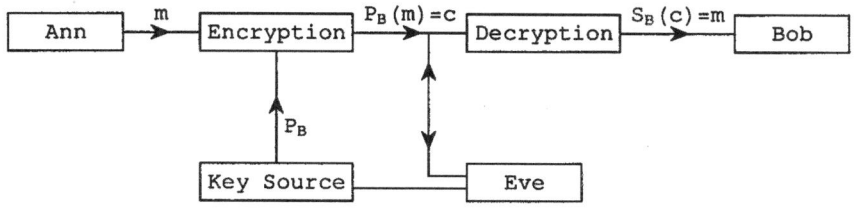

A public-key cryptosystem for encryption.

Figure 7.1

Every user U of the cryptosystem makes a pair of matching algorithms P_U and S_U (or gets them from a trustworthy authority). These algorithms operate on elements of later to be defined sets.

Algorithm P_U has to be made public by U, while algorithm S_U has to be kept secret by U. Depending on the application, these algorithms must satisfy some of following properties:

PK1 P_U and S_U are efficient algorithms, i.e. they do not need much computing time or memory space.

PK2 $S_U(P_U(m)) = m$, for every user U and for each possible message m.

PK3:   It is infeasible to find an algorithm S_U^* from P_U that satisfies $S_U^*(P_U(m)) = m$ for all m.

PK4 $P_U(S_U(m)) = m$, for every user U and for each possible message m.

PK5:   It is infeasible to find an algorithm S_U^* from P_U that satisfies $P_U^*(S_U(m)) = m$ for all m.

Properties PK3 and PK5 are not precisely formulated. Their precise meaning depends too much on the application and may vary in time.

7.1.2 Confidentiality

We assume that properties PK1, PK2, and PK3 hold.

If Alice wants to send an encrypted message m to Bob, she first looks up the public (encryption) algorithm P_B of Bob. She encrypts m by applying algorithm P_B to m. So, she sends to Bob:

$$c = P_B(m).$$

Bob recovers m from the received ciphertext c by applying his (secret) algorithm S_B to c. Indeed,

$$S_B(c) = S_B(P_B(m)) \stackrel{\text{PK2}}{=} m.$$

To make the system practical to use, property PK1 must hold. It is for the security of the system that property PK3 has to be required.

PK3 makes it possible to publish the (encryption) algorithms P_U without endangering the privacy of the transmitted messages.

We summarize the encryption scheme in the following table.

```
        Public              P_U of all users U
        Secret              S_U to all users, except U

        Properties          PK1, PK2, PK3

Encryption of m by Ann      P_B (m) = c
Decryption of c by Bob      S_B (c) = m
```

A public-key cryptosystem used for privacy.

Table 7.1

If a user U wants to change his personal key, he simply generates a new set of matching algorithms P_U and S_U satisfying PK1, PK2 and PK3 and makes P_U public. The same has to be done when a new user wants to participate in the communication system.

In [DifH76], the authors suggest to use trapdoor, one-way function for the encryption. A *one-way function* is a function $f : A \to B$ with the following properties:

F1) $f(a)$ is easy to evaluate for any $a \in A$,
F2) it is computationally infeasible to compute $f^{\leftarrow}(b)$ for almost all $b \in B$.

A *trapdoor*, one-way function is a one-way function f satisfying the further property that

F3) $f^{\leftarrow}(b)$, $b \in B$, is easy to compute given certain additional information.

Property F1 makes such a function practical to use, while property F2 makes f safe to use for encryption purposes. Property F3 makes decryption by the receiver possible.

In daily life a telephone book can be used as a one-way function; given a name one can easily find the corresponding telephone number but not the other way around. Looking up a telephone number of a person amounts to finding the name of that person. This takes $\log_2 L$ operations, if L is the number of names in the telephone guide. Finding the name if the telephone number is given means going through the whole book, name after name. The complexity is L. Property F2 is based on the exponential relation between $\log_2 L$ and L.

One-way functions f are also used to check the authenticity of a person that wants to get access to something. Each user U has his own PIN code x_U, but in a central computer only the name of U is stored together with the value $y_U = f(x_U)$.

When U wants to get access he needs to give his name and x_U. The value $f(x_U)$ will be evaluated and sent to the computer. If this values matches y_U, user U can get access, otherwise not. The advantage of this system is that the PIN codes x_U do not need to be stored in the computer. So, anybody who can read out the memory of the computer can still not determine the PIN codes.

In Chapters 8, 9, and 12 we shall discuss various proposals for trapdoor one-way functions that can be used to turn into a public-key cryptosystem. In the next chapter we shall meet a one-way function, which does not have a trapdoor.

7.1.3 Digital Signature

We assume that properties PK1, PK4, and PK5 hold.

If Alice wants to sign a message m that she wants to send to Bob, she applies her own (secret) algorithm S_A to m, so she sends

$$c = S_A(m).$$

Bob recovers m from c by applying the publicly known algorithm P_A to c. Indeed,

$$P_A(c) = P_A(S_A(m)) \overset{\text{PK4}}{=} m.$$

The value c can be used by Bob as signature for m, because, by PK5, Alice is the only person who can compute c from m, i.e. only she can make a c from a given message m such that $P_A(c) = m$.

The converse however is possible: everybody is able to find a pair (m, c) such that c carries m's signature, i.e. such that $P_A(c) = m$: simply take any c and compute $m = P_A(c)$.

So, Alice has to make sure that a randomly selected c has a negligible probability of leading to a useful message $P_A(c) = m$. This can quite easily be achieved by assuming some structure in each message m, e.g. start with the time and date.

We summarize this signature system explained above in the following table.

Public	P_U of all users U
Secret	S_U to all users, except U
Properties	PK1, PK4, PK5
Signing of m by Ann	S_A (m) = c
Verification of c by Bob	P_A (c) = m

A public-key cryptosystem used
for signing a message.

Table 7.2

Note that anybody else can also verify Alice's signature by computing $P_A(c)$, so there is no secrecy.

7.1.4   Confidentiality and Digital Signature

We assume that properties PK1, PK2, PK3, PK4, and PK5 hold.

If Alice wants to send message m in encrypted form with her own signature to Bob, she combines the techniques of Subsections 7.1.2 and 7.1.3. Thus, she uses her own secret algorithm S_A and the public algorithm P_B of Bob to send

$$c = P_B(S_A(m)).$$

Bob recovers m from c by applying P_A S_B to c. Indeed,

$$P_A(S_B(c)) = P_A(S_B(P_B(S_A(m)))) \overset{\text{PK2}}{=} P_A(S_A(m)) \overset{\text{PK4}}{=} m.$$

Although everybody can look up the public P_B, it is only Bob who can recover m from c, because only Bob knows S_B.

Bob keeps the pair $S_B(c)$, which is $S_B(P_B(S_A(m)))$, i.e. $S_A(m)$, as Alice's signature on m, just like in Subsection 7.1.3.

We summarize this in the following table.

Public	P_U of all users U
Secret	S_U to all users, except U
Properties	PK1, PK2, PK3, PK4, PK5
Ann sends	$P_B(S_A(m)) = c$
Bob computes	$P_A(S_B(c)) = m$
Bob saves	$S_B(c) = S_A(m)$

A public-key cryptosystem used
for encryption and signing.

Table 7.3

7.2 Problems

Problem 7.1

In a communication network every user U has its own public encryption algorithm P_U and secret decryption algorithm S_U. A message m from user A (for Alice) to user B (for Bob) will always be sent in the format (c, A), with $c = P_B(m)$.

The name of the sender in this message tells Bob from whom the message originates.

Bob will retrieve m from (c, A), by computing $S_B(c) = S_B(P_B(m)) = m$ (see PK2), but Bob will also automatically send $(P_A(m), B)$ back to Alice (note that $(P_A(m), B)$ has the same format as $(P_B(m), A)$). In this way, Alice knows that her message has been properly received by Bob.

a) Show how a third user E (for Eve) of the network can retrieve message m that was sent by Alice to Bob. You may assume that Eve can intercept all messages that are communicated over the network, and that Eve can also transmit her own texts, as long as they have the right format.

b) Show that communication over this network is still not safe if the protocol is such that Alice sends $P_B((P_B(m), A))$ to Bob and that Bob automatically sends $P_A((P_A(m), B))$ back to Alice.

8   Discrete Logarithm Based Systems

8.1   The Discrete Logarithm System

8.1.1   The Discrete Logarithm Problem

In [DifH76], Diffie and Hellman propose a public-key distribution system which is based on the apparent difficulty of computing logarithms over the finite field GF(p), p prime, which is also often denoted by $\mathbb{F}_p$ or $\mathbb{Z}_p$. The reader, who is not familiar with the theory of finite fields is referred to Appendix B.

Let α be a primitive element (or generator) of GF(p). So, each nonzero element c in GF(p) can be written as

$$c = \alpha^m \tag{8.1}$$

where m is unique modulo $p - 1$.

Example 8.1

In GF(7) the element $\alpha = 3$ is a primitive element, as can be checked from $3^2 \equiv 2 \,(mod\ 7)$, $3^3 \equiv 6 \,(mod\ 7)$, $3^4 \equiv 4 \,(mod\ 7)$, $3^5 \equiv 5 \,(mod\ 7)$, and $3^6 \equiv 1 \,(mod\ 7)$.

This can be done at once with

```
Mod[3^{1, 2, 3, 4, 5, 6}, 7]
```

```
{3, 2, 6, 4, 5, 1}
```

Example 8.2

In GF(197), the element $\alpha = 2$ is primitive. Such an element can be found with the Mathematica function PowerList *(for which the package* Algebra`FiniteFields *first has to be initialized). This function finds a primitive element in $\mathbb{F}_p$ and generates all its powers (starting with the 0-th). The second element in this list is the primitive element itself.*

```
<< Algebra`FiniteFields`
```

```
p = 197;
PowerList[GF[p, 1]][[2]]
```

```
{2}
```

To check that 2 is a primitive element modulo 197 is a lot easier. The multiplicative group Z_{197}^* has order 196, so each element has an order dividing 196 (see Theorem B.5).

With the function FactorInteger *one can find the different prime factors of 196.*

```
FactorInteger[196]
```

```
{{2, 2}, {7, 2}}
```

It now follows from

```
PowerMod[2, 196 / 7, 197] == 1
PowerMod[2, 196 / 2, 197] == 1
```

```
False
```

```
False
```

that the order of 2 modulo 197 does not divide $196/2$ or $196/7$, so the order must be 196.

If m is given, c can be computed from (8.1) with $2 \cdot \lceil \log_2 p \rceil$ multiplications (see [Knut81], pp. 441-466). One can realize this by creating the table $\alpha, \alpha^2, \alpha^{2^2}, \alpha^{2^3}, ..., \alpha^{2^{\lceil \log_2 p \rceil - 1}}$ (each is the square of the previous one) and multiplying elements from this table, whose exponents add up to m. To this end the binary representation of m can be used.

Example 8.3

Take $m=171$. Its binary expansion is 10101011, as follows from the Mathematica function IntegerDigits.

```
IntegerDigits[171, 2]
```

```
{1, 0, 1, 0, 1, 0, 1, 1}
```

So, now one has $\alpha^{171} = \alpha^{128} \cdot \alpha^{32} \cdot \alpha^8 \cdot \alpha^2 \cdot \alpha$.

This calculation can also be done on the fly. The leftmost 1 in the binary representation of m

stands for α. Each subsequent symbol (from the left) in the binary representation implies a squaring of the previous result, but if this symbol is a 1 also an additional multiplication by α has to be performed.

```
Clear[a];
```

$$\left(\left(\left(\left(\left(\left((a)^2\right)^2 a\right)^2\right)^2 a\right)^2\right)^2 a\right)^2 a$$

a^{171}

If one has to perform the same modular exponentiation many times, for instance on a smart card implementation, there are ways to do this with fewer multiplications.

Definition 8.1
An *addition chain* for an integer m is a sequence of integers $a_1 = 1 < a_2 < \ldots < a_{l-1} < a_l = m$, with the property that each a_k, $2 \le k \le l$, is the sum of two (not necessarily different) preceding a_i's.
The index l is called the *length* of the chain.

The way that addition chains are used for (modular) exponentiation, is clear. If $a_k = a_i + a_j$, then $\alpha^{a_k} = \alpha^{a_i}.\alpha^{a_j}$. Hence, $\alpha^m = \alpha^{a_l}$ can now be computed recursively.

It is, in general, not obvious how the shortest addition chain of an integer m can be found. See [Knut81], Section 4.6.3 and [Bos92], Chapter 4.

Example 8.4

An addition chain for $m = 15$ is the sequence 1,2,3,6,12,15.

Note that the calculation of α^{15} involves 5 multiplications with this addition chain and 6 multiplications with the binary method explained before.

In *Mathematica* the PowerMod function is a fast way to compute modular exponentiations.

```
a = 2; m = 171111111; p = 197888888;
PowerMod[a, m, p]
```

55895160

The opposite problem of finding m satisfying (8.1) from c, is not so easy. It is called the *discrete logarithm problem*, because in $\mathbf{Z}_p$ the exponent m can be written like $m = \log_\alpha c$.

In [Knut73], pp.9, 575-576, one can find an algorithm that solves the logarithm problem. It involves roughly $c_1 \sqrt{p}$ operations and $c_2 \sqrt{p}$ bits of memory space (where c_1 and c_2 are some constants). In Theorem 8.1 a more precise analysis of this algorithm will be given. Writing

$t = \log_2 p$ (and forgetting about the constants), one gets the following exponential relation between exponentiation and taking logarithms.

```
exponentiation      t
taking logarithms  2^(t/2)
```

The computational discrepancy between
exponentiation and taking logarithms

Table 8.1

8.1.2   The Diffie-Hellman Key Exchange System

We shall now describe how the discrepancy in computing time between exponentiation and taking logarithms, as depicted in Table 8.1, can be used to execute a *key exchange protocol* of a "public-key cryptography"-type. Such a protocol is a method for two parties who do not share a common secret key to agree on a common key in a secure manner.

<u>Setting up the system</u>:

1) All participants share as system parameters a prime number p and a primitive element (generator) α in GF (p).

2) Each participant P chooses an integer m_P, $1 < m_p \leq p - 2$, at random, computes $c_P = \alpha^{m_P}$ and puts c_P in the public key book. Participant P keeps m_P secret.

<u>Using the system</u>:

Let us now assume that Alice (A for short) and Bob (B) want to communicate with each other using a conventional cryptosystem, but that they have no secure channel to exchange a key. With the public key book, they can agree on the common secret key

$$k_{A,B} = \alpha^{m_A m_B}.$$

Alice can compute $k_{A,B}$ by raising the publicly known c_B of Bob to the power m_A, which only she knows herself. Indeed,

$$(c_B)^{m_A} = (\alpha^{m_B})^{m_A} = \alpha^{m_A m_B} = k_{A,B}.$$

Similarly, Bob finds $k_{A,B}$ by computing $(c_A)^{m_B}$.

If somebody else (Eve) is able to compute m_A from c_A (or m_B from c_B), she can compute the key $k_{A,B}$ just like Alice or Bob did. By taking p sufficiently large, the computation time of solving this logarithm problem will be prohibitively large. Diffie and Hellman suggest to take p about 100 bits long. A different way of finding $k_{A,B}$ from c_A and c_B does not seem to exist.

There is no obvious reason to restrict the size of the finite field to a prime number. So, from now on the size of the field can be any prime power $q = p^e$ (see Theorem B.16 or Theorem B.20).

In [Lune87], Chapter XIII, efficient algorithms to find primitive elements in finite fields are described. See also Problem B.6 and Problem B.10.

We summarize the key distribution system in Table 8.2.

system parameters	field size q primitive element α
secret key of P public key of P	m_P $c_P = \alpha^{m_P}$
common key of A and B Ann computes Bob computes	$k_{A,B} = \alpha^{m_A\, m_B}$ $(c_B)^{m_A}$ $(c_A)^{m_B}$

The Diffie-Hellman Key Exchange System

Table 8.2

Example 8.5 (Part 1)

Let $p = 197$ and $\alpha = 2$.

Alice chooses as a random secret exponent $m_A = 56$ and Bob as a random secret exponent $m_B = 111$. They compute their public key with the <u>PowerMod</u> *function.*

```
cA = PowerMod[2, 56, 197]
cB = PowerMod[2, 111, 197]
```

```
178
```

```
82
```

Alice can compute the common key with Bob by raising the publicly known c_B to the power m_A, which she only knows. She gets:

```
PowerMod[82, 56, 197]
```

```
114
```

Bob gets the same common key by raising c_A to the power m_B. Indeed, he gets:

```
PowerMod[178, 111, 197]
```

```
114
```

8.2   Other Discrete Logarithm Based Systems

8.2.1   ElGamal's Public-Key Cryptosystems

In [ElGa88], two public-key systems are described that are based on the discrete logarithm problem. One can be used for encryption purposes, the other as a signature scheme.

In both systems the transmitted text is longer than the plaintext.

□ **Setting It Up**

As system parameters, all participants share a prime number p and a generator (primitive element) α of the multiplicative group Z_p^*. The generalization to finite fields is straightforward and will be omitted.

A variation that one sees quite often is to consider Z_q^* with q prime and an element $\alpha \in Z_q^*$ of large prime order, say p, instead of taking a primitive element. Note that by Theorem B.5, p must divide $q - 1$.

Each participant P chooses an integer m_P, $1 \leq m_p \leq p - 1$, at random, computes $c_P = \alpha^{m_P} \pmod{p}$ and makes c_P public. Participant P keeps m_P secret.

As a variation, each participant can also choose his own finite field and primitive element α, instead of having them as system parameters, but there seems to be little reason to do so.

□ **ElGamal's Secrecy System**

Encryption of a message for Bob.

Suppose that Alice wants to send a private message u to Bob. The message is represented by an integer u in $\{0, 1, ..., p - 1\}$.

Alice selects a random integer r and computes $R = \alpha^r$.

Next, Alice computes $S = u.c_B^r$.

Alice sends to Bob, the pair (R, S).

Decryption by Bob.

Bob receives the pair (R, S) and can quite easily retrieve the message u with his own secret m_B with the following calculation:

$$S / R^{m_B} = u.c_B^r / \alpha^{r.m_B} = u.\alpha^{r.m_B} / \alpha^{r.m_B} = u.$$

Example 8.5 (Part 2)

We continue with Example 8.5. We have $p = 197$, $\alpha = 2$ and $c_B = 82$ as public parameters.

The number $m_B = 111$ is only known to Bob.

Suppose that Alice wants to encrypt message $u=123$ for Bob.

Let $r = 191$ be the random integer chosen by Alice (it is coprime with $p - 1$).

Alice sends the pair (R, S) computed by

```
p = 197; a = 2; cB = 82;
r = Random[Integer, {0, p-2}]
u = 123;
R = PowerMod[a, r, 197]
S = Mod[PowerMod[cB, r, 197] * u, p]
```

```
60
```

```
90
```

```
20
```

To decrypt, Bob computes $S/R^{m_B} \bmod p$ with his own secret $m_B = 111$ by means of the Mathematica functions Mod *and* PowerMod. *Note that PowerMod[a, −1, p] computes the multiplicative inverse of a modulo p (see Subsection A.3.3).*

```
mB = 111;
Mod[S * PowerMod[PowerMod[R, mB, p], -1, p], p]
```

```
123
```

An eavesdropper can not determine r from R, since we assume that taking logarithms is intractable. For that reason, this eavesdropper is not able to divide out $(c_B)^r$ from S (to obtain the secret u).

□ **ElGamal's Signature Scheme**

Signing of a message by Alice.

Suppose that Alice wants to send a signed message u to Bob. The message is again represented by an integer u in $\{0, 1, ..., p - 2\}$.

Alice selects a random integer r that is relatively prime to $p-1$ and computes $R = \alpha^r$.

Next, Alice uses her secret exponent m_A to compute S satisfying

$$u \equiv m_A R + r.S \pmod{p-1}. \tag{8.2}$$

Alice can use the extended version of Euclid's Algorithm to find S efficiently.

Alice sends to Bob the triple (u, R, S), where the pair (R, S) serves as signature on the message u.

Verification of the signature by Bob.

Bob receives the signature (R, S) together with the message u.

Bob checks this signature by verifying that

$$\alpha^u \equiv (c_A)^R R^S \pmod{p}.$$

This relation has to hold because by (8.2)

$$\alpha^u \equiv \alpha^{m_A R}.\alpha^{r.S} \equiv (\alpha^{m_A})^R.(\alpha^r)^S \equiv (c_A)^R.R^S \pmod{p}.$$

Example 8.5 (Part 3)

Continuing with Example 8.5, where we have $p = 197$, $\alpha = 2$ and $c_A = 178$ as public parameters.

The number $m_A = 56$ is only known to Alice.

Suppose that Alice wants to sign message $u=123$ for Bob.

Let $r = 97$ be the random integer chosen by Alice (it is coprime with $p-1$).

Alice computes

```
p = 197; a = 2; mA = 56;
r = 97; u = 123; S = .;
R = PowerMod[a, r, 197]
S /. Solve[{r S == u - mA * R, Modulus == p - 1}, S][[1]]
```

```
98
```

```
171
```

to find the signature $(R, S) = (98, 171)$ that she adds to her message u.

Bob checks this signature by verifying $\alpha^u \equiv (c_A)^R R^S \pmod{p}$:

```
cA = 178; R = 98; S = 171;
PowerMod[a, u, p] ==
  Mod[ PowerMod[cA, R, p] * PowerMod[R, S, p], p]
```

True

8.2.2 Further Variations

In the ElGamal scheme, the signature on a message u consists of two parts: R, being α^r with r random, and S, being a solution of $u \equiv m_A R + r.S \pmod{p-1}$ (see (8.2)). Of course one can vary this so-called *signature equation*.

The next three variations do exactly this. The reader that wants to know more about them than is presented below is referred to [MeOoV96] and [Schne96].

□ **Digital Signature Standard**

In the *Digital Signature Standard* (see [FIPS94]) the signature equation is given by:

$$r.S \equiv u + m_A.R \pmod{p-1}.$$

The system is designed by the National Security Agency (NSA) and adopted as standard by the National Institute of Standards and Technology (NIST).

DSS adds two sequences of 160 bits each to the end of a document as guarantee of its authenticity and integrity. To this end, it first compresses the document to a sequence of 160 bits by means of a cryptographically secure hash function (see Section 13.2), called the *Secure Hash Algorithm* (see [MeOoV96], §9.53 and [Schne96]).

To set up the system the following joint parameters are chosen:

i)   A prime number q whose binary representation has a word length that is divisible by 64 and lies between 512 and 1024.

ii)   A prime factor p of $q - 1$ that is 160 bits long.

iii)   A value $g = (h^{(q-1)/p} \bmod q)$, where h is less than $q - 1$, such that g is greater than 1.

Since $g^p \equiv h^{q-1} \equiv 1 \pmod{q}$ by Fermat's Theorem (A.15), it follows that the multiplicative order of g divides p. On the other p is prime, therefore, g has multiplicative order p itself (see also Theorem B.5).

Each user U chooses a secret exponent m_U, computes $c_U \equiv g^{m_U} \pmod{q}$ and makes c_U public.

When Alice wants to sign a file M, she first computes its 160 digits long hash value $h(M)$ with the Secure Hash Algorithm.

Next, she chooses a random number $r < p$ and adds as signature to M the numbers R and S, both of length 160, defined by:

$$R = ((g^r \bmod q) \bmod p),$$
$$S.r = (h(M) + m_A R \,(\mathrm{mod}\, p).$$

A receiver can check the authenticity and integrity of the received message M by evaluating:

$$w \equiv S^{-1} \, (\text{mod } p),$$
$$x \equiv h(M).w \, (\text{mod } p),$$
$$y \equiv R.w \, (\text{mod } p),$$
$$U = ((g^x.(c_A)^y \bmod q) \bmod p).$$

If $R = U$ the document will be accepted as genuine and coming from Alice. By a simple substitution one can verify that the relation $u = U$ indeed should hold.

The function of the random number r above is to hide the secret key of Alice.

□ **Schnorr's Signature Scheme**

In *Schnorr*'s signature scheme [Schno90] the signature equation (see (8.2) is given by:

$$S \equiv m_A R + r \, (\text{mod } p - 1).$$

□ **The Nyberg-Rueppel Signature Scheme**

The *Nyberg-Rueppel* signature scheme [NybR93] is slightly different from the others. Here, R is defined by

$$R = u.\alpha^r \text{ with } r \text{ random.}$$

The signature equation (see (8.2) is given by:

$$S \equiv m_A R - r \, (\text{mod } p - 1).$$

In the Nyberg-Rueppel scheme, the message u can be retrieved directly from R and S, since

$$u \equiv R.\alpha^{-r} \equiv R.\alpha^{S-m_A R} \equiv R.\alpha^S / (\alpha^{m_A})^R \equiv R.\alpha^S / c_A^{\,R} \, (\text{mod } p).$$

If u is not the hash value of a much longer other file, this feature is an advantage, because only R and S have to be sent.

8.3   How to Take Discrete Logarithms

When one has to take a logarithm in $GF(q)$, the most obvious way to reduce the workload is to factor $q - 1$ in prime power factors, compute the logarithm for each of these factors, and then combine the results with the Chinese Remainder Theorem (Thm. A.19). In Subsection 8.3.1, this method will be demonstrated for a particular technique.

As we have said before, discrete logarithm based systems are often set up in a multiplicative subgroup of $GF(q)$. This generalization does not affect the methods that will be discussed in this section.

8.3.1 The Pohlig-Hellman Algorithm

In [PohH78], Pohlig and Hellman demonstrate that discrete logarithms can be taken much faster than in $\sqrt{q}$ operations, if $q - 1$ has only small prime divisors. We shall first demonstrate this method for two special cases.

□ **Special Case:** $q - 1 = 2^n$

Examples of prime numbers that are a power of 2 plus one are given by $q = 17$, $q = 257$, and $q = 2^{16} + 1$.

```
n = 16; PrimeQ[2^n + 1]
```

```
True
```

So, let α be a primitive element in a finite field GF(q). The problem is to find m, $0 \le m \le q - 2$, satisfying (8.1) for given value of c.

Let $m_0, m_1, \ldots, m_{n-1}$ be the binary representation of the unknown m, i.e.

$$m = m_0 + m_1 2 + \ldots + m_{n-1} 2^{n-1}, \qquad m_i \in \{0, 1\}, \ 0 \le i \le n - 1.$$

Of course, it suffices to compute the unknown m_i's. Since α is a primitive element of GF(q) we know (see also Theorem B.21) that $\alpha^{q-1} = 1$ and $\alpha^i \ne 1$ for $0 < i < q - 1$.

It also follows that $\alpha^{(q-1)/2} = -1$, because the square of $\alpha^{(q-1)/2}$ is 1, while $\alpha^{(q-1)/2} \ne 1$. (We also use here that by Theorem B.15 the quadratic equation $x^2 = 1$ has ± 1 as only roots.) Hence

$$c^{(q-1)/2} = (\alpha^m)^{(q-1)/2} = \alpha^{m(q-1)/2} = \alpha^{(m_0 + m_1 2 + \ldots + m_{n-1} 2^{n-1})(q-1)/2}$$

$$\stackrel{\alpha \text{ prim.}}{=} \alpha^{m_0(q-1)/2} = \begin{cases} +1, & \text{if } m_0 = 0, \\ -1, & \text{if } m_0 = 1. \end{cases}$$

Therefore, the evaluation of $c^{(q-1)/2}$ in GF(q), which takes at most $2 \cdot \lceil \log_2 q \rceil$ multiplications, as we have seen in Subsection 8.1.1), yields m_0.

Compute $c_1 = c.\alpha^{-m_0}$. Now m_1 can be determined in the same way as above from

$$c_1^{(q-1)/4} = \alpha^{(m_1 2 + m_2 2^2 + \ldots + m_{n-1} 2^{n-1})(q-1)/4}$$

$$= \alpha^{m_1(q-1)/2} = \begin{cases} 1, & \text{if } m_1 = 0, \\ -1, & \text{if } m_1 = 1. \end{cases}$$

Compute $c_2 = c_1.\alpha^{-2 m_1} = c.\alpha^{-(m_0 + m_1 2)}$ and determine m_2 from $(c_2)^{(q-1)/8}$. Repeat this process until also m_{n-1} (and thus m) has been determined.

The above algorithm finds m from c in at most

$$n.(2. \lceil \log_2 q \rceil + 2) \approx 2. (\log_2 q)^2 \approx 2 n^2,$$

operations, where the term $+2$ comes from the evaluation of the c_i's (in the i-th step $\alpha^{-2^{i-1}}$ has to be squared and the outcome may or may not have to be multiplied to c_{i-1}).

Comparing with Table 8.1, we observe that in the current case (i.e. $q = 2^n + 1$), the discrepancy between the computational complexity of using the Diffie-Hellman scheme (one exponentiation involving $2 n$ multiplications) and breaking it ($\approx 2 n^2$ multiplications) is quadratic, which is not significant enough to make the system secure.

Remark:

Note that when $q - 1 = s . 2^t$, s odd, the t least significant bits of m can be found in exactly the same way.

Example 8.6

Consider the equation $3^m \equiv 7 \mod 17$. So, $q = 17$, $\alpha = 3$, and $c = 7$. Note that $\alpha^{-1} = 6$.

Writing $m = m_0 + 2 m_1 + 4 m_2 + 8 m_3$, we find m_0 by evaluating $c^{(q-1)/2} \mod q$.

```
PowerMod[7, 8, 17]
```

```
16
```

Since this is -1 we know that $m_0 = 1$. Compute $c_1 \equiv c/3 \equiv 6. c \equiv 8 \mod 17$. Then m_1 can be found from $c_1^{(q-1)/4} \mod q$

```
PowerMod[8, 4, 17]
```

```
16
```

Again this is -1, so $m_1 = 1$. Compute $c_2 \equiv c_1/3^2 \equiv 6^2.c_1 \equiv 16 \mod 17$. Then m_2 can be found from $c_2^{(q-1)/8} \mod q$

```
PowerMod[16, 2, 17]
```

```
1
```

Since the outcome is 1, we have $m_2 = 0$. So, $c_3 = c_2$ and m_3 can be found from $c_3^{(q-1)/16} \mod q$

```
PowerMod[16, 1, 17]
```

```
16
```

We now also have $m_3 = 1$ and thus $m = 1.2^0 + 1.2^1 + 0.2^2 + 1.2^3 = 11$. We can check this with:

```
PowerMod[3, 11, 17]
```

□ **General Case: $q - 1$ has only small prime factors**

Let $q - 1 = \prod_{i=1}^{k} p_i^{n_i}$, where the p_i's are different primes and the exponents n_i are strictly positive (see the Fundamental Theorem in Number Theory, Thm. A.6). We assume that all p_i's are small. Later we shall say precisely what we mean by that.

Instead of solving m from (8.1) directly, we shall determine

$$m^{(i)} \equiv m \,(\mathrm{mod}\ p_i^{n_i}), \quad 1 \le i \le k. \tag{8.3}$$

With the Chinese Remainder Theorem (Thm. A.19) one can compute m efficiently from these $m^{(i)}$'s.

To determine $m^{(1)}$ (the others $m^{(i)}$'s can be found in the same way) we write it in its p_1-ary representation. For the sake of convenience we drop all the sub- and superscripts referring to the $i = 1$ case.

$$m^{(1)} = m_0 + m_1 p + \ldots + m_{n-1} p^{n-1}, \quad m_l \in \{0, 1, \ldots, p-1\},\ 0 \le l \le n-1.$$

Similarly to the Special Case ($k = 1$, $p = 2$), we will find the coefficients m_i by single exponentiations.

Coefficient m_0 can be found by evaluating $c^{(q-1)/p}$. From Theorem B.21 it follows that $(c^{(q-1)/p})^p = 1$, which implies that $c^{(q-1)/p}$ is a p-th root of unity.

Define the primitive p-th root of unity ω by $\omega = \alpha^{(q-1)/p}$ and make a *table* of $1, \omega, \omega^2, \ldots, \omega^{p-1}$. Then, because $m \equiv m^{(1)} \bmod p^n$ and $m^{(1)} \equiv m_0 \bmod p$, we have

$$c^{(q-1)/p} = (\alpha^m)^{(q-1)/p} = \alpha^{m(q-1)/p} = \alpha^{m^{(1)}(q-1)/p} = \alpha^{m_0(q-1)/p} = \omega^{m_0}.$$

So, a simple table lookup of $c^{(q-1)/p}$ will yield m_0.

To determine m_1, we first compute $c_1 = c.\alpha^{-m_0}$ and then evaluate $c_1^{(q-1)/p^2}$, etc., until $m^{(1)}$ has been determined. Similar calculations have to be made to determine the other $m^{(i)}$'s.

For this algorithm, we have to make tables of the powers of the primitive p-th roots of unity for all the prime factors of $q - 1$.

The values of these factors have to be small enough to be able to store them.

Each time that we want to take a logarithm the algorithm will have to take $\sum_{i=1}^{k} n_i$ exponentiations, therefore, the algorithm involves

$$\sum_{i=1}^{k} 2. \lceil \log_2 q \rceil . n_i \approx 2. \log_2 q.(\sum_{i=1}^{k} n_i) \le 2\,(\log_2 q)^2$$

operations, if we forget about the lower order terms. Again we have a quadratic relation between using the Diffie-Hellman key-exchange system and breaking it.

□ **An Example of the Pohlig-Hellman Algorithm**

Example 8.7

Consider Equation (8.1) with q = 8101, primitive element a=6.

Note that q is a prime number, so GF(q) = $\mathbf{Z}_{8101}$.

Preliminary Calculations.

First of all we factor q − 1 and compute the multiplicative inverse of 6 modulo 8101 with the Mathematica functions `FactorInteger` *and* `PowerMod`.

```
q = 8101; a = 6;
FactorInteger[q - 1]
x = PowerMod[a, -1, q]
```

```
{{2, 2}, {3, 4}, {5, 2}}
```

```
6751
```

So, q − 1 = $2^2 \cdot 3^4 \cdot 5^2$ and $a^{-1} = 6751$.

Next we use the PowerMod function again to calculate the primitive 2-nd, 3-rd and 5-th roots of unity: $\omega_1 = 6^{(8101-1)/2} = 6^{4050}$, $\omega_2 = 6^{(8101-1)/3} = 6^{5883}$, and $\omega_3 = 6^{(8101-1)/5} = 6^{1620}$:

```
q = 8101; a = 6;
Om1 = PowerMod[a, (q - 1) / 2, q]
Om2 = PowerMod[a, (q - 1) / 3, q]
Om3 = PowerMod[a, (q - 1) / 5, q]
```

```
8100
```

```
5883
```

```
3547
```

So, $\omega_1 = 8100$, $\omega_2 = 5883$, and $\omega_3 = 3547$. With the `Table` *function we make the following three tables:*

```
q = 8101; a = 6;
Om1 = PowerMod[a, (q - 1) / 2, q];
Om2 = PowerMod[a, (q - 1) / 3, q];
Om3 = PowerMod[a, (q - 1) / 5, q];
Table[PowerMod[Om1, i, q], {i, 0, 1}]
Table[PowerMod[Om2, i, q], {i, 0, 2}]
Table[PowerMod[Om3, i, q], {i, 0, 4}]
```

```
{1, 8100}
```

```
{1, 5883, 2217}
```

```
{1, 3547, 356, 7077, 5221}
```

Hence, we have tables

$P_1 = 2$	i	0	1
	$(\omega_1)^i$	1	8100

$P_2 = 3$	i	0	1	2
	$(\omega_2)^i$	1	5883	2217

$P_3 = 5$	i	0	1	2	3	4
	$(\omega_3)^i$	1	3547	356	7077	5221

The preliminary work for the Chinese Remainder Theorem consists of solving the following three systems of linear congruence relations:

$$\begin{cases} u \equiv 1 & (\bmod 4) \\ u \equiv 0 & (\bmod 81) \\ u \equiv 0 & (\bmod 25) \end{cases}$$

$$\begin{cases} v \equiv 0 & (\bmod 4) \\ v \equiv 1 & (\bmod 81) \\ v \equiv 0 & (\bmod 25) \end{cases}$$

$$\begin{cases} w \equiv 0 & (\bmod 4) \\ w \equiv 0 & (\bmod 81) \\ w \equiv 1 & (\bmod 25) \end{cases}$$

These three systems can be solved with the Mathematica function ChineseRemainderTheorem *for which we first have to load the package* NumberTheory`NumberTheoryFunctions`

```
<<NumberTheory`NumberTheoryFunctions`
```

```
u = ChineseRemainderTheorem[{1, 0, 0}, {4, 81, 25}]
v = ChineseRemainderTheorem[{0, 1, 0}, {4, 81, 25}]
w = ChineseRemainderTheorem[{0, 0, 1}, {4, 81, 25}]
```

```
2025
```

```
6400
```

```
7776
```

So, $u \equiv 2025 \pmod{8100}$, $v \equiv 6400 \pmod{8100}$, $w \equiv 7776 \pmod{8100}$.

This concludes the preliminary work.

Solving Equation (8.1) for: $c = 7531, q = 8101$.

We first determine $m^{(i)} = m \bmod p_i^{n_i}$, $1 \leq i \leq 3$, as defined in (8.2), with the method explained above. Of course, the tables that we just made have to be consulted at each step.

First prime factor: $p_1 = 2, n_1 = 2$.

$$c = \qquad\qquad = 7531, \quad c^{(8101-1)/2} = 8100, \quad m_0 = 1,$$
$$c_1 = c.\alpha^{-1} = 8006, \quad c_1^{(8101-1)/2^2} = 1 \quad , \quad m_1 = 0.$$

Hence $m^{(1)} = 1 + 0.2^1 = 1$.

Second prime factor: $p_2 = 3, n_2 = 4$.

$$c = \qquad\qquad = 7531, \quad c^{(8101-1)/3} = 2217, \quad m_0 = 2,$$
$$c_1 = c.\alpha^{-2} = 6735, \quad c_1^{(8101-1)/3^2} = 1 \quad , \quad m_1 = 0,$$
$$c_2 = c_1 \qquad = 6735, \quad c_2^{(8101-1)/3^3} = 2217, \quad m_2 = 2,$$
$$c_3 = c_2.\alpha^{-2.3^2} = 6992, \quad c_3^{(8101-1)/3^4} = 5883, \quad m_3 = 1.$$

Hence $m^{(2)} = 2 + 0.3^1 + 2.3^2 + 1.3^3 = 47$.

Third prime factor: $p_3 = 5, n_3 = 2$.

$$c = \qquad\qquad = 7531, \quad c^{(8101-1)/5} = 5221, \quad m_0 = 4,$$
$$c_1 = c.\alpha^{-4} = 7613, \quad c_1^{(8101-1)/5^2} = 356 , \quad m_1 = 2.$$

Hence $m^{(3)} = 4 + 2.5^1 = 14$.

The final solution m is given by:

$$m \equiv u.m^{(1)} + v.m^{(2)} + w.m^{(3)} \equiv$$

```
Mod[2025 * 1 + 6400 * 47 + 7776 * 14, 8100]
```

```
6689
```

This can easily be checked.

```
PowerMod[6, 6689, 8101]
```

```
7531
```

In Mathematica, the precalculation of a, b, and c is not really necessary, because m can be computed directly from $m^{(1)}$, $m^{(2)}$, and $m^{(3)}$ with the ChineseRemainderTheorem *function:*

```
ChineseRemainderTheorem[{1, 47, 14}, {4, 81, 25}]
```

```
6689
```

If $q - 1$ has large prime factors, the dominant term in the workload of the Pohlig-Hellman algorithm will be the $\sum_{i=1}^{k} p_i$ exponentiations necessary for the generation of the tables $\{1, \omega_i, ..., \omega_i^{p_i-1}\}$, $1 \le i \le k$, and the number $\sum_{i=1}^{k} n_i$ of exponentiations, necessary to determine the $m^{(i)}$'s.

In the next subsection, we shall explain a method to take logarithms if one (or more) of the prime power factors of $q - 1$ is too large to store the tables in the Pohlig-Hellman method.

8.3.2 The Baby-Step Giant-Step Method

If one (or more) of the prime power factors of $q - 1$ is too large for the Pohlig-Hellman method, the method below can be used. It gives the user full freedom to balance the length of the table that he wants to store and the remaining workfactor.

We start with an example.

Example 8.8

Consider the equation $29^m \equiv 30 \,(mod\, 97)$ and assume that we can only store a table with 10 field elements.

We make a table of $29^i \bmod 97$ for $i = 0, 1, ..., 9$ and we compute $29^{-1} \bmod 97$ with the Mathematica functions Table, PowerMod, GridBox, *and* Transpose.

```
q = 97; a = 29;
powers = Table[{PowerMod[29, i, q], i}, {i, 0, 9}];
GridBox[Transpose[powers], RowLines -> True,
   ColumnLines -> True] // DisplayForm
x = PowerMod[a, -1, q]
```

1	29	65	42	54	14	18	37	6	77
0	1	2	3	4	5	6	7	8	9

87

We also find that $29^{-1} \equiv 87 \ (mod \ 97)$.

Writing $m = 10 \, j + i, \ 0 \le i \le 9$, *we see that* $29^m \equiv 30 \ (mod \ 97)$ *can be rewritten as* $29^i \equiv 30.29^{-10.\,j} \ (mod \ 97)$ *or as* $29^i \equiv 30.87^{10.\,j} \ (mod \ 97)$. *Since* $87^{10} \equiv 49 \ (mod \ 97)$, *we have the equivalent problem of solving* $29^i \equiv 30.49^j \ (mod \ 97), \ 0 \le i \le 9$.

We do this by trying $j = 0, 1, \ldots$ *and each time checking if* $30.49^j \ mod \ 97$ *occurs in the list of powers* $\{1, 29, 29^2, \ldots, 29^9\} \ (mod \ 97)$. *Note that* $m < 97$, *so* $j \le \lfloor 97/10 \rfloor = 9$.

To facilitate the table lookup, we sort the elements in the table of powers with the function Sort.

```
sortedpowers = Sort[powers];
GridBox[Transpose[sortedpowers],
   RowLines -> True, ColumnLines -> True] // DisplayForm
```

1	6	14	18	29	37	42	54	65	77
0	8	5	6	1	7	3	4	2	9

Next, we try $30.49^j \ mod \ 97$ *until we see the answer appear in the table above. We use the Mathematica functions,* While, MemberQ, *and* Mod. *We also print the corresponding column of the table of sorted powers (j has to be decreased by 1, because we started the numbering of j with 0).*

```
j = 0;
While[MemberQ[sortedpowers, {Mod[30 * 49^j, 97], _}] == False,
 j = j + 1];
j
Mod[30 * 49^j, 97]
```

```
4
```

```
14
```

We conclude that j = 4 and that 30.49^j mod 97 occurs in table as 14, which is 29^5 mod 97 (hence i = 5). Indeed

```
Mod[30 * 49^4, 97] == Mod[29^5, 97]
```

```
True
```

It follows that m = 10 j + i = 10.4 + 5 = 45. Indeed, 29^{45} ≡ 30 mod 97, as can be easily checked with:

```
PowerMod[29, 45, 97]
```

```
30
```

The above method will now be stated in full generality.

Theorem 8.1 **Baby-Step Giant-Step Method**
Let α be a primitive element of GF(q). Let p be a divisor of $q-1$ (not necessarily prime) and define $\omega = \alpha^{(q-1)/p}$. So, ω is a primitive p-th root of unity.
Let c be any p-th root of unity. Then, for every (trade-off value) t, $0 \le t \le 1$, one can find the exponent m, $0 \le m \le p-1$, satisfying
$$c = \beta^m$$
with an algorithm that uses
$$p^{1-t}(1 + \log_2 p^t) \qquad \text{operations,}$$
$$p^t \log_2 q \qquad \text{bits of memory space,}$$
and an initial calculation involving
$$p^t.(1 + \log_2 p^t) \text{ operations.}$$

Proof: Let $u = \lceil p^t \rceil$. We make a table of the successive powers ω^i, $0 \le i \le u-1$. This requires $u \approx p^t$ multiplications.

Next, we sort this table in $p^t \log_2 p^t$ operations, see [Knut73], pp.184. Together this explains the number of operations in the precalculation.

Each of the $u \approx p^t$ field elements in the table needs $\log_2 q$ bits of memory space. This explains the memory requirement above.

Define i and j by

$$m = j.u + i, \qquad\qquad 0 \le i < u \approx p^t.$$

Observe that

$$0 \le j \le \frac{m}{u} \le \frac{p}{u} \approx p^{1-t}.$$

Of course solving $c = \omega^m$ is equivalent to finding i and j, $0 \le i < u$, satisfying

$$\omega^i = c.\omega^{-j.u}.$$

To solve this equation, we simply compute $c.\omega^{-l.u}$, for $l = 0, 1, \ldots$ and check if the outcome appears in the table. This will happen when $l = j$, so before $l = \lceil p^{1-t} \rceil$.

For each value of l we have to perform 1 multiplication and a table look-up, which costs another $\log_2 p^t$ operations.

$\square$

For $t = 1/2$ this algorithm reduces to the $\sqrt{q}$ (both for memory and time complexity) algorithm that was mentioned at the end of Subsection 8.1.1.

The two extreme cases of the algorithm are:

$t = 0$: no table at all; all powers $1, \beta, \beta^2, \ldots$ need to be tried.

$t = 1$; complete table of $1, \beta, \beta^2, \ldots, \beta^{q-1}$ is present; only a single table look-up is needed.

Note that the product of computing time and bits of memory space in the above algorithm is more or less constant.

8.3.3   The Pollard-ρ Method

The time complexity of the Pollard-ρ Method [Poll78] is the same as that of the Baby-Step Giant-Step method explained in the previous section. The advantage lies in the minimal memory requirements.

We shall explain the Pollard-ρ Method for the special case of a multiplicative subgroup G of $GF(q)$ of prime order. So, we want to solve m, $0 \le m < p$, from the equation $c = \alpha^m$ (see (8.1)), where $\alpha \in GF(q)$ has order p, p prime, and where $c \in GF(q)$ is some given p-th root of unity. Note that p divides $q - 1$ by Theorem B.5..

Example 8.9 (Part 1)

To avoid calculations in a finite field, we take for q the prime number 4679. Note that $q - 1 = 2 \times 2339$.

Further we observe that 11 is a primitive element of GF(4679) and thus that $\alpha = 11^{(q-1)/2339} = 11^2 = 121$ is the generator of a multiplicative subgroup of order 2339. All these calculations can be easily checked with the Mathematica functions PrimeQ, FactorInteger, PowerMod *and the function MultiplicativeOrder*

```
MultiplicativeOrder[a_, n_] := If[GCD[a, n] == 1,
    Divisors[ EulerPhi[n] ] //.
    {x_, y___} -> If[PowerMod[a, x, n] == 1, x, {y}] ];
```

that was introduced in Subsection B.4.1, but which is a standard function in Mathematica 4.

```
q = 4679;
PrimeQ[q]
FactorInteger[q - 1]
MultiplicativeOrder[11, q]
PowerMod[11, 2, q]
MultiplicativeOrder[121, q]
```

```
True
```

```
{{2, 1}, {2339, 1}}
```

```
4678
```

```
121
```

```
2339
```

Further on, we shall continue with this example, when we want to solve the equation

$$121^m \equiv 3435 \,(mod\,4679).$$

Note that this equation must have a solution, since 3435 is indeed a 2339-th root of unity in GF(4679). Indeed, all 2339-th roots of unity are a zero of $x^{2339} - 1$ and by Theorem B.15 there are no other zeros of this polynomial.

```
PowerMod[3435, 2339, 4679]
```

```
1
```

In order to solve $c = \alpha^m$, we partition the multiplicative subgroup G of $GF(q)$ of order p, in three subsets G_i, $i = 0, 1, 2$, as follows:

$$x \in G_i \qquad \Longleftrightarrow \qquad x \equiv i \,(\mathrm{mod}\,3).$$

We define a sequence $\{x_i\}_{i \geq 0}$ in $GF(q)$ recursively by $x_0 = 1$ and

$$x_{i+1} = f(x_i) = \begin{cases} (x_i^2 \bmod q), & \text{if } x_i \in G_0, \\ (c.x_i \bmod q), & \text{if } x_i \in G_1, \\ (\alpha.x_i \bmod q), & \text{if } x_i \in G_2. \end{cases} \tag{8.4}$$

With the sequence $\{x_i\}_{i \geq 0}$ we associate two other sequences $\{a_i\}_{i \geq 0}$ and $\{b_i\}_{i \geq 0}$ in such a way that for all $i \geq 0$

$$x_i = \alpha^{a_i} c^{b_i}.$$

To this end, take $a_0 = b_0 = 0$ and use the recursions

$$a_{i+1} = \begin{cases} (2\,a_i \bmod p), & \text{if } x_i \in G_0, \\ a_i, & \text{if } x_i \in G_1, \\ (a_i + 1 \bmod p, & \text{if } x_i \in G_2. \end{cases}$$

$$b_{i+1} = \begin{cases} (2\,b_i \bmod p), & \text{if } x_i \in G_0, \\ (b_i + 1 \bmod p), & \text{if } x_i \in G_1, \\ b_i, & \text{if } x_i \in G_2. \end{cases}$$

Note that by induction

$$x_{i+1} = x_i^2 = (\alpha^{a_i} c^{b_i})^2 = \alpha^{2\,a_i} c^{2\,b_i} = \alpha^{a_{i+1}} c^{b_{i+1}}, \text{ if } x_i \in G_0,$$
$$x_{i+1} = c.x_i = c.\alpha^{a_i} c^{b_i} = \alpha^{a_i+1} c^{b_i} = \alpha^{a_{i+1}} c^{b_{i+1}}, \text{ if } x_i \in G_1,$$
$$x_{i+1} = \alpha.x_i = \alpha.\alpha^{a_i} c^{b_i} = \alpha^{a_i} c^{b_i+1} = \alpha^{a_{i+1}} c^{b_{i+1}}, \text{ if } x_i \in G_2.$$

As soon as we have two distinct indices i and j with $x_i = x_j$ we are done, because this would imply that $\alpha^{a_i} c^{b_i} = \alpha^{a_j} c^{b_j}$ and thus that $\alpha^{a_i - a_j} = c^{b_j - b_i}$. Provided that $b_i \neq b_j$, we have found the solution $m \equiv (a_j - a_i)/(b_i - b_j) \,(\mathrm{mod}\,p)$.

If $b_i = b_j$, which happens with negligible probability, we put $c' = c.\alpha$ and solve $c' = \alpha^{m'}$, where $m' = m + 1$.

To find indices i and j with $x_i = x_j$, we follow *Floyd's cycle-finding algorithm*: find an index i such that $x_i = x_{2i}$ (so, take $j = 2i$).

To this end, we start with the pair (x_1, x_2), calculate (x_2, x_4), then (x_3, x_6), and so on, each time calculating (x_{i+1}, x_{2i+2}) from the previously calculated (x_i, x_{2i}) by the defining rules $x_{i+1} = f(x_i)$ and $x_{2i+2} = f^2(x_{2i})$ In this way, huge storage requirements can be avoided.

Example 8.9 (Part 2)

We continue with Example 8.9. Hence, we have $q = 4679$, $\alpha = 121$, an element of (prime) order $p = 2339$, and $c = 3435$. I.e. we have the equation:

$$121^m \equiv 3435 \,(\mathrm{mod}\,4679).$$

The recurrence relation for the $\{x_i\}_{i \geq 0}$ sequence can be evaluated by means of the <u>Which</u> *and* <u>Mod</u> *functions.*

```
RecX[x_, alp_, c_, q_] := Which[ Mod[x, 3] == 0, Mod[x², q],
    Mod[x, 3] == 1, Mod[c * x, q], Mod[x, 3] == 2, Mod[alp * x, q] ]
```

The smallest index i, $i \geq 1$, satisfying $x_i = x_{2i}$ can quite easily be found with the help of the <u>While</u> *function.*

```
alp = 121; c = 3435; q = 4679;
x1 = RecX[1, alp, c, q];
x2 = RecX[x1, alp, c, q]; i = 1;
While[x1 != x2, x1 = RecX[x1, alp, c, q];
  x2 = RecX[RecX[x2, alp, c, q], alp, c, q]; i = i + 1];
i
```

```
76
```

So, $x_{76} = x_{152}$ *and* $m \equiv (a_{152} - a_{76})/(b_{76} - b_{152}) \pmod{2339}$. *However, above we did not update the values of the sequences a_i and b_i. We will do that now.*

```
RecurrDef[{x_, a_, b_}] := Which[
        Mod[x, 3] == 0, {Mod[x², q], Mod[2 a, p], Mod[2 b, p]},
        Mod[x, 3] == 1, {Mod[c * x, q], a, Mod[b + 1, p]},
        Mod[x, 3] == 2, {Mod[alp * x, q], Mod[a + 1, p], b}]
```

```
alp = 121; c = 3435; q = 4679; p = 2339;
x1 = 1; a1 = 0; b1 = 0;
x2 = 1; a2 = 0; b2 = 0;
{x1, a1, b1} = RecurrDef[{x1, a1, b1}]; i = 1;
{x2, a2, b2} = RecurrDef[RecurrDef[{x2, a2, b2}]];
While[x1 != x2, {x1, a1, b1} = RecurrDef[{x1, a1, b1}];
        {x2, a2, b2} = RecurrDef[RecurrDef[{x2, a2, b2}]];
 i = i + 1];
Print["i=", i]
Print["x₁=", x1, ", a₁=", a1, ", b₁=", b1];
Print["x₂.₁=", x2, ", a₂.₁=", a2, ", b₂.₁=", b2];
```

i=76

x_i=492, a_i=84, b_i=2191

$x_{2.i}$=492, $a_{2.i}$=286, $b_{2.i}$=915

Indeed, the relation $\alpha^{a_i} c^{b_i}$ gives the same value for $i = 76$ and $i = 2 \times 76$:

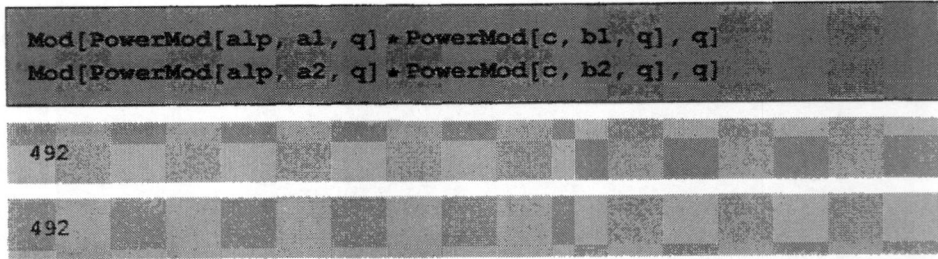

```
Mod[PowerMod[alp, a1, q] * PowerMod[c, b1, q], q]
Mod[PowerMod[alp, a2, q] * PowerMod[c, b2, q], q]
```

492

492

The solution m of $121^m \equiv 3435 \,(mod\,4679)$ can now be determined from $m \equiv (286 - 84)/(2191 - 915)\,(mod\,2339)$.

```
m = Mod[(a2 - a1) * PowerMod[b1 - b2, -1, p], p]
```

1111

That $m = 1111$ is indeed the solution can be checked with

```
PowerMod[alp, 1111, q] == c
```

True

The ρ in the name of this algorithm reflects the shape of the $\{x_i\}_{i\geq0}$-sequence: after a while it starts cycling around. The memory requirements of Floyd's cycle finding algorithm are indeed minimal. The expected running time is $\sqrt{p}$. For further details, the reader is referred to [Poll78].

8.3.4   The Index-Calculus Method

□ **General Discussion**

To describe the index-calculus method in general we consider a cyclic group G of order N generated by an element g. So, $G = \{e, g, g^2, \ldots, g^{N-1}\}$ and $g^N = e$.

In this setting we want to solve m from $g^m = h$ (see (8.1)) for a given $h \in G$.

The basic idea of the index-calculus method consists of the following steps:

1) Select an appropriate subset S of G with the property that a large proportion of the elements of G can be expressed as a product of elements of S in an efficient way. This set S is called the *factor base*. An element $g \in G$ that can be expressed as a product of elements of S is called *smooth* with respect to S. Let k be the size of S. In the next two steps each element in S will be written as a power of g.

2) Find a sufficiently large collection I of exponents i with the property that each g^i, $i \in I$, can be expressed efficiently as a product of elements of S, say $g^i = s_{1i,1}^u \, s_{2i,2}^u \ldots s_{ki,k}^u$. Taking the $\log_g$ of both hands, we get a set of linear congruence relations

$$i \equiv u_{i,1} \log_g s_1 + u_{i,2} \log_g s_2 + \ldots + u_{i,k} \log_g s_k \pmod{N}, \qquad i \in I.$$

3) Treating the numbers $\log_g s_j$, $1 \leq j \leq k$, as unknowns, solve the above system of linear congruence relations (for this, the system of linear congruence relations has to have rank k and the set I will have to be sufficiently large).

4) Pick a random exponent r and try to express $g^r h$ as a product of elements of S. As soon as this has happened, say $g^r . h = s_{11}^v \, s_{22}^v \ldots s_{kk}^v$, we again take the $\log_g$ of both hands and get

$$r + m \equiv v_1 \log_g s_1 + v_2 \log_g s_2 + \ldots + v_k \log_g s_k \pmod{N}.$$

Since the values of each $\log_g s_i$ has already been determined in Step 3 and r was chosen, m can be determined from this congruence relation.

Note that Steps 2 and 3 aim to solve the logarithm problem for all the elements in the factor base. Step 4 tries to reduce the current logarithm problem to the factor base elements.

It may be clear that the optimal size of the factor base S is a compromise between manageable storage requirements and the probability that a random element in G (namely $g^r h$) can be expressed as a product of elements of S.

In general, there are two (related) unresolved problems in the above approach.

- How can one determine a good factor base?

- How does one express an element in G as product of elements of S?

In the next subsubsections we demonstrate the above method for two special cases where more can be said about the above two questions.

Complexity

There are many variations of the index-calculus method. Typically, their complexity grows subexponential in $\log_2 N$, while the methods described in Subsections 8.3.1, 8.3.2, and 8.3.3 are all exponential in $\log_2 N$.

□ $\mathbb{Z}_p^*$, i.e. the Multiplicative Group of GF(p)

In this case, $G = \{1, 2, ..., p-1\}$. Let g be a generator of this group.

Choice of the factor base S: the first k prime numbers, $p_1, p_2, ..., p_k$.

If k is sufficiently big, a large proportion of the elements in G can be expressed as product of powers of these k primes, i.e. they will be smooth with respect to S.

Technique to express an element in G as product of elements of S: divide the element by the p_i's.

Complexity

Adleman in [Adle79] analyzes this technique in detail and arrives at a complexity of

$$\exp^{C\sqrt{\ln p \ln\ln p}}$$

for some constant C.

Example 8.10

Consider $\mathbb{Z}_{541}^*$ with primitive element $g = 2$. That 541 is prime and that 2 is a primitive element can be checked with the Mathematica functions PrimeQ, FactorInteger, *and* PowerMod. *Indeed, the order of 2 divides $|\mathbb{Z}_{541}^*| = 540$ by Theorem B.5, therefore, we only have to check that $2^{(p-1)/d} \neq 1 \pmod{541}$ for the divisors of $p = 541$.*

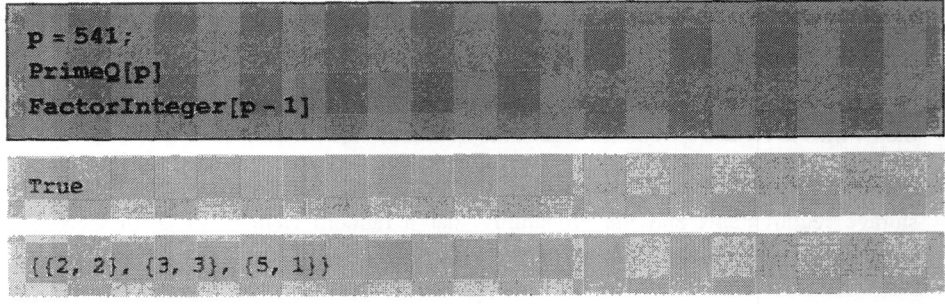

```
p = 541;
PrimeQ[p]
FactorInteger[p - 1]
```

```
True
```

```
{{2, 2}, {3, 3}, {5, 1}}
```

```
PowerMod[2, (541 - 1) / 2, p]
PowerMod[2, (541 - 1) / 3, p]
PowerMod[2, (541 - 1) / 5, p]
```

```
540
```

```
129
```

```
48
```

As factor base S we take the set of the first five prime numbers, which can be generated with the Mathematica functions `Prime` *and* `Table`.

```
Table[Prime[i], {i, 1, 5}]
```

```
{2, 3, 5, 7, 11}
```

We want to write each of the elements in this factor base as a power of g = 2, i.e. we want to solve the logarithm problem for the elements in the factor base. To this end, we try to find powers of g = 2 in Z_{541}^* that can be expressed as product of elements in {2, 3, 5, 7, 11}. *For this, we can use the Mathematica function* `FactorInteger` *and* `PowerMod`. *When trying*

```
p = 541;
try = PowerMod[2, 102, p]
FactorInteger[try]
```

```
136
```

```
{{2, 3}, {17, 1}}
```

we see that we have no complete factorization in {2, 3, 5, 7, 11}.

After some trial and error we did find the elements 2^{14}, 2^{81}, 2^{207}, 2^{214}, and 2^{300} achieving our goal.

```
p = 541;
FactorInteger[PowerMod[2, 14, p]]
FactorInteger[PowerMod[2, 81, p]]
FactorInteger[PowerMod[2, 207, p]]
FactorInteger[PowerMod[2, 214, p]]
FactorInteger[PowerMod[2, 300, p]]
```

```
{{2, 1}, {7, 1}, {11, 1}}
```

```
{{2, 1}, {3, 1}, {7, 2}}
```

```
{{5, 2}, {11, 1}}
```

```
{{5, 1}, {7, 1}}
```

```
{{2, 5}, {11, 1}}
```

Writing $m_1 = \log_2 2$, $m_2 = \log_2 3$, $m_3 = \log_2 5$, $m_4 = \log_2 7$, $m_5 = \log_2 11$ *and taking the logarithms on both sides gives five linear congruence relations in* m_1, m_2, ..., m_5.

For example, $2^{207} \equiv 5^2 . 11^1 \bmod 541$ *can be rewritten as*

$$2^{207} \equiv 2^{2. \log_2 5} \, 2^{1. \log_2 11} \equiv 2^{2 m_3} \, 2^{m_5} \bmod 541.$$

Taking $\log_2$ *on both sides gives the congruence relation*

$$207 \equiv 2 m_3 + m_5 \bmod 540.$$

So, we have:

$$14 \equiv m_1 + m_4 + m_5 \pmod{540},$$
$$81 \equiv m_1 + m_2 + 2 m_4 \pmod{540},$$
$$207 \equiv 2 m_3 + m_5 \pmod{540},$$
$$214 \equiv m_3 + m_4 \pmod{540},$$
$$300 \equiv 5 m_1 + m_5 \pmod{540},$$

The above system of linear congruence relations can be solved with the Solve *function:*

```
m1 = .; m2 = .; m3 = .; m4 = .; m5 = .;
Solve[{m1 + m4 + m5 == 14 , m1 + m2 + 2 * m4 == 81,
   2 * m3 + m5 == 207, m3 + m4 == 214, 5 m1 + m5 == 300,
   Modulus == 540}, {m1, m2, m3, m4, m5}]
```

```
{{Modulus → 540, m2 → 104, m3 → 496, m1 → 1, m4 → 258, m5 → 295}}
```

So, we know that

$$m_1 = \log_2 2 = 1, \qquad m_2 = \log_2 3 = 104, \quad m_3 = \log_2 5 = 496, m_4 = \log_2 7 = 258,$$
$$m_5 = \log_2 11 = 295$$

or, equivalently

$$2^1 \equiv 2 \bmod 541, \qquad 2^{104} \equiv 3 \bmod 541, \qquad 2^{496} \equiv 5 \bmod 541, \; 2^{258} \equiv 7 \bmod 541,$$
$$2^{295} \equiv 11 \bmod 541.$$

If the above linear congruence relations are not linearly independent one has to replace some equations by others until they are linearly independent.

Let us now find a solution of $2^m \equiv 345 \, (\bmod \, 541)$.

From

```
FactorInteger[345]
FactorInteger[Mod[2² 345, 541]]
FactorInteger[Mod[2¹⁰⁰ 345, 541]]
FactorInteger[Mod[2¹³ 345, 541]]
```

```
{{3, 1}, {5, 1}, {23, 1}}
```

```
{{2, 1}, {149, 1}}
```

```
{{3, 2}, {41, 1}}
```

```
{{2, 3}, {7, 1}}
```

we see that 345 can not be expressed as product of elements of S, nor can $2^2 \times 345$ and $2^{100} \times 345$, but $2^{13} \times 345 = 2^3 \, 7^1$ in GF(541).

We conclude that

$$13 + m \equiv 3 . m_1 + 1 . m_4 \equiv 3 \times 1 + 258 \equiv 261 \, (\bmod \, 540),$$

therefore, the solution of $2^m \equiv 345 \, (\bmod \, 541)$ is given by

$$m \equiv 248 \, (\bmod \, 540).$$

This can easily be checked with

```
PowerMod[2, 248, 541]
```

345

Because of the small parameters, we can find out explicitly how many elements in {1, 2, ..., 540} can be expressed as product of elements of S. We use the Mathematica functions *Select*, *Flatten*, *Table*, *Sort*, and *Length* and make use of the fact that the exponent of 2 is at most $\lfloor \log_2 541 \rfloor = 9$, the exponent of 3 is at most $\lfloor \log_3 541 \rfloor = 5$, etc., in any number less than 541.

```
BaseProd = Select[
    Flatten[ Table[ 2^{i1} 3^{i2} 5^{i3} 7^{i4} 11^{i5},
        {i1, 0, Log[2, 541]},
        {i2, 0, Log[3, 541]},
        {i3, 0, Log[5, 541]},
        {i4, 0, Log[7, 541]},
        {i5, 0, Log[11, 541]}] ],
            # < 541 &] // Sort
Length[BaseProd]
```

```
{1, 2, 3, 4, 5, 6, 7, 8, 9, 10, 11, 12, 14, 15, 16, 18, 20, 21, 22,
  24, 25, 27, 28, 30, 32, 33, 35, 36, 40, 42, 44, 45, 48, 49, 50,
  54, 55, 56, 60, 63, 64, 66, 70, 72, 75, 77, 80, 81, 84, 88, 90,
  96, 98, 99, 100, 105, 108, 110, 112, 120, 121, 125, 126, 128,
  132, 135, 140, 144, 147, 150, 154, 160, 162, 165, 168, 175, 176,
  180, 189, 192, 196, 198, 200, 210, 216, 220, 224, 225, 231, 240,
  242, 243, 245, 250, 252, 256, 264, 270, 275, 280, 288, 294, 297,
  300, 308, 315, 320, 324, 330, 336, 343, 350, 352, 360, 363, 375,
  378, 384, 385, 392, 396, 400, 405, 420, 432, 440, 441, 448, 450,
  462, 480, 484, 486, 490, 495, 500, 504, 512, 525, 528, 539, 540}
```

142

Therefore, about a quarter of all elements in G can be expressed as product of elements of S. That means that on the average it takes four trials (choices of r) before $g^r h$ can expressed as a product of elements of {2, 3, 5, 7, 11}.

□ **GF(2^n)**

All elements in GF(2^n) can be represented by means of binary polynomials of degree $< n$ in x modulo an irreducible polynomial $f(x)$ (see Theorem B.16). One writes GF(2^n) = GF(2)$[x]/(f(x))$.

Let the polynomial $\alpha = \alpha(x)$ denote a primitive element of GF(2^n). Then GF(2^n) can also be represented by binary polynomials of degree $< n$ modulo the minimal polynomial $p(x)$ of α. It follows that α is a primitive element in GF(2)$[\alpha]/(p(\alpha))$, i.e. x is a primitive element in GF(2)$[x]/(p(x))$.

See Example B.6, where $f(x) = x^4 + x^3 + x^2 + x + 1$ defines GF(2^4) and where $\alpha(x) = 1 + x$ is a primitive element of GF(2^4) = GF(2)$[x]/(x^4 + x^3 + x^2 + x + 1)$. This element α is a zero of the primitive polynomial $p(x) = x^4 + x^3 + 1$. In GF(2)$[x]/(x^4 + x^3 + 1)$ the element x is a primitive element

Equation (8.1), that we want so solve, can be reformulated as:

for every polynomial $c(x)$ of degree $< n$, find the exponent m, $0 \leq m \leq 2^n - 2$, such that $x^m \equiv c(x) \pmod{p(x)}$.

As choice of the factor base S we take all binary, irreducible polynomials of degree $\leq \sigma$, say $p_1(x)$, $p_2(x)$, ..., $p_k(x)$. (The number of such polynomials is given by Theorem B.17).

As a <u>technique</u> to express an element in GF(2^n) as a product of elements of S, we simply divide the element by the polynomials $p_i(x)$.

A polynomial $u(x)$ that can be expressed as a product of elements of S is called smooth with respect to S.

<u>Complexity</u>

Coppersmith [Copp84] analyzes this algorithm and finds as asymptotic running time

$$\exp^{C \sqrt[3]{(\ln n)(\ln \ln n)^2}}$$

Later, further improvements have been found with names like *number field sieve* and *function field sieve* (see [AdDM93], [Adle94], and [HelR83]).

For an excellent survey on the discrete logarithm problem we refer the reader to [Odly85].

Example 8.11

We want to take a logarithm in GF(2^{10}). To represent GF(2^{10}) properly and to find a primitive element in it, we look for a primitive polynomial of degree 10. We do this with the Mathematica function `FieldIrreducible` *for which the package* `Algebra`FiniteFields`` *has to be read first.*

```
<< Algebra`FiniteFields`
```

```
fld = GF[2, 10];
FieldIrreducible[fld, x]
```

```
1 + x^7 + x^10
```

So, we take $GF(2^{10}) = GF(2)[x]/(x^{10} + x^7 + 1)$ which has x as primitive element. Equation (8.1) now reads like:

find m such that $x^m \equiv c(x) \pmod{x^{10} + x^7 + 1}$.

As factor base S we shall take the set of all irreducible polynomials of degree ≤ 4.

The reader may remember that all binary, irreducible polynomials of degree d appear in the factorization of $x^{2^d} - x$ (see Theorem B.35).

```
Clear[x];
Factor[x^2^3 - x, Modulus -> 2]
Factor[x^2^4 - x, Modulus -> 2]
```

```
x (1 + x) (1 + x + x^3) (1 + x^2 + x^3)
```

```
x (1 + x) (1 + x + x^2) (1 + x + x^4) (1 + x^3 + x^4) (1 + x + x^2 + x^3 + x^4)
```

Hence, as factor base S we have:

$$p_1(x) = x, \qquad p_2(x) = 1 + x,$$
$$p_3(x) = 1 + x + x^2, \qquad p_4(x) = 1 + x + x^3,$$
$$p_5(x) = 1 + x^2 + x^3, \qquad p_6(x) = 1 + x + x^2 + x^3 + x^4,$$
$$p_7(x) = 1 + x + x^4, \qquad p_8(x) = 1 + x^3 + x^4.$$

We want to write each of the elements in this factor base as a power of x, i.e. we want to solve the logarithm problem for the elements in the factor base. To this end, we try to find powers of x in $GF(2)[x]/(x^{10} + x^7 + 1)$ that can be expressed as a product of the polynomials $p_j(x)$, $1 \leq j \leq 8$. We use the Mathematica function _Factor_ and _PolynomialMod_.

```
attempt = PolynomialMod[x^85, {x^10 + x^7 + 1, 2}]
Factor[attempt, Modulus -> 2]
```

```
1 + x + x^2 + x^3 + x^4 + x^5 + x^6 + x^9
```

```
(1 + x)^2 (1 + x + x^2) (1 + x^2 + x^3 + x^4 + x^5)
```

We conclude that x^{85} is not smooth with respect to our factor base S. After some trial and error we find the following list of smooth powers of x:

```
Factor[PolynomialMod[x, {x^10 + x^7 + 1, 2}], Modulus -> 2]
Factor[PolynomialMod[x^86, {x^10 + x^7 + 1, 2}], Modulus -> 2]
Factor[PolynomialMod[x^140, {x^10 + x^7 + 1, 2}], Modulus -> 2]
Factor[PolynomialMod[x^211, {x^10 + x^7 + 1, 2}], Modulus -> 2]
Factor[PolynomialMod[x^319, {x^10 + x^7 + 1, 2}], Modulus -> 2]
Factor[PolynomialMod[x^457, {x^10 + x^7 + 1, 2}], Modulus -> 2]
Factor[PolynomialMod[x^605, {x^10 + x^7 + 1, 2}], Modulus -> 2]
Factor[PolynomialMod[x^787, {x^10 + x^7 + 1, 2}], Modulus -> 2]
```

x

$(1 + x + x^3) (1 + x^2 + x^3)$

$x^2 (1 + x + x^2)^2$

$(1 + x)^5 (1 + x + x^2 + x^3 + x^4)$

$(1 + x) (1 + x^3 + x^4) (1 + x + x^2 + x^3 + x^4)$

$(1 + x + x^2) (1 + x + x^3) (1 + x + x^4)$

$(1 + x) (1 + x^2 + x^3) (1 + x + x^4)$

$(1 + x + x^3) (1 + x^2 + x^3)^2$

Writing $p_i(x) \equiv x^{m_i} (mod\ x^{10} + x^7 + 1)$, these relations give rise to eight linear congruence relations. For instance, the last equation gives

$$x^{787} \equiv (1 + x + x^3)(1 + x^2 + x^3)^2 \equiv (x^{m_4})(x^{m_5})^2 \equiv x^{m_4 + 2 m_5} \ (mod\ x^{10} + x^7 + 1).$$

Taking the logarithm on both sides gives the linear congruence relations

$$787 \equiv m_4 + 2 m_5 \ (mod\ 1023),$$

since 1023 is the multiplicative order of the primitive element x. In this way, the eight relations above can be rewritten as

$$1 \equiv m_1 \ (mod\ 1023),$$
$$86 \equiv m_4 + m_5 \ (mod\ 1023),$$

$$140 \equiv 2\,m_1 + 2\,m_3 \pmod{1023},$$
$$211 \equiv 5\,m_2 + m_6 \pmod{1023},$$
$$319 \equiv m_2 + m_6 + m_8 \pmod{1023},$$
$$457 \equiv m_3 + m_4 + m_7 \pmod{1023},$$
$$605 \equiv m_2 + m_5 + m_7 \pmod{1023},$$
$$787 \equiv m_4 + 2\,m_5 \pmod{1023}.$$

This forms a system of congruence relations that can be solved with the Mathematica function `Solve`.

```
Clear[m1, m2, m3, m4, m5, m6, m7, m8];
Solve[{m1 == 1 , m4 + m5 == 86, 2 m1 + 2 m3 == 140,
    5 m2 + m6 == 211, m2 + m6 + m8 == 319, m3 + m4 + m7 == 457,
    m2 + m5 + m7 == 605, m4 + 2 m5 == 787, Modulus == 1023},
   {m1, m2, m3, m4, m5, m6, m7, m8}]
```

```
{{Modulus -> 1023, m8 -> 827, m1 -> 1, m3 -> 69,
   m6 -> 591, m7 -> 1003, m2 -> 947, m4 -> 408, m5 -> 701}}
```

So, we know that $m_1 = 1$, $m_2 = 947$, $m_3 = 69$, $m_4 = 408$, $m_5 = 701$, $m_6 = 591$, $m_7 = 1003$, *and* $m_8 = 827$.

If the linear congruence relations are not linearly independent one has to replace some equations by others until they are linearly independent.

Let us now find a solution of $x^m \equiv 1 + x + x^6 + x^9 \pmod{x^{10} + x^7 + 1}$.

From

```
Factor[
   PolynomialMod[1 + x + x^6 + x^9, {x^10 + x^7 + 1, 2}], Modulus -> 2]
Factor[PolynomialMod[x^50 (1 + x + x^6 + x^9), {x^10 + x^7 + 1, 2}],
   Modulus -> 2]
```

$$(1 + x)^2 (1 + x + x^2 + x^3 + x^4 + x^5 + x^7)$$

$$(1 + x + x^2)^2 (1 + x + x^4)$$

we see that $1 + x + x^6 + x^9$ *can not be written as product of polynomials in* S, *but* $x^{50}(1 + x + x^6 + x^9)$ *can.*

We conclude that $50 + m \equiv 2\,m_3 + m_7 \equiv 2 \times 69 + 1003 \equiv 118 \pmod{1023}$, *so the solution of* $x^m \equiv 1 + x + x^6 + x^9 \pmod{x^{10} + x^7 + 1}$ *is given by*

$$m \equiv 68 \pmod{1023}.$$

This can be checked by

```
PolynomialMod[x^68, {x^10 + x^7 + 1, 2}]
```

```
1 + x + x^6 + x^9
```

8.4   Problems

Problem 8.1^M
Users A and B want to use the Diffie-Hellman system to fix a common key over a public channel. They use GF(p), with $p = 541$ and primitive element $\alpha=2$.
User B makes $c_B = 123$ public. If $m_A = 432$, what will be the common key $k_{A,B}$ that A and B use for their communication?

Problem 8.2
Users A and B want to use the Diffie-Hellman system to fix a common key over a public channel. They use $\mathbb{F}_2[x]/(x^{10} + x^3 + 1)$ as representation of GF(2^{10}). User B makes $c_B = 0100010100$ public, which stands for the field element $x + x^5 + x^7$. If $m_A = 2$, what will be the common key that A and B use for their communication?

Problem 8.3
Demonstrate the Special Case version of the Pohlig-Helmann algorithm, that computes logarithms in finite fields of size $q = 2^n + 1$, by evaluating $\log_3(142)$ in GF(257).

Problem 8.4^M
Check that 953 is a prime number and that 3 is a generator of $\mathbb{Z}^*_{953}$. Find the three least significant bits of the solution m of the congruence relation $3^m \equiv 726 \bmod 953$.
(See the remark in the discussion of the special case $q - 1 = 2^n$ in Subsection 8.3.1.)

Problem 8.5
Compute $\log_3(135)$ in GF(353) with the Pohlig-Hellman algorithm.

Problem 8.6^M
Find a solution of $\log_{44} 55$ in GF(197) by means of the Baby-Step Giant-Step method, when only 15 field elements can be stored.

Problem 8.7^M
Check that $\alpha = 662$ is a primitive 2003-th root of unity in GF(4007) (note that 4007 is a prime number). Let G be the multiplicative subgroup G of order 2003 in GF(4007) generated by α. Check that 2124 is an element of G.
Determine $\log_{662} 2124$ by the Pollard-ρ method.

Problem 8.8^M
Check that $g = 996$ is a generator of the multiplicative group $\mathbb{Z}^*_{4007}$. Set up the index-calculus method with a factor base of size 6 and determine $\log_{996} 1111$.

Problem 8.9[M]
Solve the equation $x^m \equiv 1 + x^3 + x^9 \pmod{x^{10} + x^3 + 1}$ in the setting of Example 8.11.

Problem 8.10[M]
What is the probability that a random element $x^m \pmod{x^{10} + x^3 + 1}$ is smooth with respect to the set of irreducible, binary polynomials of degree ≤ 10 (see Example 8.11).

9 RSA Based Systems

9.1 The RSA System

In 1978 R.L. Rivest, A. Shamir and L. Adleman [RivSA78] proposed a public key cryptosystem that has become known as the RSA system. It makes use of the following three facts:

1) Exponentiation modulo a composite number n, i.e. computing c from $c \equiv m^e \pmod{n}$ for given m and e, is a relatively simple operation (see Subsection 8.1.1).

2) The opposite problem of taking roots modulo a large, composite number n, i.e. computing m from $c \equiv m^e \pmod{n}$ (which can be written as $m \equiv \sqrt[e]{c} \pmod{n}$) for given c and e, is, in general, believed to be intractable.

3) If the prime factorization of n is known, the problem of taking roots modulo n is feasible.

9.1.1 Some Mathematics

From Appendix A we quote Theorem A.14 and the definition of Euler's Totient function (Def. A.6):

> **Theorem 9.1** *Euler*
> Let a and n be integers. Then
>
> $$\gcd(a, n) = 1 \implies a^{\varphi(n)} \equiv 1 \pmod{n}, \tag{9.1}$$
>
> where Euler's Totient Function $\varphi(n)$ counts the number of integers in between 1 and n that are coprime with n. The function $\varphi(n)$ can be computed from the relation:
>
> $$\varphi(n) = n \prod_{p|n,\ p\ \text{prime}} \left(1 - \frac{1}{p}\right). \tag{9.2}$$

The reader can check the above in any example with the *Mathematica* functions GCD and EulerPhi.

```
n = 1999; a = 1234;
GCD[a, n]
ph = EulerPhi[n]
PowerMod[a, ph, n]
```

```
1
```

```
1998
```

```
1
```

9.1.2 Setting Up the System

□ **Step 1** Computing the Modulus n_U

Each user U of the system chooses two different large prime numbers, say p_U and q_U. In the original proposal the suggested length was about 100 digits.

Let $n_U = p_U \, q_U$. It follows from (9.2) that

$$\varphi(n_U) = n_U\left(1 - \tfrac{1}{p_U}\right)\left(1 - \tfrac{1}{q_U}\right) = (p_U - 1)(q_U - 1). \tag{9.3}$$

This can also be seen directly. The n integers in between 1 and $n_U = p_U \, q_U$ are all coprime with n_U except for the q_U multiples of p_U (namely $p_U, 2.\, p_U, 3.\, p_U, \ldots, q_U.p_U$) and the p_U multiples of q_U (namely $q_U, 2.\, q_U, 3\, q_U, \ldots, p_U.q_U$) In this counting, one should realize that the number $p_U \, q_U$ has been subtracted once too often.

Example 9.1 (Part 1)

To keep this example manageable participant Bob will keep his primes reasonably small. He makes use of the Mathematica functions `Prime` *and* `EulerPhi`.

```
pB = Prime[1200]
qB = Prime[1250]
nB = pB * qB
phiB = EulerPhi[nB]
```

```
9733
```

```
10177
```

```
99052741
```

```
99032832
```

□ **Step 2** **Computing the Exponents e_U and d_U**

User U chooses an integer e_U, $1 < e_U < \varphi(n_U)$, with $\gcd(e_U, \varphi(n_U)) = 1$. User U computes the unique integer d_U, satisfying

$$e_U \, d_U \equiv 1 \,(\mathrm{mod}\,\varphi(n_U)), \qquad 1 < d_U < \varphi(n_U). \tag{9.4}$$

For instance, U can use Euclid's Algorithm (see Section A.2) to find d_U in less than $\log_f \varphi(n_U)$ operations (Theorem A.9) with $f = \left(1 + \sqrt{5}\right)/2$.

Example 9.1 (Part 2)

The random choice of e_B and the computation of d_B can be made with the Mathematica functions Random, While, *and* ExtendedGCD.

```
eB = Random[Integer, {1, nB}];
While[GCD[eB, phiB] != 1,
                        eB = Random[Integer, {1, nB}]];
eB
ExtendedGCD[eB, phiB]
```

```
81119923
```

```
{1, {17089915, -13998717}}
```

So, Bob has $e_B = 81119923$ and $d_B = 17089915$. This can be checked by the Mod *calculation:*

```
dB = 17089915;
Mod[eB * dB, phiB]
```

```
1
```

□ **Step 3** **Making Public:** e_U **and** n_U

Each user U makes e_U and n_U public, but keeps d_U secret. The primes numbers p_U and q_U no longer play a role. User U may use them to reduce the complexity of his calculations as we shall see later on. They may not be made public by U.

9.1.3 RSA for Privacy

If user A, say Alice, wants to send a secret message to Bob (user B) she represents her message in any standardized way by a number m, $0 < m < n_B$. Next, Alice looks up the public exponent e_B of Bob. She will send the ciphertext c computed from

$$c \equiv m^{e_B} \pmod{n_B}.$$

Bob can recover m from c by raising it to the power d_B which he only knows. Indeed, for some integer l one has

$$c^{d_B} \equiv (m^{e_B})^{d_B} \equiv m^{e_B\,d_B} \overset{(9.4)}{\equiv} m^{1+l.\varphi(n_B)} \equiv m.(m^{\varphi(n_B)})^l \overset{(9.1)}{\equiv} m \pmod{n_B}. \tag{9.5}$$

when $\gcd(m, n_B) = 1$. In Problem 9.2 the reader is invited to verify that the system also works when $\gcd(m, n_B) \neq 1$.

We summarize the RSA secrecy system in the next table.

public	e_U and n_U of all users U
secret	d_U of user U
property	$e_U\,d_U \equiv 1 \pmod{\varphi(n_U)}$
message to Bob	$0 < m < n_B$
encryption by A	$c \equiv m^{e_B} \pmod{n_B}$
decryption by B	$c^{d_B} \equiv m \pmod{n_B}$

The RSA System for Privacy

Table 9.1

The public and secret exponents in the RSA system are traditionally called e_U and d_U to denote the encryption resp. decryption functions that they have in this subsection.

Example 9.1 (Part 3)

We continue with the parameters of Example 9.1, so $n_B = 99052741$, $e_B = 81119923$, and $d_B = 17089915$. The encryption $c \equiv m^{e_B} \pmod{n_B}$ of message $m = 12345678$ leads with the Mathematica function PowerMod *to*

```
nB = 99052741; eB = 81119923; dB = 17089915;
m = 12345678;
c = PowerMod[m, eB, nB]
```

```
38447790
```

Bob decrypts this by computing $c^{d_B} \pmod{n_B}$, which gives m.

```
PowerMod[c, dB, nB]
```

```
12345678
```

It is possible to reduce the work factor of the decryption process by means of the Chinese Remainder Theorem (Thm. A.19). Indeed, since Bob knows the factorization of n into $p \times q$, he can do the following.

Bob precomputes integers a and b mod n, satisfying

$$\begin{cases} a \equiv 1 & (\text{mod} \quad p) \\ a \equiv 0 & (\text{mod} \quad q) \end{cases}$$

$$\begin{cases} b \equiv 0 & (\text{mod} \quad p) \\ b \equiv 1 & (\text{mod} \quad q) \end{cases}$$

Next, Bob computes $m_1 \equiv c_1^d \pmod{p}$ and $m_2 \equiv c_2^d \pmod{q}$, where $c_1 = (c \bmod p)$ and $c_2 = (c \bmod q)$. Note that all these calculations take place modulo the integers p and q that are typically half the length of n. By the Chinese Remainder Theorem, $m = (c^d \bmod n)$ is now given by $m_1.a + m_2.b \pmod{n}$.

There is even an extra bonus in this approach. The exponent d in the calculations of m_1 and m_2 can be reduced modulo $p-1$, resp. $q-1$, by Fermat's Theorem (Thm. A.15). Indeed, $m_1 \equiv c^d \equiv c^{d_1} \bmod p$, with $d_1 = (d \bmod p)$ and a similar statement is true for the mod q calculations.

Altogether, this way of computing $c^d \bmod n$ reduces the workload by a factor of about 4.

Example 9.1 (Part 4)

We continue with the parameters of Example 9.1, so $p_B = 9733$, $q_B = 10177$, $n_B = 99052741$, $e_B = 81119923$, and $d_B = 17089915$. To compute the solutions to

$$\begin{cases} a \equiv 1 & (\text{mod} \quad 9733) \\ a \equiv 0 & (\text{mod} \quad 10177) \end{cases}$$

$$\begin{cases} b \equiv 0 & (\text{mod} \quad 9733) \\ b \equiv 1 & (\text{mod} \quad 10177) \end{cases}$$

we load the Mathematica package `NumberTheory`NumberTheoryFunctions``

```
<<NumberTheory`NumberTheoryFunctions`
```

and find a and b with the function ChineseRemainderTheorem.

```
a = ChineseRemainderTheorem[{1, 0}, {9733, 10177}]
b = ChineseRemainderTheorem[{0, 1}, {9733, 10177}]
```

```
45287650
```

```
53765092
```

Next, we calculate $m_1 \equiv c^{d_1} \pmod{p}$ *and* $m_2 \equiv c^d \equiv c^{d_2} \pmod{q}$. *We get*

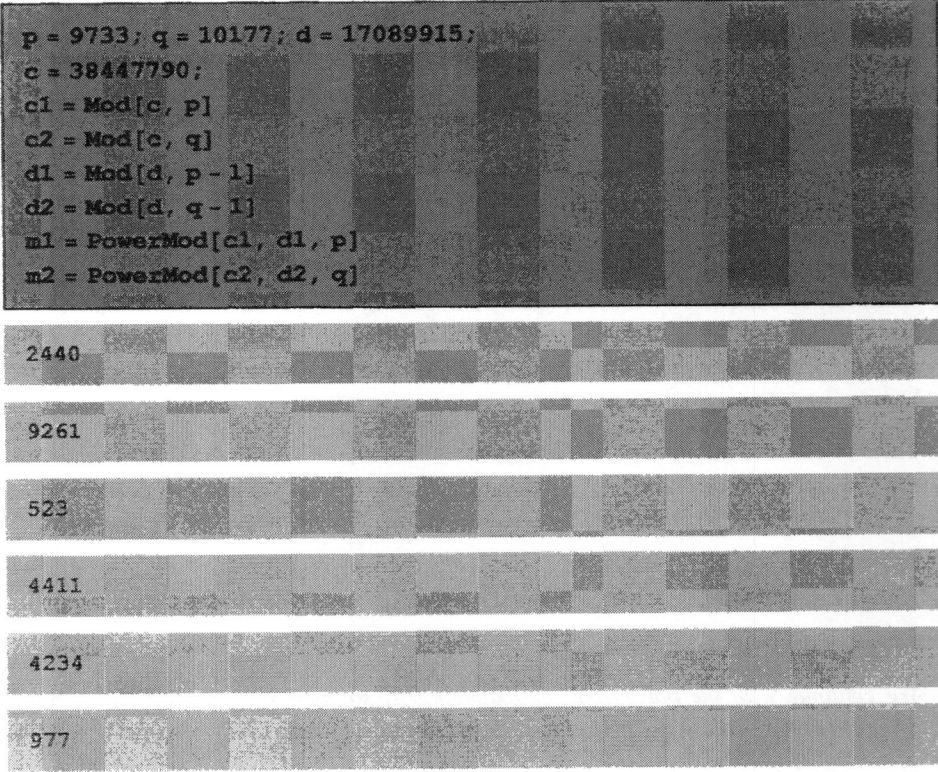

```
p = 9733; q = 10177; d = 17089915;
c = 38447790;
c1 = Mod[c, p]
c2 = Mod[c, q]
d1 = Mod[d, p - 1]
d2 = Mod[d, q - 1]
m1 = PowerMod[c1, d1, p]
m2 = PowerMod[c2, d2, q]
```

```
2440
```

```
9261
```

```
523
```

```
4411
```

```
4234
```

```
977
```

The result of the decryption process is now given by $m_1.a + m_2.b \bmod n$ *and coincides with our earlier decryption process.*

```
n = 99052741;
Mod[m1 * a + m2 * b, n]
```

```
12345678
```

9.1.4 RSA for Signatures

The RSA system can equally be used to sign messages. To sign a message m, $0 < m < n_B$, Bob will compute $c = (m^{d_B} \bmod n_B)$.

The receiver of c, say Alice, can easily retrieve the original message from $c^{e_B} \pmod{n_B}$, because Bob's parameters e_B and n_B are public. To check this we repeat (9.5) (with a minor variation):

$$c^{e_B} \equiv (m^{d_B})^{e_B} \equiv m^{e_B d_B} \overset{(9.4)}{\equiv} m^{1+l.\varphi(n_B)} \equiv m.(m^{\varphi(n_B)})^l \overset{(9.1)}{\equiv} m \pmod{n_B}. \tag{9.6}$$

for all m with $\gcd(m, n_B) = 1$. The relation $c^{e_B} \equiv m \pmod{n_B}$ also holds when $\gcd(m, n_B) \neq 1$. In Problem 9.2 the reader is asked to prove this.

Alice should keep c as Bob's signature on m. Only Bob can have made c out of m, because he is the only one knowing d_B. The reader is advised to reread the discussion above Table 7.2.

public	e_U and n_U of all users U
secret	d_U of user U
property	$e_U d_U \equiv 1 \pmod{\varphi(n_U)}$
message of Bob	$0 < m < n_B$
signing by B	$c \equiv m^{d_B} \pmod{n_B}$
verification by A	$c^{e_B} \equiv m \pmod{n_B}$
signature	the pair (m, c)

The RSA System for Signing

Table 9.2

Example 9.1 (Part 5)

Bob signs message $m = 11111111$ by computing $c \equiv m^{d_B} \pmod{n_B}$.

```
m = 11111111;
c = PowerMod[m, dB, nB]
```

```
74138899
```

Alice verifies this by computing c^{e_B} (mod n_B), which gives m.

9.1.5   RSA for Privacy and Signing

Suppose that Alice wants to sign a confidential message m to Bob. The solution described in Subsection 7.1.4 , namely Alice first signs m with her secret key and then encrypts the result with Bob's public key, can not always be applied directly in the RSA-case.

To see this, we observe that Alice would like to send

$$c = (m^{d_A} \pmod{n_A})^{e_B} \pmod{n_B}. \tag{9.7}$$

However, this mapping is not one-to-one if $n_A > n_B$. For instance, the messages $m = 1$ and $m = (1 + n_B)^{e_A}$ will both be mapped to $c = 1$.

Since Alice and Bob do not want to share their prime numbers, we must have $n_A < n_B$. In this case, Bob can recover m as follows:

$$(c^{d_B} \pmod{n_B})^{e_A} \pmod{n_A} = m.$$

To verify this, combine (9.5) with (9.6).

Of course, there now is the problem of what to do when Bob wants to sign a confidential message to Alice. A simple solution is to have every user U make two sets of parameters, one with its modulus smaller than some threshold T and the other with its modulus larger than T. In this setting, the sender uses his own smaller modulus for the signature and the receivers larger modulus for the encryption.

public	e_{Ui} and n_{Ui} of all users U, $i = 1, 2$
secret	d_{Ui} of user U, $i = 1, 2$
properties	$e_{Ui} d_{Ui} \equiv 1 \pmod{\varphi(n_{Ui})}$
	$n_{U1} < T < n_{U2}$
message from Alice to Bob	$0 < m < n_{A1}$
Alice sends	$c \equiv ((m^{d_{A1}} \bmod n_{A1})^{e_{B2}} \bmod n_{B2})$
Bob computes	$((c^{d_{B2}} \bmod n_{B2})^{e_{A1}} \bmod n_{A1}) = m$
Bob keeps as signature	m and $(c^{d_{B2}} \bmod n_{B2})$ which is equal to $(m^{d_{A1}} \bmod n_{A1})$

RSA for privacy and signing

Table 9.3

If there is an argument between Alice and Bob, they will go to an arbitrator. This arbitrator is given the pair m and $u = (c^{d_{B2}} \bmod n_{B2})$ by Bob. As an integer, the latter is equal to $(m^{d_{A1}} \bmod n_{A1})$, since

$$(c^{d_{B2}} \bmod n_{B2}) \overset{(9.7)}{=} \left(((m^{d_{A1}} \bmod n_{A1})^{e_{B2}} \bmod n_{B2})^{d_{B2}} \bmod n_{B2}\right) \overset{(9.5)}{=} (m^{d_{A1}} \bmod n_{A1}).$$

Just like in Subsection 9.1.4, the arbitrator now checks if $u^{e_{A1}} \equiv m \pmod{n_{A1}}$.

If this is the case, the message m came indeed from Alice, if not, u will not be considered as Alice's signature on m.

Note that the arbitrator does not need to know the secret exponents of Alice or Bob to make his decision. Therefore, Alice and Bob can continue to use their original set of parameters.

9.2   The Security of RSA: Some Factorization Algorithms

9.2.1   What the Cryptanalist Can Do

Suppose that an eavesdropper, say Eve, gets hold of a secret message $c = m^{e_B} \pmod{n_B}$ for Bob. Once Eve knows the secret exponent d_B of Bob, she can compute m from the ciphertext c in exactly the same way as Bob can, namely by computing $c^{d_B} \pmod{n_B}$ (see (9.5)).

To determine d_B from the public exponent e_B and the relation $e_B.d_B \equiv 1 \pmod{\varphi(n_B)}$ (see (9.4)) is easy for Eve as soon as she knows $\varphi(n_B)$: just like Bob did when he set up the system, she will use Euclid's Algorithm.

To find $\varphi(n_B) = p_B.q_B$ (see (9.3)) from the publicly known modulus n_B, Eve will have to find the factorization of n_B.

At the time of the introduction of RSA, Schroeppel (not published) had a modification of a factorization algorithm by Morrison and Brillhart [MorB75]. It involved

$e^{\sqrt{\ln n \ln \ln n}}$ operations

In the next table we have made use of the *Mathematica* functions `TableForm`, `Table`, `Exp`, `Sqrt`, `Log`, and `N` to give an impression of the growth of the above expression.

```
TableForm[    Table[
  {k, N[Exp[ Sqrt[Log[10^k] Log[Log[10^k]]]], 3]},
  {k, 25, 250, 25}],         TableHeadings ->
  {{}, {"length in digits", "complexity"}},
  TableAlignments -> {Center}]
```

length in digits	complexity
25	4.3×10^{6}
50	1.42×10^{10}
75	8.99×10^{12}
100	2.34×10^{15}
125	3.41×10^{17}
150	3.26×10^{19}
175	2.25×10^{21}
200	1.2×10^{23}
225	5.17×10^{24}
250	1.86×10^{26}

As one can see, if n is about 200 digits long, the above cryptanalysis is clearly not tractable. On the other hand, much larger numbers have been factored than was thought to be possible at the time that the original RSA scheme was proposed (at the time of the printing the record stood at 512 bits numbers). For this reason, one now sees proposals for implementations of RSA with a much larger modulus.

An example of a fast modern factorization algorithm can be found in [LensH86]. Other methods will be discussed in Section 9.2.3. There does exist special factorization algorithms that run faster if n is of a special form. We shall discuss one of these methods in the next subsection.

Up to now, there seems to be no way of breaking the RSA system other than by factoring the modulus n. There is no formal proof however that these two problems are equivalent. In Section 9.5 we shall discuss a variant of the RSA system for which it can be shown that breaking it is equivalent to factoring its modulus.

A drawback of having to choose large moduli is that the execution of a single exponentiation takes more time than one may like, especially when one wants to encrypt a long file. Quite often in such a situation one shall use a hybrid system: a symmetric system with secret key k is used for encryption of the data and the RSA scheme is used to send this key securely to the receiver (using the public parameters of the receiver).

When generating p and q it is a bad idea to first generate p and then try out $p+2$, $p+4$, ... for primality. One really wants $p-q$ to be large. Indeed, if a cryptanalist can guess $p-q$, for instance by checking all likely values, it follows from

$$4n = 4p.q = (p+q)^2 - (p-q)^2$$

that $p+q$ also can be determined. From these two linear relations p and q can be found, which implies that the system has been broken.

Example 9.2

Let $n = 5007958289$. Guessing that $q - p = 200$, we get $p + q$ from

$$n = 5007958289; \quad \sqrt{4n + 200^2}$$

$$141534$$

From $p + q = \sqrt{4n + 200^2}$ and $q - p = 200$, we get that $q = \left(\sqrt{4n + 200^2} + 200 \right) / 2$.

$$q = \left(\sqrt{4\,n + 200^2} + 200 \right) \Big/ 2$$

$$p = q - 200$$

70867

70667

p * q == n

True

We conclude that $|p - q|$ has to be large. A way to do this is to take q more than $p + \sqrt{p}$.

In the literature one can also find a few attacks on the RSA system, that have a probability of success which is not significantly more than the probability that a randomly chosen integer a smaller than n has a non-trivial factor in common with n. This factor would then be p or q. The probability that the latter happens can be evaluated with the Euler Totient function $\varphi(n)$ and is given by

$$\frac{n - \varphi(n)}{n} \overset{(9.3)}{=} \frac{p.q - (p-1)(q-1)}{p.q} = \frac{p+q-1}{p.q} \approx \frac{1}{p},$$

assuming that $p < q$. That one should not take p too small will follow from the factorization algorithm that we shall discuss in the next subsection.

Because the "attacks" mentioned above have such a small probability of success, we choose not to discuss them here. Some of the problems at the end of this chapter are based on them.

9.2.2   A Factorization Algorithm for a Special Class of Integers

We shall now briefly discuss a factorization algorithm that runs faster than the general factorization algorithms that we shall address later under the assumption that at least one of the prime factors of n, say p, has the property that $p - 1$ only contains small prime factors.

□ **Pollard's $p - 1$ Method**

In [Poll75], Pollard describes a way to factor n in $\sqrt{p}$ steps, where p is the smallest prime divisor of n. This explains why we have to take p and q both large.

The assumption in Pollard's $p - 1$ method is that in the factorization of n at least one of the two factors, say p, has the property that $p - 1$ has only small prime factors. To be more precise, an integer is said to be *smooth* (see also Subsection 8.3.4) with respect to S if all its prime factors are less than or equal to S. We shall assume that $p - 1$ is smooth with respect to some integer S.

Example 9.3

The prime number $p = 70877$ has the property that $p - 1$ is smooth with respect to $S = 50$, as one can check with the Mathematica function <u>FactorInteger</u> *and* <u>PrimeQ</u>.

```
p = 70877; PrimeQ[p]
FactorInteger[p - 1]
```

```
True
```

```
{{2, 2}, {13, 1}, {29, 1}, {47, 1}}
```

For each prime number r, $r \leq S$, the largest power of r that is still less than or equal to n can be determined from

$$r^i \leq n \quad \text{, or, equivalently,} \quad i \leq \log_r n.$$

Define R by

$$R = \prod_{p \leq S,\, p \text{ prime}} p^{\lfloor \log_r n \rfloor} \tag{9.8}$$

Example 9.4 (Part 1)

Consider the number $n = 6700892281$ and assume that at least on of its factors, say p, is smooth with respect to $S = 50$. It follows from

```
Prime[15]
Prime[16]
```

```
47
```

```
53
```

that there are 15 primes less than or equal to $S = 50$. So, R can be calculated from (9.8) with the Mathematica functions <u>Prime</u>, <u>Log</u>, *and* <u>Floor</u> *as follows*

```
n = 6700892281; R = ∏_{i=1}^{15} (Prime[i])^Floor[Log[Prime[i],n]]
```

404956718036087157154810988735114505168715463893514902450607270767
022142824248137349465019194031679620397545778700300894863360000000
00000000

To see the exponents of the primes up to 50 (out of curiosity), we give

```
FactorInteger[R]
```

```
{{2, 32}, {3, 20}, {5, 14}, {7, 11}, {11, 9}, {13, 8}, {17, 7}, {19, 7},
 {23, 7}, {29, 6}, {31, 6}, {37, 6}, {41, 6}, {43, 6}, {47, 5}}
```

If $p - 1$ is smooth with respect to S, each prime power r^i that divides $p - 1$, will also be a factor of R, since i will be at most $\lfloor \log_r n \rfloor$. It follows that $(p - 1)$ divides R.

We know from Fermat's Theorem (Thm. A.15) that any integer a, $1 \le a < p$, will satisfy $a^{p-1} \equiv 1 \pmod{p}$. Since $(p - 1) \mid R$, also $a^R \equiv 1 \pmod{p}$.

Now take a random integer a, $2 \le a < n$, and check if $\gcd(a, n) = 1$. If this gcd is not 1, we have found a factor of n and we are done.

If $\gcd(a, n) = 1$ it follows from $a^R \equiv 1 \pmod{p}$ that $p \mid (a^R - 1)$. Since it is very unlikely that also $a^R \equiv 1 \pmod{q}$, we shall almost certainly find a factor of n (namely p) from $\gcd(a^R - 1, n)$. Note that a^R does not have to be evaluated for this calculation, the value of $a^R \pmod{n}$ suffices.

Example 9.4 (Part 2)

To find a factor of $n = 6709248019$ we pick a random a in between 2 and $n - 1$ and compute the gcd of $a^R - 1$ with n by means of the Mathematica functions Random, PowerMod, *and* GCD.

```
a = Random[Integer, {2, n}]
GCD[PowerMod[a, R, n] - 1, n]
```

```
3922094384
```

```
81919
```

It follows that $p = 81919$ is a factor of n. The other factor follows from $n / p = 81799$. Note that if q is also smooth with respect to S, we would have found n as outcome of the gcd calculation.

We summarize Pollard's $p - 1$ method in the following table.

> **input** : integer n.
> **select** a smoothness parameter S.
> **calculate** R from (9.8).
> **select** a random a, $2 \le a < n$.

```
compute d = gcd (aᴿ - 1, n).
if 1 < d < n then d is a factor of n
              else STOP or select a new random a
```

Pollard's $p - 1$ Method to Factor n

Figure 9.1

To make Pollard's $p - 1$ method infeasible, one often chooses so-called *safe primes* when setting up the RSA system. These strong primes are primes p of the form $p = 2p' + 1$, where p' is a (large) prime. In this case, $p - 1$ has just one small factor.

9.2.3 General Factorization Algorithms

□ The Pollard-ϱ Method

Let p be an unknown prime factor of the integer n that we want to factor. Now look at the sequence $a_0, a_1, \ldots$, defined recursively by

$$a_0 = 1,$$
$$a_{i+1} \equiv a_i^2 + 1 \pmod{p}, i \geq 0.$$

Suppose that we have found indices u and v with $v > u$ and $a_u \equiv a_v \pmod{p}$. Then clearly $\gcd(a_v - a_u, n)$ is divisible by p and very likely this gcd is equal to p.

Of course, p is not known, so we replace the above recursion relation by

$$a_0 = 1,$$
$$a_{i+1} \equiv a_i^2 + 1 \pmod{n}, i \geq 0. \tag{9.9}$$

Since $p \mid n$ we will find the factor p from $\gcd(a_v - a_u, n)$ for the same values of u and v (the probability that other large factors of n divide this gcd is negligible).

Instead of having to store all previously computed values of a_i, $i \geq 0$, we use Floyd's cycle-finding algorithm to find an index k such that $a_{2k} = a_k$ and then we take $u = k$ and $v = 2k$. The idea is simply that one starts with a_1 and a_2 and recursively determines the pair (a_i, a_{2i}) from $(a_{i-1}, a_{2(i-1)})$.

The above is summarized in the following figure.

```
input : integer n.
put a = 1, b = 2.
do a ← (a² + 1) mod n,
   b ← (((b² + 1) mod n)² + 1) mod n
until d = gcd (b - a, n) > 1
if d < n then d is a factor of n
        else STOP
```

Pollard's ϱ Method to Factor n

Figure 9.2

Example 9.5

To find a factor of n = 9032411471 with the above method we use the Mathematica functions While, Mod, *and* GCD *functions.*

```
n = 168149075693;
a = 1; b = 2; d = GCD[b - a, n];
While[d == 1,      a = Mod[a^2 + 1, n];
 b = Mod[(Mod[b^2 + 1, n])^2 + 1, n]; d = GCD[b - a, n]]
d
```

```
350377
```

So, 350377 is a factor of n = 168149075693. The quotient n/p is 479909, which happens to be a prime too, as can easily be checked with the function PrimeQ.

```
a = n / 350377
PrimeQ[a]
```

```
479909
```

```
True
```

□ **Random Square Factoring Methods**

This method and the next one are related to the Index-Calculus Method discussed in Subsection 8.3.4. The reader may want to read the introduction there first, but that will not necessary for the understanding of the discussion here. We assume that n is a composite odd integer.

The method consists of the following four steps.

Step 1:

Construct the set $S = \{p_1, p_2, ..., p_k\}$ consisting of the first k prime numbers, so $p_1 = 2$, $p_2 = 3$, etc. The set S will be called the factor base.

Step 2:

Find sufficiently many pairs (a_i, b_i) such that

$$a_i^2 \equiv b_i \pmod{n} \tag{9.10}$$

and such that b_i is smooth with respect to S, i.e. b_i factors completely into elements of the factor base S, say

$$b_i = \prod_{j=1}^{k} p_j^{u_{i,j}}, \qquad \text{with } u_{i,j} \geq 0.$$

Put $\underline{u}_i = (u_{1,1}, u_{1,2}, \ldots, u_{1,k})$. Pairs (a_i, b_i) satisfying property (9.10) can be found by trying random choices of a_i. An alternative is to use any suitable recursion relation that generates candidates for a_i. For instance, after trying $a_i = a$ one may want to try $a_i = ((a^2 + 1) \bmod n)$.

Step 3:

Find a collection of b_i's whose product is a perfect square. Quite clearly, only the parity of the $u_{i,j}$'s matters in this condition, so let us put $v_{i,j} = (u_{i,j} \bmod 2)$ and $\underline{v}_i = (v_{1,1}, v_{1,2}, \ldots, v_{1,k})$. We write $\underline{v}_i \equiv \underline{u}_i \pmod 2$.

Since any $k + 1$ vectors $\underline{v}_i$ (all of length k) must be linearly dependent over $\mathbb{Z}_2$, there must be a non-trivial linear combination adding up to $\underline{0}$. Such a linear combination can be found very efficiently with standard methods from linear algebra.

Let I denote the subset of $\{1, 2, \ldots, k\}$ with $\sum_{i \in I} \underline{v}_i \equiv \underline{0} \pmod 2$. Set

$$x = \prod_{i \in I} a_i \qquad \text{and} \qquad y = (\prod_{i \in I} b_i)^{1/2}.$$

Step 4:

It follows from (9.10) that $x^2 \equiv y^2 \pmod n$, i.e. n divides $(x - y)(x + y)$. Assume that $x \not\equiv \pm y \pmod n$ (the probability that this happens is at least 1/2 as we shall see in a moment and as will be demonstrated more extensively in Subsection 9.5.1 for the case that n is the product of two different primes). Then $x - y$ must be divisible by a non-trivial divisor of n. In other words, $\gcd(x - y, n)$ yields a non-trivial factor of n.

If $\gcd(x - y, n) = n$ one has to try to find another perfect square, either by another linear dependency between the $\underline{v}_i$'s or by exchanging one of the pairs (a_i, b_i) for a new one.

Consider the congruence relation $x^2 \equiv y^2 \pmod n$ where y is assumed to have a given fixed value that is coprime with n. Further, let p^a be any factor in the prime power decomposition of n (see Theorem A.6). Then $x^2 \equiv y^2 \pmod{p^a}$ has just two solutions, namely $x \equiv \pm y \pmod{p^a}$. Indeed, for $a = 1$ this follows from Theorem B.15. For $a > 1$, we still have that p^a must divide either $x - y$ or $x + y$, because if $p \mid (x - y)$ and $p \mid (x + y)$ then $p \mid 2y$, but $p \nmid y$ (since n is odd, also p will be odd). We conclude that $x \equiv \pm y \pmod{p^a}$ also when $a > 1$.

It now follows directly from the Chinese Remainder Theorem (Thm. A.19) that relation $x^2 \equiv y^2 \pmod n$ has 2^l solutions, where l is the number of different prime numbers dividing n. Only two of these 2^l, $l \geq 2$, solutions are given by $x \equiv \pm y \pmod n$, therefore, the probability that $\gcd(x - y, n)$ yields a non-trivial factor of n is at least $(2^l - 2)/2^l \geq 2/4 = 1/2$.

```
input : integer n.
make factor base S = {p₁, …, p_k}
find pairs (aᵢ, bᵢ) with
```

a_i random, $\quad a_i^2 \equiv b_i \pmod{n}$, $\quad b_i$ smooth w.r.t. S

find index set I such that $\prod_{i \in I} b_i$ is a perfect square

put $\quad x = \prod_{i \in I} a_i$, $\quad y = \sqrt{\prod_{i \in I} b_i}$

put $d = \gcd(x - y, n)$

if $d < n$ **then** d is a factor of n
         else retry with other I

Factoring by Random Squares

Figure 9.3

Example 9.6

Suppose that we try to factor $n = 1271$ with the above method. We first make the factor base consisting of the first 8 primes by means of the Mathematica functions Table *and* Prime.

```
S = Table[Prime[i], {i, 1, 8}]
```

```
{2, 3, 5, 7, 11, 13, 17, 19}
```

Next, we use the function Random *to generate a random* a, $1 \le a \le n$, *and the function* FactorInteger *to factor* $b \equiv a^2 \pmod{n}$.

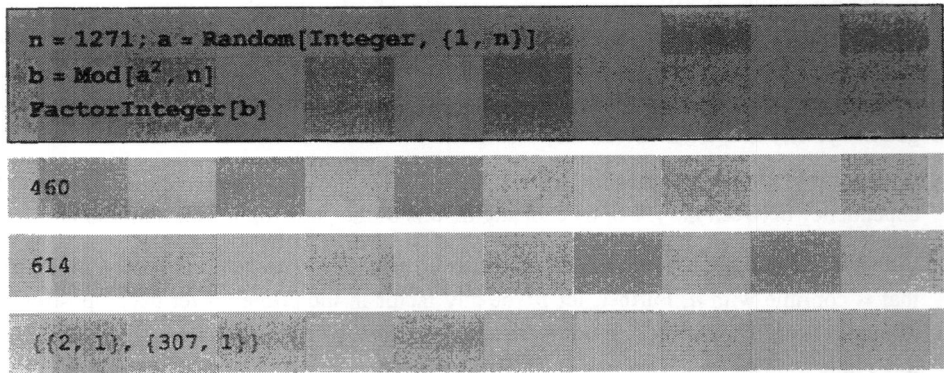

```
n = 1271; a = Random[Integer, {1, n}]
b = Mod[a^2, n]
FactorInteger[b]
```

```
460
```

```
614
```

```
{{2, 1}, {307, 1}}
```

Unfortunately, $b = 614$ is not smooth with respect to S, but after some trial and error we found the following nine smooth numbers (they are put in a list called a).

```
n = 1271;
a = {583, 879, 1137, 421, 727, 1034, 1051, 107, 1111};
b = Mod[a^2, n];
TableForm[Table[ {a[[i]], b[[i]],
    Times @@ Superscript @@@ FactorInteger[ b[[i]] ] },
   {i, 1, Length[a]} ], TableHeadings ->
   {{}, {"a", "a^2 mod n", "factors"}},
  TableAlignments -> {Left}]
```

a	a^2 mod n	factors
583	532	$2^2 \, 7^1 \, 19^1$
879	1144	$2^3 \, 11^1 \, 13^1$
1137	162	$2^1 \, 3^4$
421	572	$2^2 \, 11^1 \, 13^1$
727	1064	$2^3 \, 7^1 \, 19^1$
1034	245	$5^1 \, 7^2$
1051	102	$2^1 \, 3^1 \, 17^1$
107	10	$2^1 \, 5^1$
1111	180	$2^2 \, 3^2 \, 5^1$

The exponents in the factorization of the b_i's are given by the vectors $\underline{u}_i$, that form the rows of the matrix U below. The vectors $\underline{v}_i$ are the modulo 2 reductions of the $\underline{u}_i$'s. They form the rows of the matrix V below.

For instance, $b_1 = 532 = 2^2 . 7 . 19$ gives $\underline{u}_1 = \{2, 0, 0, 1, 0, 0, 0, 1\}$ and $\underline{v}_1 = \{0, 0, 0, 1, 0, 0, 0, 1\}$. These two rows are the first row of the matrices U resp. V below. We use the function MatrixForm to display them.

```
U = {{2, 0, 0, 1, 0, 0, 0, 1}, {3, 0, 0, 0, 1, 1, 0, 0},
    {1, 4, 0, 0, 0, 0, 0, 0}, {2, 0, 0, 0, 1, 1, 0, 0},
    {3, 0, 0, 1, 0, 0, 0, 1}, {0, 0, 1, 2, 0, 0, 0, 0},
    {1, 1, 0, 0, 0, 0, 1, 0}, {1, 0, 1, 0, 0, 0, 0, 0},
    {2, 2, 1, 0, 0, 0, 0, 0}};
V = Mod[U, 2];
MatrixForm[U]
MatrixForm[V]
```

$$\begin{pmatrix}
2 & 0 & 0 & 1 & 0 & 0 & 0 & 1 \\
3 & 0 & 0 & 0 & 1 & 1 & 0 & 0 \\
1 & 4 & 0 & 0 & 0 & 0 & 0 & 0 \\
2 & 0 & 0 & 0 & 1 & 1 & 0 & 0 \\
3 & 0 & 0 & 1 & 0 & 0 & 0 & 1 \\
0 & 0 & 1 & 2 & 0 & 0 & 0 & 0 \\
1 & 1 & 0 & 0 & 0 & 0 & 1 & 0 \\
1 & 0 & 1 & 0 & 0 & 0 & 0 & 0 \\
2 & 2 & 1 & 0 & 0 & 0 & 0 & 0
\end{pmatrix}$$

$$\begin{pmatrix}
0 & 0 & 0 & 1 & 0 & 0 & 0 & 1 \\
1 & 0 & 0 & 0 & 1 & 1 & 0 & 0 \\
1 & 0 & 0 & 0 & 0 & 0 & 0 & 0 \\
0 & 0 & 0 & 0 & 1 & 1 & 0 & 0 \\
1 & 0 & 0 & 1 & 0 & 0 & 0 & 1 \\
0 & 0 & 1 & 0 & 0 & 0 & 0 & 0 \\
1 & 1 & 0 & 0 & 0 & 0 & 1 & 0 \\
1 & 0 & 1 & 0 & 0 & 0 & 0 & 0 \\
0 & 0 & 1 & 0 & 0 & 0 & 0 & 0
\end{pmatrix}$$

To find a non-trivial linear combination of the rows of V adding up to the all-zero vector modulo 2, we use the NullSpace *and* Transpose *functions.*

```
NullSpace[Transpose[V], Modulus -> 2]
```

```
{{0, 0, 0, 0, 0, 1, 0, 0, 1}, {0, 0, 1, 0, 0, 1, 0, 1, 0},
 {1, 0, 1, 0, 1, 0, 0, 0, 0}, {0, 1, 1, 1, 0, 0, 0, 0, 0}}
```

We see that the first of the above linear dependencies between rows of V reflect two identical rows, but the third one does give an index set I that can be used, namely I = {1, 3, 5}.

It leads to the values $x = a_1 a_3 a_5$ and $y = \sqrt{b_1 b_3 b_5}$

```
x = a[[1]] * a[[3]] * a[[5]]
y = (b[[1]] * b[[3]] * b[[5]])^{1/2}
GCD[x - y, n]
```

```
481907217
```

```
9576
```

```
41
```

We conclude that $p = 41$ is a factor of $n = 1271$. Indeed $1271 = 31 \times 41$.

□ **Quadratic Sieve**

The complexity of this method is given by

$$e^{1.923.. \, (\ln n)^{1/3} \, (\ln \ln n)^{2/3}} \text{ operations.}$$

As with the previous methods, we shall not explain all details of this factorization technique. Let n be the number that we want to factor.

To start we need a so-called factor base S, which means that S is a list of k primes (which k primes will be determined later).

Let $r = \lfloor \sqrt{n} \rfloor$ and let the polynomial $f(x)$ be defined by

$$f(x) = (x + r)^2 - n = x^2 + 2\,r.x + r^2 - n.$$

Note that $r^2 \le n < (r + 1)^2$, so $0 \le n - r^2 < 2r + 1 \le 2\sqrt{n} + 1$. It follows that if x is small in absolute value, then also $f(x)$ will be small (when compared to n).

For $x = 0, \pm 1, \pm 2, \ldots$ define a by $a = x + r$ and test $b = (x + r)^2 - n$ for smoothness with respect to S, i.e. test if all prime factors of b are in S. If so, we save the pair (a, b) in a list of pairs (a_i, b_i) with this property.

Note that $a_i^2 \equiv (x + r)^2 \equiv b_i \pmod{n}$, just as in equation (9.10).

If a prime p divides b_i, then $p \,|\, ((x + r)^2 - n)$ for some known value of x. This means that $n \equiv (x + r)^2 \pmod{p}$ and thus that n is a quadratic residue (QR) mod p. This means that the only prime factors that will appear in the factorization of any of the b_i's will have Jacobi symbol $(n / p) = 1$.

So, we let the factor basis S consist of the k smallest p_j, $1 \le j \le k$, with the property that $(n / p_j) = 1$. We also add -1 and 2 to S, because the b_i's may be negative and/or even.

Now that we know how to construct a list of pairs (a_i, b_i), satisfying

$$a_i^2 \equiv b_i \pmod{n},$$

b_i is smooth with respect to S,

we can continue with Step 3 in the algorithm described in the previous subsubsection.

We summarize the quadratic sieve method in the following figure.

input :   integer n.
make factor base $S = \{-1, 2, p_1, ..., p_k\}$ with $(n / p_j) = 1$
find pairs (a_i, b_i) with $a_i - \lfloor \sqrt{n} \rfloor$ small,
        $a_i^2 \equiv b_i \pmod{n}$, and b_i smooth $w.r.t.$ S
find index set I such that $\prod_{i \in I} b_i$ a perfect square
put $x = \prod_{i \in I} a_i$, $y = \sqrt{(\prod_{i \in I} b_i)}$
put $d = \gcd (x - y, n)$
if $d < n$ **then** d is a factor of n
             else retry with other I

Quadratic Sieve Factoring Algorithm

Figure 9.4

We shall only give an example of the first two steps of the quadratic sieve method.

Example 9.7

Let $n = 661643$. To make a factor base with 10 primes, we use the Mathematica functions `While`, `Length`, `JacobiSymbol`, `Prime`, *and* `AppendTo`.

```
n = 661643; k = 10;
SS = {-1, 2}; i = 2;
While[Length[SS] - 2 < k,
    If[JacobiSymbol[n, Prime[i]] == 1,
       AppendTo[SS, Prime[i]]]; i = i + 1];
SS
```

```
{-1, 2, 11, 19, 23, 31, 37, 47, 53, 59, 79, 89}
```

To try out if any of $f(-5)$, $f(-4)$, ..., $f(5)$ is smooth with respect to S we use the functions `TableForm`, `Table`, *and* `FactorInteger`:

```
n = 661643; Clear[x, f];
r = ⌊√n⌋; m = 5;
f[x_] := (x + r)^2 - n;
TableForm[        Table[ {r + i, f[i],
    FactorInteger[f[i]] // OutputForm},        {i, -m, m}]]
```

```
808    -8779    {{-1, 1}, {8779, 1}}
809    -7162    {{-1, 1}, {2, 1}, {3581, 1}}
810    -5543    {{-1, 1}, {23, 1}, {241, 1}}
811    -3922    {{-1, 1}, {2, 1}, {37, 1}, {53, 1}}
812    -2299    {{-1, 1}, {11, 2}, {19, 1}}
813    -674     {{-1, 1}, {2, 1}, {337, 1}}
814    953      {{953, 1}}
815    2582     {{2, 1}, {1291, 1}}
816    4213     {{11, 1}, {383, 1}}
817    5846     {{2, 1}, {37, 1}, {79, 1}}
818    7481     {{7481, 1}}
```

We see that we have only found three pairs (a_i, b_i), *namely* $(811, -3922)$, $(812, -2299)$, *and* $(817, 5846)$.

So, we need to try a larger range of values. We leave the rest of this example as an exercise to the reader (see Problem 9.7).

9.3   Some Unsafe Modes for RSA

9.3.1   A Small Public Exponent

We shall discuss here two particular dangers described in [Håst88] (see also [CoppFPR96]). The first one is the situation that more people have chosen the same (small) public exponent and that a sender wants to transmit the same message to all of them. The second danger is when a sender wants to transmit several mathematically related messages to the same receiver, who happens to have a small public exponent.

Both dangers may appear farfetched to the reader, but since exponentiations modulo large numbers are still rather cumbersome, it remains very appealing in practical situations to select small public exponents.

□ **Sending the Same Message to More Receivers Who All Have the Same Small Public Exponent**

Suppose that Alice wants to send the same secret message m to Bob, Chuck, and Dennis. Let the public modulus of these three people be given by the numbers n_B, n_C, and n_D. Now assume that they all happen to have the same public exponent $e = 3$. The messages that Alice will transmit are

$$\begin{aligned}
c_B &\equiv m^3 \ (\text{mod } n_B) \quad \text{for Bob,} \\
c_C &\equiv m^3 \ (\text{mod } n_C) \quad \text{for Chuck,} \\
c_D &\equiv m^3 \ (\text{mod } n_D) \quad \text{for Dennis.}
\end{aligned} \tag{9.11}$$

Almost certainly the three moduli will be coprime (otherwise at least two of moduli are compromised in a trivial way). The eavesdropper Eve, who intercepts c_B, c_C, and c_D can use the Chinese Remainder Theorem (Thm. A.19) to determine $m^3 \pmod{n_B\, n_C\, n_D}$ from (9.11).

Since it can be assumed that $m < \min \{n_B,\ n_C,\ n_D\}$, also $m^3 < n_B\, n_C\, n_D$ holds. So, the above means that Eve in fact has found the integer m^3. To compute m is now straightforward.

Example 9.8

Suppose that $n_B = 137703491$, $n_C = 144660611$, and $n_D = 149897933$. Let the three intercepted messages be given by $c_B = 124100785$, $c_C = 85594143$, and $c_D = 148609330$.

To solve the system of linear congruence relations

$$m^3 \equiv c_B \ (mod\ n_B);\ m^3 \equiv c_C \ (mod\ n_C);\ m^3 \equiv c_D \ (mod\ n_D),$$

with known right hand sides and known moduli, we use the Mathematica function ChineseRemainderTheorem. *To this end we first have to load the package* NumberTheory`NumberTheoryFunctions`.

```
<<NumberTheory`NumberTheoryFunctions`
```

```
nB = 137703491; nC = 144660611; nD = 149897933;
cB = 124100785; cC = 85594143; cD = 148609330;
mCubed = ChineseRemainderTheorem[{cB, cC, cD}, {nB, nC, nD}]
```

```
1881563525396008211918161
```

We conclude that $m^3 \equiv 1881563525396008211918161 \ (mod\ n_B\, n_C\, n_D)$. Since $m^3 < n_B\, n_C\, n_D$, we even have

$$m^3 = 1881563525396008211918161.$$

To find m is now easy.

```
m = (mCubed)^(1/3)
```

```
123454321
```

That this outcome is correct can easily be checked by means of the Mod *function.*

```
Mod[m^3, nB] == cB
Mod[m^3, nC] == cC
Mod[m^3, nD] == cD
```

```
True
```

```
True
```

```
True
```

□ Sending Related Messages to a Receiver with Small Public Exponent

Alice wants to send two secret messages, say m_1 and m_2 to Bob, who happens to have a public exponent e_B that is rather small. Let n_B be Bob's modulus. Now, assume that the two messages of Alice are related in a linear way, say $m_2 = a.m_1 + b$, where a and b are in Z_{n_B} and assume further that eavesdropper Eve knows this linear relation.

Coppersmith et al. [CoppFPR96] describe two surprising methods for Eve to recover the plaintext m.

Direct Method

We shall first describe this method for the case $e = 3$.

Let the encryptions of m_1 and m_2 be denoted by c_1, resp. c_2. So, $c_1 \equiv m_1^3 \pmod{n_B}$ and $c_2 \equiv (a.m_1 + b)^3 \pmod{n_B}$. Then

$$\frac{b(c_2 + 2a^3 c_1 - b^3)}{a(c_2 - a^3 c_1 + 2b^3)} \equiv \frac{3a^3 bm_1^3 + 3a^2 b^2 m_1^2 + 3ab^3 m_1}{3a^3 bm_1^2 + 3a^2 b^2 m_1 + 3ab^3} \equiv m_1 \pmod{n_B}. \qquad (9.12)$$

With the *Mathematica* function <u>Simplify</u> one can verify these calculations as follows

```
Clear[a, b, c1, c2, m1, m2];
Simplify[ b (c2 + 2 a^3 c1 - b^3) //. {c1 -> m1^3, c2 -> (a*m1+b)^3}]
          a (c2 - a^3 c1 + 2 b^3)
```

```
m1
```

A particular simple case is given by $m_1 = m$ and $m_2 = m + 1$, i.e. $a = b = 1$. Then (9.12) reduces to

$$\frac{(m+1)^3 + 2m^3 - 1}{(m+1)^3 - m^3 + 2} \equiv \frac{3m^3 + 3m^2 + 3m}{3m^2 + 3m + 3} \equiv m \pmod{n_B}$$

Example 9.9

Suppose that $n_B = 477310661$ and that the messages m_1 and m_2 are related by $m_2 \equiv 3\, m_1 + 5 \pmod{n_B}$. So, $a = 3$ and $b = 5$. Let $c_1 = 477310661$ and $c_2 = 5908795$. Then m_1 can be computed with the Mathematica functions Mod *and* Solve *as follows*

```
Clear[c1, c2, f, g, m1, m2, a, b];
n = 477310661;
c1 = 5908795; c2 = 374480016;
a = 3; b = 5;
f = Mod[b (c2 + 2 a^3 c1 - b^3), n];
g = Mod[a (c2 - a^3 c1 + 2 b^3), n];
Solve[{f == g*m1, Modulus == n}, m1]
```

```
{{Modulus → 477310661, m1 → 321321321}}
```

So, we have found $m_1 = 321321321$. That this is indeed the solution can be verified quite easily as follows

```
m1 = 321321321;
m2 = Mod[3*m1 + 5, n]
PowerMod[m1, 3, n] == c1
PowerMod[m2, 3, n] == c2
```

```
9342646
```

```
True
```

```
True
```

If $a = b = 1$ and $e_B > 3$, a method like the above still exists. In fact, it can be shown [CoppFPR96] that polynomials $P(m)$ and $Q(m)$ exist such that each of them can be expressed as rational polynomials in $c_1 \equiv m^e \pmod{n_B}$ and $c_2 \equiv (m+1)^e \pmod{n_B}$ and such that $Q(m) = m.P(m)$. For $e_b = 5$ these polynomials are given by

$$P(m) = c_2^3 + 2\,c_1\,c_2^2 - 4\,c_1^2\,c_2 + c_1^3 - 2\,c_2^2 + 9\,c_1\,c_2 + 8\,c_1^2 + c_2 - 2\,c_1,$$

$$Q(m) = 9\,c_1\,c_2^2 - 9\,c_1^2.$$

Again, one can check this with

```
Clear[c1, c2, m];
P = c2^3 + 2 c1 * c2^2 - 4 c1^2 c2 + c1^3 - 2 c2^2 +
   9 c1 * c2 + 8 c1^2 + c2 - 2 c1; Q = 9 c1 * c2^2 - 9 c1^2;
Expand[P //. {c1 -> m^3, c2 -> (m + 1)^3}]
Expand[Q //. {c1 -> m^3, c2 -> (m + 1)^3}]
Simplify[Q/P //. {c1 -> m^3, c2 -> (m + 1)^3}]
```

$$9 m^2 + 54 m^3 + 135 m^4 + 171 m^5 + 135 m^6 + 54 m^7 + 9 m^8$$

$$9 m^3 + 54 m^4 + 135 m^5 + 171 m^6 + 135 m^7 + 54 m^8 + 9 m^9$$

$$m$$

To find such a solution, write $P = \sum_{i+j \le e} p_{i,j} c_2^i c_1^j$ and $Q = \sum_{i+j \le e} q_{i,j} c_2^i c_1^j$. Next, substitute $c_2 = (m + 1)^e$ and $c_1 = m^e$ in P and Q to obtain two polynomials in m of degree $\le e^2$. Now, equate the coefficients of m in $Q(m) = m.P(m)$. This gives $2 ((e + 1) + e + \dots + 2 + 1) = 2 \binom{e+2}{2} = (e + 2)(e + 1)$ linear equations in the coefficients of P and Q. So, there is in fact a large solution space.

Since the number of terms in $P(m)$ and $Q(m)$ grows quadratic in e the above approach will still be rather cumbersome for larger values of e.

Method through GCD calculation

For arbitrary values of e there is a more direct way to determine m_1 and m_2 from c_1 and c_2, when they satisfy a polynomial relation that is known to the eavesdropper. Suppose that $m_2 \equiv f(m_1) \pmod{n_B}$. The idea is to compute the gcd of $z^e - c_1$ and $(f(z))^e - c_2$. Indeed, since m_1 is a zero of both polynomials, it follows that both are divisible by $z - m_1$. As a consequence, also the gcd will contain this factor. Almost certainly the gcd will not contain any other factors.

We shall demonstrate this idea with an example.

Example 9.10

Let $e_B = 5$, $n_B = 466883$. Further suppose that the message m_1 and m_2 are related by $m_2 = 2 m_1 + 3$ and that they are encrypted into $c_1 = 66575$, resp. $c_2 = 387933$. We want to compute $gcd(z^5 - 66575, (2 z + 3)^5 - 387933) \mod 466883$. In general, this can not be done since n_B is not prime. Also Mathematica can not do this directly. We shall simply follow the polynomial version of Euclid's Algorithm step for step. Problems may arise, when numbers appear that are not coprime with n. This happens rarely and is not bad at all. Indeed, one almost always finds in this way a non-trivial factor of n, so the system will be broken!

In the first step we calculate $f_1 = (2z + 3)^5 - 387933$ and $f_2 = z^5 - 66575$ and then divide f_1 by f_2. We use the Mathematica functions PolynomialMod and Expand.

```
n = 466883;
c1 = 66575; c2 = 387933;
f1 = Expand[(2 z + 3)^5 - c2]
f2 = z^5 - c1
f3 = PolynomialMod[f1 - 32 f2, n]
```

$$-387690 + 810 z + 1080 z^2 + 720 z^3 + 240 z^4 + 32 z^5$$

$$-66575 + z^5$$

$$342061 + 810 z + 1080 z^2 + 720 z^3 + 240 z^4$$

To keep the division process more manageable, we normalize f_3 by multiplying it with the multiplicative inverse of its leading coefficient (mod n_B). We use the Mathematica function PowerMod.

```
InverseLeadCoeff = PowerMod[240, -1, n]
f3 = PolynomialMod[InverseLeadCoeff * f3, n]
```

$$258731$$

$$376877 + 408526 z + 233446 z^2 + 3 z^3 + z^4$$

We continue with this division process until $f_k = 0$ for some k. The gcd will be given by f_{k-1}.

```
f4 = PolynomialMod[f2 - f3 * (z + 466880), n]
```

$$130290 + 381818 z + 291812 z^2 + 233446 z^3$$

```
InverseLeadCoeff = PowerMod[233446, -1, n]
f4 = PolynomialMod[InverseLeadCoeff * f4, n]
```

$$103752$$

```
184581 + 292352 z + 116723 z^2 + z^3
```

```
f5 = PolynomialMod[f3 - f4 * (z + 350163), n]
```

```
355162 + 4681 z + 203714 z^2
```

```
InverseLeadCoeff = PowerMod[203714, -1, n]
f5 = PolynomialMod[InverseLeadCoeff * f5, n]
```

```
349909
```

```
397084 + 98465 z + z^2
```

```
f6 = PolynomialMod[f4 - f5 * (z + 18258), n]
```

```
451016 + 87731 z
```

```
InverseLeadCoeff = PowerMod[87731, -1, n]
f6 = PolynomialMod[InverseLeadCoeff * f6, n]
```

```
132235
```

```
466340 + z
```

```
f7 = PolynomialMod[f5 - f6 * (z + 99008), n]
```

```
0
```

We conclude that k = 7 and that

$$gcd(z^5 - 66575, (2z + 3)^5 - 387933) \equiv z + 466340 \equiv z - 543 \pmod{466883}.$$

Therefore, the secret message m is 543. One can check this with the Mathematica function
PowerMod.

```
m = 543;
PowerMod[m, 5, n] == c1
PowerMod[2 m + 3, 5, n] == c2
```

```
True
```

```
True
```

The above approach of finding m by computing a gcd is still practical for e up to 32 bits long ([CoppFPR96]).

9.3.2   A Small Secret Exponent; Wiener's Attack

Wiener [Wien90] shows that it is unsafe to use the RSA system with a small secret exponent d, where "small" means something like $\sqrt{n}$. This observation is of importance, because often one is inclined to reduce the work load of the exponentiation, by choosing a small exponent. For instance, if a smart card is used to sign messages (see Subsection 9.1.3), it will have to compute exponentiations $c^d \pmod n$. If the card has limited computing power, a relatively small value of d (of course not so small that d can be found by exhaustive search) would be handy.

We first show that we can replace (9.4) by the slightly stronger relation

$$e.d \equiv 1 \ (\text{mod} \ \text{lcm}(p-1, q-1)),$$

where lcm denote the least common multiple. We remark that $p-1$ and $q-1$ both divide $\phi(n)$ and so does $\text{lcm}(p-1, q-1)$. Now note that for a correct functioning of the RSA system, one only needs that $e.d \equiv 1 \pmod{p-1}$ and $e.d \equiv 1 \pmod{q-1}$. The reason is that these two congruences are sufficient to prove that (9.5) and (9.6) hold modulo p resp. modulo q. From the Chinese Remainder Theorem it then follows that (9.5) and (9.6) also hold modulo n. We conclude that it is sufficient that $e.d \equiv 1 \ (\text{mod} \ \text{lcm}(p-1, q-1))$.

The subsequent cryptanalysis will deal with this most general case. It is the cryptanalist's aim to find d satisfying this relation (and also p and q). The above congruence can be rewritten as

$$e.d = 1 + K.\text{lcm}(p-1, q-1) = 1 + \frac{K}{G}(p-1)(q-1),$$

where $G = \gcd(p-1, q-1)$. If K and G have a factor in common, the above relation may be further simplified to

$$e.d = 1 + \frac{k}{g}(p-1)(q-1), \qquad \text{with } \gcd(k, g) = 1. \qquad (9.13)$$

One should realize that often G (and thus also g) will be very small. In a typical RSA system, p and q will be safe primes, meaning that $p - 1 = 2.p'$ and $q - 1 = 2.q'$, with p' and q' prime. So, in this case $G = 2$ and $g = 1$ or 2.

Let us rewrite (9.13) by dividing both hands by $d.n$ ($= d.p.q$) and rearranging the terms:

$$\frac{k}{d.g} = \frac{e}{n} + \frac{k}{d.g}\left(\frac{1}{p} + \frac{1}{q} - \frac{1}{n}\right) - \frac{1}{d.n}. \tag{9.14}$$

What we like to show is that $k/(d.g)$ is a convergent of the continued fraction of the known rational e/n. Since these continued fractions are easy to compute, it is then possible to find the secret exponent d (and k and g).

> **Theorem 9.2**
> Assume that $p \sim q \sim \sqrt{n}$, $e \sim n$, and $2g < d$.
> Then $k \sim (g.d)$ and the numbers d, k, g, p, and q can be found from the continued fraction of e/n for secret exponents d up to $n^{1/4}$.

Remark 1:

We shall be a little sloppy with the use of the $\sim$ symbol. What we mean with $a \sim b$ is something like "a and b have the same order of magnitude".

Remark 2:

We already discussed the likelihood that g is small. If d is selected as a small integer, the value of e will be like that of a random number in the range $\{1, 2, \ldots, \text{lcm}(p - 1, q - 1)\}$, so also the assumption $e \sim n$ is very reasonable. The same holds for $p \sim q \sim \sqrt{n}$ (see the discussion around Example 9.2).

Remark 3:

Relation (9.14) implies that $\frac{k}{d.g} > \frac{e}{n}$, therefore, it suffices to check only the odd convergents of e/n.

Proof of Theorem 9.2:

If $e \sim n$ then $k \sim g.d$ by (9.14), since the other terms there all tend to zero. It further follows from (9.14) that

$$\left|\frac{k}{d.g} - \frac{e}{n}\right| = \left|\frac{k}{d.g}\left(\frac{1}{p} + \frac{1}{q} - \frac{1}{n}\right) - \frac{1}{d.n}\right| \le \frac{k+g}{d.g.n} + \frac{k}{d.g}\left(\frac{1}{p} + \frac{1}{q}\right)$$

$$\sim \frac{d+1}{d.n} + \frac{1}{\sqrt{n}} \sim \frac{1}{\sqrt{n}}.$$

Since $2g.d < d^2 < n^{1/2}$, we conclude that

$$\left|\frac{k}{d.g} - \frac{e}{n}\right| \le \frac{1}{\sqrt{n}} < \frac{1}{2(d.g)^2}.$$

It follows from Theorem A.35 that the rational number $k/(d.g)$ will appear as a convergent in the continued fraction of e/n. Since $\gcd(k, g) = 1$ and since (9.13) also implies that $\gcd(k, d) = 1$, it follows from Corollary A.32 that k and $d.g$ will be obtained from one of the convergents. Because

g is very small, we can find g and d with a small trial and error effort.

From (9.13) one can now compute $(p-1)(q-1)$ and since $p.q$ is known, one can also find the factorization of n into p and q.

□

Example 9.11

Consider $n = 9998000099$ and $e = 6203014673$. Let us compute the successive convergents of e/n. We first load the Mathematica package NumberTheory`ContinuedFractions` *and then we can use the functions* ContinuedFraction *and* FromContinuedFraction *(use Normal in Mathematica 3).*

```
<<NumberTheory`ContinuedFractions`
```

```
n = 9998000099; e = 6203014673;
FromContinuedFraction[ContinuedFraction[e / n, 2]]
FromContinuedFraction[ContinuedFraction[e / n, 4]]
FromContinuedFraction[ContinuedFraction[e / n, 6]]
FromContinuedFraction[ContinuedFraction[e / n, 8]]
```

$$1$$

$$\frac{2}{3}$$

$$\frac{5}{8}$$

$$\frac{18}{29}$$

Let us check why the last one does not lead to d (the other cases are even simpler). Writing $18/29 = k/(d.g)$ leads to $k = 18$, $g = 1$, and $d = 29$. An easy argument to show that this is not the right value of d is an encryption followed by a decryption, not resulting into the original message. We use the function PowerMod.

```
m = 123; d = 29;
c = PowerMod[m, e, n];
PowerMod[c, d, n] == m
```

```
False
```

Let us try the next convergent.

```
FromContinuedFraction[ContinuedFraction[e/n, 10]]
```

$$\frac{85}{137}$$

Writing $85/137 = k/(d.g)$ *leads to* $k = 85$, $g = 1$, *and* $d = 137$. *From* *(9.13)* *we get* $(p-1)(q-1) = 9993745862$.

```
k = 85; g = 1; d = 137;
(e*d - 1) g / k
```

```
9997800120
```

Together with $n = p.q = 9998000099$ *we get* $p + q - 1 = p.q - (p-1)(q-1) =$

```
n - (e*d - 1) g / k
```

```
199979
```

So, p *and* q *are the roots of* $(x-p)(x-q) = x^2 - 199980x + 9998000099$. *They can be found with the function* $\underline{Solve}$

```
Clear[x];
Solve[x^2 - 199980 x + 9998000099 == 0, {x}]
```

```
{{x -> 99989}, {x -> 99991}}
```

Indeed, $99989 \times 99991 = n$.

```
99989*99991 == n
```

```
True
```

9.3.3 Some Physical Attacks

Clearly physical attacks on cryptographic implementations are beyond the scope of this introduction. Nevertheless, two such attacks will be mentioned briefly, because of their relation to theory that we have explained here.

□ **Timing Attack**

Suppose that RSA is implemented on a hardware device (like a smartcard), and that the secret exponentiation ($m \rightarrow (m^d \bmod n)$ or $c \rightarrow (c^d \bmod n)$) in the RSA process follows a computational scheme of the type explained in Subsection 8.1.1, i.e. any method that consists of repeated squarings and/or multiplications. See for instance Example 8.1.3.

It is further assumed in this attack (see [Koch96]) that an observer can measure the electromagnetic radiation or power consumption of the device and can clock the length of the various calculations. Typically, a multiplication takes longer than a simple squaring operation.

In this way, the attacker can determine the particular sequence of squarings and multiplications that the program went through. Based on the outcome, he can simply compute the secret exponent d stored on the card.

For instance, if the measurements give Sq.Sq.M.Sq.Sq.M.Sq.Sq.M.Sq.M, where Sq stands for Squaring and M for Multiplying, we get the exponent from

```
Clear[a];
```

$$\left(\left(\left(\left(\left(\left((a)^2\right)^2 a\right)^2\right)^2 a\right)^2\right)^2 a\right)^2 a$$

$$a^{171}$$

□ **The "Microwave" Attack**

Suppose again that RSA is implemented on a hardware device (say a smartcard), but now assume that the secret exponentiation ($m \rightarrow (m^d \bmod n)$ or $c \rightarrow (c^d \bmod n)$) in the RSA process makes use of the Chinese Remainder Theorem (Thm. A.19). See for instance Example 9.1, Part 4. So, we assume that two independent exponentiations take place on this device: one modulo p and one modulo q, where $n = p.q$.

Now suppose that this RSA implementation is used to sign data (this is the simplest version of the attack, cfr. [LensA96] and [BoDML97]). So, typically, the attacker presents a message m to the smart card and would normally expect $c = (m^d \bmod n)$ back. However, the attacker submits the smart card, when it is making its calculations, to the right kind of radiation ("just put it in a microwave" is an oversimplification of this attack) and hopes that in one of the two exponentiations an incorrect calculation will be made.

For instance, the smart card calculates $c_1 = (m^d \bmod p)$ correctly, but gets a wrong value for c_2, i.e. $c_2' \neq (m^d \bmod q)$. The reader should remember that in the smart card values a and b are stored satisfying

$$\begin{cases} a \equiv 1 & (\bmod \ p) \\ a \equiv 0 & (\bmod \ q) \end{cases}$$

$$\begin{cases} b \equiv 0 & (\bmod \ p) \\ b \equiv 1 & (\bmod \ q) \end{cases}$$

So, the card will output $c' = (a.c_1 + b.c_2' \bmod n)$. Now note that since $b \equiv 0 \,(\bmod \ p)$ and $a \equiv 0 \,(\bmod \ q)$

$$c - c' \equiv a.c_1 - a.c_1 \equiv 0 \,(\bmod \ p),$$

$$c - c' \equiv b.c_2 - b.c_2' \equiv b(c_2 - c_2') \not\equiv 0 \ (\bmod \ q).$$

It follows that $\gcd(c - c', n)$ gives a non-trivial factorization of n.

It depends on the application whether the attacker can let the card give the correct value of c too, for instance by having the card sign m again without introducing any radiation). A way around this problem is to let the attacker select a message c, compute $m = (c^e \bmod n)$ with the public exponent e and submit m when attacking the card. In this way, the correct value of c is already known beforehand.

Example 9.1 (Part 6)

We continue with the parameters of Example 9.1, so $p_B = 9733$, $q_B = 10177$, $n_B = 99052741$, $e_B = 81119923$, and $d_B = 17089915$.

Further, $a = 45287650$ and $b = 53765092$ (see Ex. 9.1, Part 4).

When, $m = 12345678$, the correct value of c is given by

```
n = 99052741; e = 81119923;
c = 11111111;
m = PowerMod[c, e, n]
```

```
24307114
```

So, when signing $m = 24307114$ the card should produce $c = 11111111$.

In his calculations the card computes numbers c_1 and c_2 and gets c as follows:

```
p = 9733; q = 10177;
d = 17089915; d1 = Mod[d, p - 1]; d2 = Mod[d, q - 1];
m1 = Mod[m, p]; m2 = Mod[m, q];
a = 45287650; b = 53765092;
c1 = PowerMod[m1, d1, p];
c2 = PowerMod[m2, d2, q];
c = Mod[a * c1 + b * c2, n]
```

```
11111111
```

However, when c_1 is calculated incorrectly due to radiation, say $c_1' = 8765$, the card will produce an incorrect value c' for $c = 11111111$ and the gcd of the difference of these two numbers with n will yield a factor of n.

```
c1Prime = 8765;
cPr = Mod[a * c1Prime + b * c2, n]
GCD[c - cPr, n]
```

```
92608527
```

```
10177
```

The number 10177 is indeed one of the two factors of n.

9.4   How to Generate Large Prime Numbers; Some Primality Tests

9.4.1   Trying Random Numbers

To make the RSA system practical, one needs an efficient way to generate very long prime numbers. The following pseudo-algorithm describes a probabilistic way of how this can be done.

> **Algorithm 9.3   Method to generate an *l*-digits long prime number**
>
> Step 1: Write down a random, odd integer u of l digits long.
>
> Step 2: Test the candidate u for primality.
>   If u is not prime, go back to Step 1, otherwise STOP.

In the next two paragraphs we shall discuss several ways to test an integer u for primality. The first two tests do not give an absolute guarantee that u is prime, but the probability that a composite number u meets the test can be made arbitrary small. The second test (of which only an outline will be given in Section 9.3.3) can guarantee the primality, but it is much slower. For other tests we refer the reader to [Knut81], Section 4.5.4.

Example 9.12

In Mathematica one can use the functions `Random`, `PrimeQ`, *and* `While` *to simulate the above algorithm. Note that the parity of u is not tested below (this is not an essential part of the above algorithm anyway).*

```
u = 1; l = 3;
att = 0;
While[PrimeQ[u] == False, att = att + 1;
                          u = Random[Integer, {10^{l-1}, 10^l}]];
Print["prime number is ", u]
Print[att, " attempt(s)"]
```

```
prime number is 907

7 attempt(s)
```

How often does one expect to have to go through Steps 1 and 2 in the above "algorithm" before obtaining a prime? To answer this question we have to know the fraction of the prime numbers in the set of odd, l-digit numbers. To this end we quote the Prime Number Theorem (Th. A.2).

Theorem 9.4
Let $\pi(x)$ count the number of primes less than or equal to x (see Definition A.1). Then
$$\lim_{x \to \infty} \frac{\pi(x)}{x/\ln x} = 1$$

With the Prime Number Theorem one can quite easily obtain an approximation of the fraction of odd, l-digit numbers that are prime. One gets

$$\frac{\pi(10^l) - \pi(10^{l-1})}{(10^l - 10^{l-1})/2} \overset{P.N.T.}{\approx} \frac{\frac{10^l}{\ln 10^l} - \frac{10^{l-1}}{\ln 10^{l-1}}}{(10^l - 10^{l-1})/2} = \frac{2(9\,l-10)}{9.l.(l-1).\ln 10} \approx \frac{2}{l.\ln 10}$$

For instance, with $l = 100$, one gets

```
l = 100;
EstimateProb[l_] =
               2 (9 * l - 10) / (9 * l * (l - 1) * Log[10]);
N[EstimateProb[100], 3]
```

```
0.00868
```

Since the reciprocal of this number is about 115, we estimate that the expected number runs in the prime generation algorithm above will be 115.

9.4.2   Probabilistic Primality Tests

□ **The Solovay and Strassen Primality Test**

Let p be a prime number. We recall from Definition A.9 that an integer u with $p \nmid u$ (read: p does not divide u), is called a *quadratic residue* (*QR*) modulo p, if the equation

$$x^2 \equiv u \,(\mathrm{mod}\, p),$$

has an integer solution. If $p \nmid u$ and this congruence relation does not have an integer solution, u will be called a *quadratic non-residue* modulo p (*NQR*). The well known Legendre symbol (u / p) (see Definition A.10) is defined by

$$\left(\tfrac{u}{p}\right) = \begin{cases} +1 & \text{if } u \text{ is a quadratic residue mod } p, \\ -1 & \text{if } u \text{ is a quadratic nonresidue mod } p, \\ \ \ 0 & \text{if } p \text{ divides } u. \end{cases}$$

The Jacobi symbol $(\tfrac{u}{m})$ (see Definition A.11) generalizes the Legendre symbol to all odd integers m. Let $m = \Pi_i \, (p_i)^{e_i}$ where the p_i's are (not necessarily distinct) odd primes. Then, $(\tfrac{u}{m})$ is defined by

$$\left(\tfrac{u}{m}\right) = \prod_i \left(\tfrac{u}{p_i}\right)^{e_i}$$

In Section A.4, the reader can find all kinds of properties of the Legendre symbol and the Jacobi symbol. These properties culminate in an extremely efficient algorithm to compute the values of these symbols. An example can be found there. In *Mathematica*, both symbols can be computed with the `JacobiSymbol` function:

```
u = 12703; m = 16361; JacobiSymbol[u, m]
```

```
1
```

As a matter of fact, since m in the example above, is a prime number, it is quite easy to compute a "square-root" of u. For a discussion of how this can be done, we refer the reader to Section 9.5. In *Mathematica* one can simply use the `Solve` function.

```
Clear[x];
Solve[{x^2 == 12703, Modulus == 16361}, x]
```

```
{{Modulus -> 16361, x -> 7008}, {Modulus -> 16361, x -> 9353}}
```

Indeed, $(\pm 7008)^2 \equiv 12703 \pmod{16361}$, as can be checked with the `PowerMod` function.

```
PowerMod[7008, 2, 16361]
```

```
12703
```

To find a solution of the equation $x^2 \equiv u \pmod{m}$ for composite integers m is, in general, a very difficult problem and intractable for large values of m (see [Pera86] for a discussion of this problem).

If m is the product of different primes and this factorization is known (!), one can find the square root of u by finding the square root of u modulo all the prime factors of m and then combine the result by means of the Chinese Remainder Theorem. In Section 9.5, this method will be demonstrated. When m has higher prime powers in its factorization, matters get much more complicated.

Let p be a prime number, $p > 2$. We recall from Theorem A.23 that for all integers u:

$$\left(\tfrac{u}{p}\right) \equiv u^{(p-1)/2} \pmod{p}. \tag{9.15}$$

The Solovay and Strassen Algorithm [SolS77] relies on the following theorem.

Theorem 9.5
Let m be an odd integer and let G be defined by

$$G = \left\{ 0 \le u < m \mid \gcd(u, m) = 1 \text{ and } \left(\tfrac{u}{m}\right) \equiv u^{(m-1)/2} \pmod{m} \right\}$$

Then

$$|G| = m - 1 \qquad \text{if } m \text{ is prime.} \tag{9.16}$$

$$|G| \le (m-1)/2 \qquad \text{if } m \text{ is not a prime,} \tag{9.17}$$

Proof: If m is prime, every integer $0 < u < m$ satisfies (9.15), and has gcd 1 with m, so $|G| = m - 1$ in this case.

So, we now consider the case that m is not a prime number. Clearly, G is a subgroup of the multiplicative group

$$\mathbf{Z}_m^* = \{ 0 \le u < m \mid \gcd(u, m) = 1 \}.$$

It follows (from Theorem B.5) that the cardinality of G divides that of $\mathbf{Z}_m^*$. So, if $G \ne \mathbf{Z}_m^*$ we can conclude that $|G| \le |\mathbf{Z}_m^*|/2 = \varphi(m)/2 \le (m-1)/2$. This would prove the theorem. We conclude, that it suffices to prove the existence of an element u in $\mathbf{Z}_m^*$ with $\left(\frac{u}{m}\right) \not\equiv u^{(m-1)/2} \pmod{m}$.

We distinguish two cases. In [SolS77], the authors omit to consider the case that m is a square. In the proof below, which is due to J.W. Nienhuys (private communication), Case 1 will cover this possibility.

Case 1: The number m is divisible by at least the square of some prime number. We write $m = p^r.s$ with p an odd prime, $r \ge 2$, and $\gcd(p, s) = 1$.

Let u be a solution of the system simultaneous congruence relations:

$$u \equiv 1 + p \pmod{p^r}, \tag{9.18}$$

$$u \equiv 1 \pmod{s}. \tag{9.19}$$

By the Chinese Remainder Theorem (Thm. A.19) such a solution u exists and is unique modulo m. Clearly, $\gcd(u, p^r) = \gcd(u, s) = 1$, so $\gcd(u, m) = 1$, i.e. $u \in \mathbf{Z}_m^*$.

It follows from (9.18), the binomial theorem, and an argument similar to the proof of Theorem B.26 that $u^m \equiv (1 + p)^m \equiv 1 \pmod{p^r}$. By (9.19) we also have that $u^m \equiv 1 \pmod{s}$. By the Chinese Remainder Theorem we now have that $u^m \equiv 1 \pmod{m}$.

Since $u \not\equiv 1 \pmod{m}$ by (9.18), it also follows that $u^{m-1} \not\equiv 1 \pmod{m}$. This in turn implies that $u^{(m-1)/2} \not\equiv \pm 1 \pmod{m}$, which implies that u can not satisfy (9.15). We conclude that this element u is a member of $\mathbf{Z}_m^*$, but not of G.

Case 2: m is the product of s distinct prime numbers, say $m = p_1 p_2 \ldots p_s$, with $s \ge 2$.

Let a be a quadratic non-residue modulo p_1. By the Chinese Remainder Theorem there is a unique integer u modulo m satisfying the system simultaneous congruence relations

$$u \equiv a \pmod{p_1}, \tag{9.20}$$

$$u \equiv 1 \pmod{p_i}, \ 2 \le i \le s. \tag{9.21}$$

Clearly, $\gcd(u, p_i) = 1$ for $1 \le i \le s$, so $u \in \mathbf{Z}_m^*$. To show that $u \notin G$, we need to show that (9.15) does not hold.

Since $u \equiv 1 \pmod{p_i}$, $2 \le i \le s$, it follows that $\left(\frac{u}{p_i}\right) = 1$ for these indices. But $\left(\frac{u}{p_1}\right) = \left(\frac{a}{p_1}\right) = -1$, because a is NQR. From the definition of the Jacobi symbol (Def. A.11) it follows that $\left(\frac{u}{m}\right) = -1$. In particular this implies that $\left(\frac{u}{m}\right) \equiv -1 \pmod{p_i}$ for any $2 \le i \le s$.

On the other hand, (9.21) implies that $u^{(m-1)/2} \equiv 1 \pmod{p_i}$ for any $2 \le i \le s$. Hence

$$(u/m) \not\equiv u^{(m-1)/2} \pmod{p_i}$$

for any i, $2 \le i \le s$, and a fortiori (9.15) does not hold.

$\square$

We can now describe the Solovay and Strassen Algorithm.

```
Algorithm 9.6              Solovay and Strassen primality test
    input    odd integer m (candidate)
             security parameter k
    initialize prime=True; i=1;
    while prime and i ≤ k do
        begin
            select a random integer u, 1 < u < m;
            if gcd(u, m) ≠ 1 or (u/m) ≢ u^((m-1)/2) (mod m) then prime=False;
            i=i+1;
        end
    output prime
```

In the algorithm above, k can be any positive integer. The probability that k independently and randomly selected elements u will pass the two tests, given in Algorithm 9.6, while m is not prime, is less than or equal to 2^{-k} by Theorem 9.5. By taking k sufficiently large, the probability that a non-prime number survives the above algorithm can be made arbitrary small.

See however the Miller-Rabin test in the next subsubsection, where we have 4^{-k} as probability that a composite number is not detected after k tests.

Example 9.13

To test if the odd number m = 1234563 is prime we use the Mathematica functions GCD, JacobiSymbol, PowerMod, *and* Mod:

```
m = 1234563;
u = 1212121;
GCD[u, m] == 1
Mod[JacobiSymbol[u, m] - PowerMod[u, (m - 1) / 2, m], m] == 0
```

```
True
```

```
False
```

The reader is invited to test $m = 104729$ for primality.

□ **Miller-Rabin Test**

The Miller-Rabin test [Mill76], [Rabi80a] is based on the fact (see Theorem B.14) that the equation $x^2 \equiv 1 \pmod{p}$ has only two solutions: $x \equiv \pm 1 \pmod{p}$.

So, let m be an odd integer that we want to test for primality. Assuming for a moment that m is in fact prime, we have by Fermat's Theorem (Thm. A.15) that any integer a with $\gcd(a, m) = 1$

satisfies $a^{m-1} \equiv 1 \pmod{m}$.

Since $m - 1$ is even, it follows that $a^{(m-1)/2} \equiv \pm 1 \pmod{m}$. If $a^{(m-1)/2}$ happens to be $+1$ and $(m-1)/2$ is even, we can repeat the argument, so in this case we conclude that $a^{(m-1)/4} \equiv \pm 1 \pmod{m}$, etc. In this way, one can prove the following lemma.

> **Lemma 9.7**
> Let p be a prime and write $p - 1 = a.2^f$, with a odd. Let u be an integer in between 1 and $p - 1$. Then
> either $\quad u^a \equiv 1 \pmod{p}$
> or $\qquad u^{a.2^i} \equiv -1 \pmod{p}$ for some $0 \le i < f$.

To test an odd integer m for primality we proceed as follows. First we write $m - 1 = a.2^f$, with a odd. Next we pick a random integer u, $2 \le u < m$, and compute from left to right u^a, $u^{a.2}$, ..., $u^{a.2^f}$. As soon as one of these numbers is not in $\{-1, 1\}$, while the next one is $+1$, or if $u^{a.2^f} \not\equiv 1 \pmod{m}$ we may conclude that m is composite and we can stop.

We repeat the test k times, where k is a security parameter, that will be discussed in a moment.

Let m be an integer and let u be such that $u^{a.2^j} \equiv 1 \pmod{m}$, $j \ge 1$, while $u^{a.2^{j-1}} \not\equiv \pm 1 \pmod{m}$. Then u is called a *strong witness* to the compositeness of m. It gives a proof that m is composite.

On the other hand, let m be composite and let u be an integer that satisfies $u^a \equiv 1 \pmod{m}$ or $u^{a.2^j} \equiv -1 \pmod{m}$ for some $0 \le j \le f - 1$, then this u is called a *strong liar* (to the primality) of m.

For an efficient primality test we want composite numbers to have as few strong liars as possible.

> **Algorithm 9.8**   *Miller-Rabin* primality test
> **input**   odd integer m (candidate)
>   security parameter k
> **initialize** prime=True; i=1;
> **write** $m - 1 = a.2^f$, a odd.
> **while** prime and $i \le k$ **do**
>   **begin**
>   select a random integer u, $1 < u < m - 1$;
>   **compute** $x \equiv (u^a \bmod m)$
>   **if** $x \not\equiv \pm 1 \pmod{m}$ **then**
>   **begin**   **put** $j = 1$
>   **while** $x \not\equiv \pm 1 \pmod{m}$ **and** $j \le f - 1$
>   **do**   **begin** $x \leftarrow (x^2 \bmod m)$
>   **if** $x \equiv 1 \pmod{m}$ **then** prime=False
> $j \leftarrow j + 1$
>   **end**
>   **if** $x \not\equiv -1 \pmod{m}$ **then** prime=False
>   **end**
> $i = i + 1$;
>   **end**
> **output** prime

Example 9.14

Let $m = 7933$. Then $m - 1 = 1983.2^2$. Let us pick a random u and compute $u^{1983.2^i}$ for $i = 0, 1, 2$. We use the Mathematica functions <u>While</u> and <u>EvenQ</u> to write $m - 1$ as $a.2^f$ and use <u>Random</u>, <u>PowerMod</u>, <u>Print</u>, and <u>Do</u> for the actual test.

```
m = 7933;
f = 0; a = m - 1; While[EvenQ[a], f = f + 1; a = a / 2];
{a, f}
u = Random[Integer, {1, m - 2}]
x = PowerMod[u, a, m];
Do[{Print[x], x = Mod[x^2, m]}, {i, 0, f}]
```

```
{1983, 2}
```

```
4225
```

```
7932
```

```
1
```

```
1
```

We see that no matter how often we run this, we shall always get $(+1, +1, +1)$ or $(-1, +1, +1)$, or $(*, -1, +1)$.

Example 9.15

Let $m = 429$. A strong witness of the compositeness of m is given by the choice $u = 34$, as we can see below.

```
m = 429;
f = 0; a = m - 1; While[EvenQ[a], f = f + 1; a = a / 2];
{a, f}
u = 34
x = PowerMod[u, a, m];
Do[{Print[x], x = Mod[x^2, m]}, {i, 0, f}]
```

```
{107, 2}
```

```
34
```

```
265
```

```
298
```

```
1
```

What remains to be done is to give an estimate of the fraction of strong liars modulo a composite number. The next theorem says that this fraction is at most 1/4. This means that the probability that a composite number will not be detected after k runs of the Miller-Rabin test is at most $(1/4)^k$. This compares very favorably with the Solovay and Strassen primality test where this probability can only be upperbounded by $(1/2)^k$.

> **Theorem 9.9**
> Let m be a composite number, $m \neq 9$. Then the number of strong liars in between 1 and $m - 1$ is at most $\varphi(m)/4$, where φ denotes Euler's totient function.
> In other words: the probability that after k runs Algorithm 9.8 has not established the compositeness of a non-prime m is at most 4^{-k}.

The proof of Theorem 9.8 (see [Moni80] or [Rabi80a]) is very technical and does not give further insight to the reader of this introduction.

If $m = 9$, $\varphi(m)/4$ will be $6/4$, which is less than the two "strong liars" -1 and $+1$.

9.4.3   A Deterministic Primality Test

Primality tests that prove in a deterministic way that a certain is prime or not are of course much slower than probabilistic algorithms of the type discussed in the previous subsection.

We shall now explain the idea behind the deterministic primality test of H. Cohen and H.W. Lenstra jr. [CohL82]. This test is an improvement of [AdPR83]. We shall not give a complete description of this test. That would involve too much advanced and deep number theory. We closely follow the excellent introductory article by Lenstra [LensH83].

We start by quoting Fermat's Theorem (Thm. A.15).

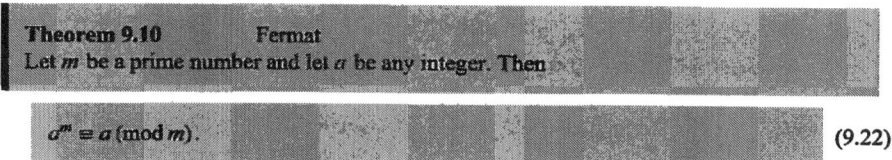

Theorem 9.10   Fermat
Let m be a prime number and let a be any integer. Then

$$a^m \equiv a \,(\mathrm{mod}\, m). \tag{9.22}$$

Let m be an integer that we want to test for primality. A single integer a that does not satisfy (9.22), proves that m is not a prime number.

Unfortunately, the opposite is not true. For instance, $m = 561$ satisfies (9.22), while $m = 3 \times 11 \times 17$. To see this we first compute $\mathrm{lcm}(\varphi(3),\, \varphi(11),\, \varphi(17))$ $\overset{\text{Thm.}A.17}{=} \mathrm{lcm}(2,\, 10,\, 16) = 80$. Let a be coprime with 561. It follows from Euler's Theorem (Thm. A.14) that a^{80} is congruent to 1 modulo each of the three prime divisors of 561. The Chinese Remainder Theorem (Thm. A.19) now implies that $a^{80} \equiv 1 \,(\mathrm{mod}\, 561)$. Hence, $a^{561} \equiv a.(a^{80})^7 \equiv a \,(\mathrm{mod}\, 561)$.

For the values of a that have a factor in common with 561, (9.22) can be proved in a similar way.

The reader may want to verify the above with the *Mathematica* functions `FactorInteger` and `PowerMod`:

```
m = 561; FactorInteger[m]
a = 543;
PowerMod[a, m, m] == a
```

```
{{3, 1}, {11, 1}, {17, 1}}
```

```
True
```

Composite integers m with the property that $a^{m-1} \equiv 1 \,(\mathrm{mod}\, m)$, for all a with $\gcd(a, m) = 1$, are commonly called *Carmichael numbers*.

The converse of a slightly stronger statement than Theorem 9.10 does hold however. In the sequel, (a/m) denotes, as usual, the Jacobi symbol.

Theorem 9.11
An odd integer m is prime if and only if for all integers a

$$\gcd(a, m) = 1 \implies a^{(m-1)/2} \equiv (a/m) \,(\mathrm{mod}\, m).$$

Proof: That the relation above holds for prime numbers was already remarked on in (9.15). The converse was first proved by Lehmer [Lehm76], but it also follows directly Theorem 9.5.

□

The above theorem is of course not a very efficient primality test for numbers that are more than 100 digits long. Lenstra offers the following "attractive" alternative.

Theorem 9.12
An odd integer m is prime if and only if every divisor d of m is a power of m.

Proof: This statement is completely trivial, since $d = 1 = m^0$ and $d = m = m^1$ are the only divisors of a prime number m. All other numbers in between 1 and m can not be written as power of m.

□

Clearly it is not this theorem that we want to use as a primality test, but a variation of it does turn out to be very powerful. We shall show that under certain conditions every divisor of m looks a little bit like a power of m.

Theorem 9.13
Let m be an integer m that is coprime with 6. Assume further that

$$(u/m) \equiv u^{(m-1)/2} \pmod{m} \qquad \text{for } u = -1, 2, \text{ and } 3, \tag{9.23}$$

$$a^{(m-1)/2} \equiv -1 \pmod{m} \text{ for some integer } a. \tag{9.24}$$

Then, for each d dividing m

$$d \equiv m^j \pmod{24} \qquad \text{for some non-negative integer.} \tag{9.25}$$

In fact, (9.19) can be strengthened to

$$d \equiv m^j \pmod{24} \qquad \text{for } j = 0 \text{ or } 1. \tag{9.26}$$

Condition (9.24) can not be omitted in the theorem above. Indeed, $m = 1729 = 7 \times 13 \times 19$ does satisfy (9.23), but does not satisfy (9.25). Note that $m \equiv 1 \pmod{24}$, therefore, no power of m will ever be equal to one of the prime divisors of m.

All these statements can be checked with the *Mathematica* functions `FactorInteger`, `JacobSymbol`, `PowerMod`, and `Mod`:

```
m = 1729; FactorInteger[m]
Mod[m, 24]
Mod[JacobiSymbol[-1, m] - PowerMod[-1, (m - 1) / 2, m], m] == 0
Mod[JacobiSymbol[2, m] - PowerMod[2, (m - 1) / 2, m], m] == 0
Mod[JacobiSymbol[3, m] - PowerMod[3, (m - 1) / 2, m], m] == 0
```

```
{{7, 1}, {13, 1}, {19, 1}}
```

```
1
```

```
True
```

```
True
```

```
True
```

Before we prove Theorem 9.13, we shall illustrate how it can be used to test the primality of integers m, $24 < m < 24^2$. After the proof we shall discuss generalizations of Theorem 9.13, that yield efficient primality tests for larger values of m.

Algorithm 9.14	(Cohen and Lenstra limited primality test)
input	m, $24 < m < 24^2$.
initialize	prime=True,
test 1:	if $\gcd(m, 6) \neq 1$ then prime=False
test 2:	if $(u / m) \not\equiv u^{(m-1)/2} \pmod{m}$ for $u = -1, 2,$ or 3
	then prime=False
test 3:	find an integer a with $a^{(m-1)/2} \equiv -1 \pmod{m}$;
	if no such integer a exists then prime=False
test 4:	compute $d = (m \bmod 24)$.
	if $d > 1$ and $d \mid m$ then prime=False
output prime	

Proof: The first matter to be addressed is Test 3. If m is prime, the probability that a random $1 < a < m$ satisfies (9.24) is 1/2 by Theorem A.23 and Theorem A.20. So, in two tries one can expect to find an integer a satisfying (9.24). If no such integer a exists, m is not prime.

More can be said about this step. Assuming the Extended Riemann Hypothesis one can even prove that (9.24) has a solution a, $1 < a < 2 (\log m)^2$, if m is prime. (See also [Pera86].)

If m meets the first three tests, we know from Theorem 9.13 that each divisor d of m must be congruent to 1 or m modulo 24. Since $m < 24^2$, we may assume that $d < 24$ (otherwise consider n/d instead of d). It follows that d is in fact *equal* to 1 or to $(m \bmod 24)$.

The possibility that $d = (m \bmod 24)$, $d > 1$, is ruled out by Test 4. It follows that this divisor d must be equal to 1. We conclude that m is prime.

□

To be able to prove Theorem 9.13 we need the following lemmas. The first gives a necessary and sufficient condition for two integers m_1 and m_2, both having gcd 1 with 6, to be congruent to each other modulo 24.

Lemma 9.15
Let m_1 and m_2 be two integers, both coprime with 6. Then

$$m_1 \equiv m_2 \ (\bmod\, 24) \quad \Longleftrightarrow \quad (u/m_1) = (u/m_2) \text{ for } u = -1, 2, \text{ and } 3.$$

Proof: There are eight integers m, $1 \le m \le 24$, that are coprime with 6, namely 1, 5, 7, 11, 13, 17, 19 and 23. For each of these values m we calculate the values (u/m) for $u = -1, 2$, and 3 by means of Corollary A.24, Theorem A.25, resp. Theorem A.27 or with the *Mathematica* functions `JacobSymbol`, which can be applied at once to a whole list of numbers.

```
m = {1, 5, 7, 11, 13, 17, 19, 23};
JacobiSymbol[-1, m]
JacobiSymbol[2, m]
JacobiSymbol[3, m]
```

```
{1, 1, -1, -1, 1, 1, -1, -1}
```

```
{1, -1, 1, -1, -1, 1, -1, 1}
```

```
{1, -1, -1, 1, 1, -1, -1, 1}
```

It is easy to verify that the matrix with these three vectors as rows has the property that all columns are different. This shows that the three values (u/m), $u = -1, 2, 3$, uniquely define m from $\{1, 5, 7, 11, 13, 17, 19, 23\}$.

□

For example, by looking at the second column, we see that $m = 5$ is uniquely defined in $\{1, 5, 7, 11, 13, 17, 19, 23\}$ by the three values $(-1, m) = 1$, $(2/m) = -1$, and $(3/m) = -1$.

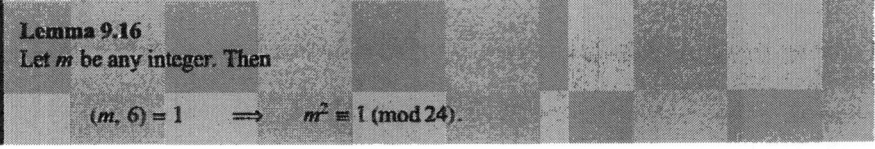

Lemma 9.16
Let m be any integer. Then

$$(m, 6) = 1 \quad \Longrightarrow \quad m^2 \equiv 1 \ (\bmod\, 24).$$

Proof: Since m is not divisible by 3, it follows that $m^2 \equiv 1 \pmod{3}$. Similarly, since m is odd, it follows that $m^2 \equiv 1 \pmod{8}$. To see this, write $m = 2.n + 1$. Then $m^2 = (2.n + 1)^2 = 4n(n + 1) + 1$.

Since, 3 and 8 are coprime, the statement follows from the Chinese Remainder Theorem.

$\square$

Of course, we could have checked the above lemma with the *Mathematica* function <u>Mod</u> as follows

```
m = {1, 5, 7, 11, 13, 17, 19, 23}
Mod[m^2, 24]
```

```
{1, 5, 7, 11, 13, 17, 19, 23}
```

```
{1, 1, 1, 1, 1, 1, 1, 1}
```

We are now ready to prove Theorem 9.13.

Proof of Theorem 9.13:

It is a direct consequence of condition $\gcd(m, 6) = 1$ and Lemma 9.16 that each exponent j in (9.25) can be reduced modulo 2. This shows that (9.25) can be replaced by (9.26)

Next, note that it suffices to prove (9.25) for prime divisors d of m only. Write $m - 1 = f.2^k$ and $d - 1 = g.2^l$, where f and g are odd and where $k > 0, l > 0$.

We shall first prove that $l \geq k$ and then use Lemma 9.15 to show that either $d \equiv n^0 \pmod{24}$ or $d \equiv n^1 \pmod{24}$.

Raise both sides in condition (9.24) to the power g and reduce the result modulo d. Since $d \mid m$ and g is odd, one obtains

$$a^{f.g.2^{k-1}} \equiv (-1)^g \equiv -1 \pmod{d}.$$

Since we assume that d is prime and since a can not have a factor in common with d or m, it follows from Fermat's Theorem (Thm. A.15) that

$$a^{f.g.2^l} \equiv a^{f(d-1)} \equiv 1^f \equiv 1 \pmod{d}.$$

We conclude from these two congruence relations that

$$k - 1 < l.$$

Now consider $u \in \{-1, 2, 3\}$. Since g is odd and $d \mid m$, we have

$$u^{f.g.2^{k-1}} \equiv u^{g(m-1)/2} \overset{(9.23)}{\equiv} (u/m)^g \equiv (u/m) \pmod{d}.$$

On the other hand (again because d is prime), we have

$$u^{f.g.2^{l-1}} \equiv u^{f(d-1)/2} \overset{(9.15)}{\equiv} (u/d)^f \equiv (u/d) \pmod{d}.$$

It follows from the two last congruence relations that for $i = -1, 2, 3$

$$(u/d) = (u/m)^{2^{l-k}}.$$ (9.27)

Note that we have replaced the congruence relation above by an equality sign. We can do this, because both hands have value -1 or 1.

If $l = k$, relation (9.27) and Lemma 9.15 together imply that $d \equiv m \equiv m^1 \pmod{24}$.

On the other hand, if $l > k$, the right hand side of (9.27) is equal to 1, which is also $(u/1)$. So, Lemma 9.15 yields that $d \equiv 1 \equiv n^0 \pmod{24}$.

$\square$

Crucial in the application of Theorem 9.13 is the fact that we can replace (9.25) by (9.26). Because of this, only one condition needed to be tested in the fourth step of Algorithm 9.14. The reason that (9.25) could be replaced by (9.26) (see Lemma 9.16) is the fact that

$$\gcd(n, 24) \implies n^2 \equiv 1 \pmod{24}.$$

Theorem 9.13 can only prove the primality of integers m, $24 < m < 24^2$. For larger values of m one needs generalizations of Theorem 9.13. As may be expected, the exponent in Lemma 9.16 will have to be increased in these generalizations. An example of such a generalization would be

$$\gcd(m, 65520) = 1 \implies m^{12} \equiv 1 \pmod{65520}.$$

In order to test 100-digit numbers for primality, one uses

$$\gcd(m, s) = 1 \implies m^{5040} \equiv 1 \pmod{65520}.$$

where s is the 53-digit number

$$2^6 \times 3^3 \times 5^2 \times 7^2 \times 11 \times 13 \times 17 \times 19 \times 31 \times 37 \times 41 \times 43 \times 61 \times 71$$
$$\times 73 \times 113 \times 127 \times 181 \times 211 \times 241 \times 281 \times 337 \times 421 \times 631 \times 1009 \times 2521.$$

Note that $\sqrt{m} < s$, if m has not more than 100 digits. A rough outline of the primality test of a 100-digit number is as follows.

Algorithm 9.17 (Cohen and Lenstra: outline of primality test)
- **input** $m < 10^{100}$
- **initialize**   prime=True.
- **test 1:**   if $\gcd(m, s) \neq 1$ then prime=False
- **test 2:**   if m fails any of 67 congruence relations like (9.23) then prime=False
- **test 3:**   compute $d = (n^i \bmod s)$, for $i = 1, 2, ..., 5039$, if any of these d divide m then prime=False
- **output** prime

If m is composite, the algorithm above will sometimes yield a factor of m. The probability that this will happen however, is very small. In most cases that m is composite, the algorithm will terminate in Step 2 and one does not obtain a factor of m. The algorithm above can be adapted to test larger integers for primality. The expected running time is

$$(\ln n)^{c \ln \ln n}$$

where c is some constant.

9.5 The Rabin Variant

In Subsection 9.2.1, it was mentioned that no other general method of breaking RSA is known than by factoring n. In [Rabi79], *Rabin* proposes a variant of the RSA system, whose cryptanalysis can be proved to be equivalent to the factorization of n.

9.5.1 The Encryption Function

In the RSA system, each user U had to select a public exponent e_u with $\gcd(e_U, n_U) = 1$ (see (9.2)). In Rabin's variant, all users U take the same exponent

$$e_U = 2. \tag{9.28}$$

We remind the reader of the discussion in Subsection 9.3.1.

Since $\gcd(2, \varphi(n_U)) = 2$, because both $p_U - 1$ and $q_U - 1$ are even, encryption is no longer a one-to-one mapping. Indeed, if $c \equiv m^2 \pmod{n_U}$, with $\gcd(c, n_U) = 1$ and $n_U = p_U q_U$, it follows that the congruence relation $x^2 \equiv c \pmod{p_U}$ has two solutions, namely $\pm m \pmod{p_U}$ and, similarly, the congruence relation $x^2 \equiv c \pmod{q_U}$ will have the two solutions $\pm m \pmod{q_U}$. By the Chinese Remainder Theorem (Thm. A.19), the congruence relation

$$x^2 \equiv c \pmod{n_U} \tag{9.29}$$

has four solutions modulo n_U. What happens if $\gcd(c, n_U) \neq 1$ is an easy exercise for the reader (see Problem 9.5).

Example 9.16 (Part 1)

Consider the encryption of the message $m = 12345678$ modulo the modulus $n = 9733 \times 10177 = 99052741$ (we use the Mathematica functions `Prime` *and* `PowerMod`*).*

```
pB = Prime[1200];
qB = Prime[1250];
nB = pB * qB
m = 12345678;
PowerMod[m, 2, nB]
```

```
99052741
```

```
43962531
```

To find the four messages that are mapped to the same ciphertext, we have to combine the four systems of linear congruence relations $x \equiv \pm m \pmod{p}$ and $x \equiv \pm m \pmod{q}$ with the Chinese Remainder Theorem. We have to load the package NumberTheory`NumberTheoryFunctions` to be able to use the function ChineseRemainderTheorem.

```
<<NumberTheory`NumberTheoryFunctions`
```

```
m1 = ChineseRemainderTheorem[
    {12345678, 12345678}, {9733, 10177}]
m2 = ChineseRemainderTheorem[
    {-12345678, 12345678}, {9733, 10177}]
m3 = ChineseRemainderTheorem[
    {12345678, -12345678}, {9733, 10177}]
m4 = ChineseRemainderTheorem[
    {-12345678, -12345678}, {9733, 10177}]
```

```
12345678
```

```
48738630
```

```
50314111
```

```
86707063
```

To check this we calculate

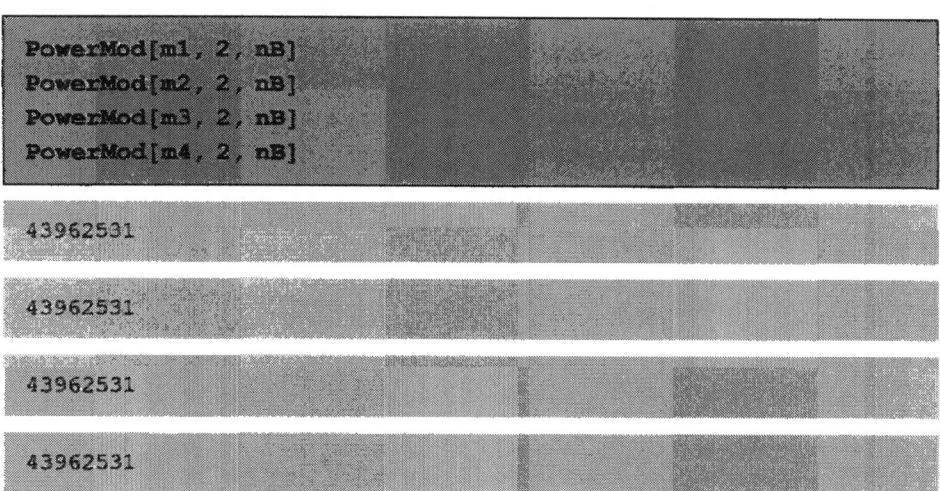

```
PowerMod[m1, 2, nB]
PowerMod[m2, 2, nB]
PowerMod[m3, 2, nB]
PowerMod[m4, 2, nB]
```

43962531

43962531

43962531

43962531

We note that the image space of the encryption function is not the whole set $\{0, 1, \ldots, n_U\}$. As a consequence, this variant by Rabin can not be used in a straightforward way as a signature scheme. (See the related Fiat-Shamir protocol in Chapter 14.)

9.5.2 Decryption

□ **Precomputation**

How does one decrypt a message $c \equiv m^2 \pmod{n}$ in the Rabin variant of the RSA system? As explained earlier in this section, we do this with the Chinese Remainder Theorem. As precalculation, one computes integers a and b satisfying

$$a \equiv 1 \pmod{p_U} \qquad \text{and} \qquad a \equiv 0 \pmod{q_U}, \qquad\qquad (9.30)$$

$$b \equiv 0 \pmod{p_U} \qquad \text{and} \qquad b \equiv 1 \pmod{q_U}. \qquad\qquad (9.31)$$

The solutions a and b can easily be found as follows; for instance, to find a, we obtain $a = l.q_U$ from the second congruence relation and substitute this in the first congruence relation. One gets the congruence relation $l.q_U \equiv 1 \pmod{p_U}$, which can be solved with the extended version of Euclid's Algorithm, (Alg. A.8). See also Example A.3.

These systems of congruence relations can also be solved directly with the *Mathematica* function ChineseRemainderTheorem for which the package NumberTheory`NumberTheoryFunctions` has to be loaded first.

Example 9.16 (Part 2)

Continuing with the parameters of Example 9.16, we need to solve

$$a \equiv 1 \ (mod\ 9733) \quad and \quad a \equiv 0 \ (mod\ 10177),$$

$b \equiv 0 \,(mod\,9733) \quad and \quad b \equiv 1 \,(mod\,10177).$

```
<<NumberTheory`NumberTheoryFunctions
```

```
a = ChineseRemainderTheorem[{1, 0}, {9733, 10177}]
b = ChineseRemainderTheorem[{0, 1}, {9733, 10177}]
```

```
45287650
```

```
53765092
```

So, a = 45287650 and b = 53765092.

□ Finding a Square Root Modulo a Prime Number

Next, one has to solve the congruence relation $x^2 \equiv c \,(mod\, p_U)$ (and, similarly, $x^2 \equiv c \,(mod\, q_U)$). If $c = 0$ the solution is obvious, so, let us assume that $c \not\equiv 0 \,(mod\, p_U)$.

For notational reasons we omit the subscript U from now on. It turns out that an immediate technique to find x is not always possible. We consider three cases.

<u>Case 1</u>: $p \equiv 3 \,(mod\, 4)$

If c is the square of some element m in $\mathbb{Z}_p$ (such a c is called a quadratic residue modulo p; see Section A.4), the two solutions of $x^2 \equiv c \,(mod\, p)$ are given by $\pm c^{(p+1)/4}$. Indeed, if we square this expression we get from Fermat's theorem:

$$(\pm c^{(p+1)/4})^2 \equiv c^{(p+1)/2} \equiv c.c^{(p-1)/2} \equiv c.m^{p-1} \overset{\text{Thm. } A.15}{\equiv} c \,(mod\, p).$$

Example 9.17

Consider the prime $p = 3571$ which is congruent to 3 modulo 4. The number $c = 2868$ is a quadratic residue modulo p as can be checked with the Legendre symbol. To verify all these assertions we use the Mathematica functions Prime, Mod, *and* JacobiSymbol.

```
p = Prime[500]
Mod[p, 4] == 3
c = 2868;
JacobiSymbol[c, p] == 1
```

```
3571
```

```
True
```

```
True
```

The solution of $x^2 \equiv 2868 \pmod{p}$ is given by $m \equiv \pm 2868^{(p+1)/4} \equiv \pm 3234 \pmod{3571}$.

To verify this we use the Mathematica function <u>PowerMod</u>.

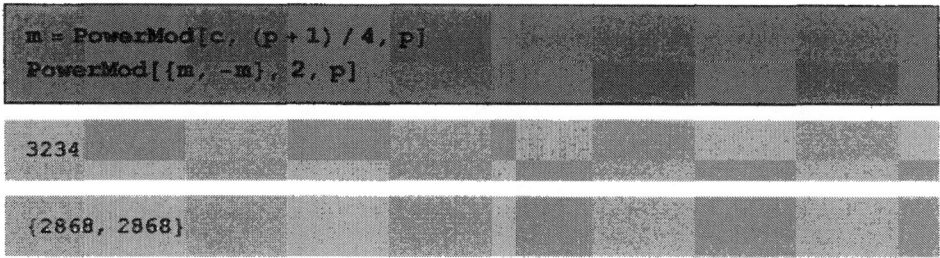

```
m = PowerMod[c, (p + 1) / 4, p]
PowerMod[{m, -m}, 2, p]
```

```
3234
```

```
{2868, 2868}
```

<u>Case 2:</u> $p \equiv 5 \pmod{8}$

With a slight refinement of the method used above it can be shown that the solution of $x^2 \equiv c \pmod{p}$ in this case is given by $\pm c^{(p+3)/8}$ if $c^{(p-1)/4} \equiv 1 \pmod{p}$ and by $\pm 2 . c . (4 . c)^{(p-5)/8}$ if $c^{(p-1)/4} \equiv -1 \pmod{p}$.

See Problem 9.14, which addresses this case.

Example 9.18

Consider the prime $p = 3581$ which is congruent to 5 modulo 8. The number $c = 2177$ is a quadratic residue modulo p as can be checked with the Legendre symbol, which is a special case of the Jacobi symbol.

```
p = Prime[501]
Mod[p, 8] == 5
c = 2177;
JacobiSymbol[c, p] == 1
```

```
3581
```

```
True
```

```
True
```

The solution of $x^2 \equiv 2177\,(mod\ p)$ is given by $m \equiv \pm 2177^{(p+1)/4} \equiv \pm 3100\,(mod\ 3581)$ because $c^{(p-1)/4} \equiv 1\,(mod\ p)$ (otherwise the answer would be $\pm 2.\,c.(4.\,c)^{(p-5)/8}$).

```
If[PowerMod[c, (p - 1) / 4, p] == 1,
            m = PowerMod[c, (p + 3) / 8, p],
            m = Mod[2 c * PowerMod[4 c, (p - 5) / 8, p] p]]
PowerMod[{m, -m}, 2, p]
```

```
3100
```

```
{2177, 2177}
```

<u>Case 3</u>: $p \equiv 1\,(mod\ 8)$

A fast deterministic algorithm to solve this congruence relation does not exist. We follow [Rabi79].

In Section A.4 we have introduced QR as the set of quadratic residues modulo p and NQR as the set of quadratic non-residues modulo p.

Let r and s denote the two solutions $\pm m$ of the congruence relation $x^2 \equiv c\,(mod\ p)$. Then $r + u$ and $s + u$ are the two solutions of $(x - u)^2 - c \equiv 0\,(mod\ p)$. In other words,

$$(x - u)^2 - c = (x - (r + u))\,(x - (s + u)) \tag{9.32}$$

over the finite field $\mathbf{Z}_p\ (=GF(p))$.

Since $r \not\equiv s\,(mod\ p)$, it follows that the field element $(r + u)/(s + u)$ will never take on value 1. Since the mapping $u \to (r + u)/(s + u)$ is one-to-one for $u \in \mathbf{Z}_p, u \neq -s$, we conclude that

$$\{(r + u)/(s + u) \mid u \in \mathbf{Z}_p \setminus \{-s\}\} = \mathbf{Z}_p \setminus \{1\}. \tag{9.33}$$

The reader may want to verify this by means of the *Mathematica* functions `Table`, `Mod`, `PowerMod`, and `Union`.

```
p = 19; s = 9;
r = p - s;
S1 = Table[Mod[(r + u) * PowerMod[(s + u), -1, p], p],
        {u, 0, r - 1}]
S2 = Table[Mod[(r + u) * PowerMod[(s + u), -1, p], p],
        {u, r + 1, p - 1}]
S = Union[S1, S2]
```

```
{18, 3, 8, 9, 4, 16, 15, 7, 10, 0}
```

```
{2, 11, 14, 6, 5, 17, 12, 13}
```

```
{0, 2, 3, 4, 5, 6, 7, 8, 9, 10, 11, 12, 13, 14, 15, 16, 17, 18}
```

It follows from (9.33) and Theorem A.20 that for half of the admissible values of u the element $(r + u)/(s + u)$ will be in QR $\bigcup \{0\}$ and for the other half it will be in NQR. In the first case, either $u = -r$ or (by Theorem A.21) both $r + u$ and $s + u$ will be an element of QR or they will both be in NQR. In the latter case, exactly one of them will be in QR and the other will be in NQR.

A property of quadratic residues modulo a prime number that we shall need later on is given by (A.16):

$$x^{(p-1)/2} - 1 = \prod_{u \text{ is QR}} (x - u).$$

Example 9.19

As an example, consider the QR's mod 11. We introduce a new function:

```
ListQuadRes[p_] :=
  Select[Range[p], JacobiSymbol[#1, p] == 1 &]
```

```
p = 11;
ListQuadRes[p]
```

```
{1, 3, 4, 5, 9}
```

So, the QR's modulo 11 are given by: 1, 3, 4, 5, and 9. We now compute with the Mathematica function PolynomialMod:

```
PolynomialMod[(x - 1) (x - 3) (x - 4) (x - 5) (x - 9), 11]
```

```
10 + x^5
```

This is indeed equal to $x^5 - 1$ modulo 11.

It follows from the above discussion, in particular from (9.33) and (A.16), that for a randomly chosen u, $u \in \mathbf{Z}_p \setminus \{-s\}$,

$$\gcd((x - u)^2 - c, \; x(x^{(p-1)/2} - 1)) \qquad (\bmod p) \tag{9.34}$$

will be

$$
\begin{array}{lll}
x - u - r, & \text{if } u + r \in \text{QR} \bigcup \{0\} & \text{and } u + s \in \text{NQR}, \\
x - u - s, & \text{if } u + r \in \text{NQR} & \text{and } u + s \in \text{QR} \bigcup \{0\}, \\
1, & \text{if } u + r \in \text{NQR} & \text{and } u + s \text{ is NQR}, \\
(x - u)^2 - c, & \text{if } u + r \in \text{QR} \bigcup \{0\} & \text{and } u + s \in \text{QR} \bigcup \{0\}.
\end{array}
$$

The counting arguments above imply that with probability $\frac{(p-1)/2}{p-1} = \frac{1}{2}$ one of the first two possibilities will occur. So, with probability 1/2 we have a non-trivial factor of $(x - u)^2 - c$. Since u is known, one also has found the value of r or s.

Note that in the extremely unlikely, remaining case, namely if $u = -s$, expression $(x - u)^2 - c$ will reduce to $x^2 + 2s.x$. So, the gcd in (9.34) will contain a factor x and the other factor will yield the solution s.

An example of the above method will be given later.

The expected number of u's that one has to try in this algorithm before finding a solution of $x^2 \equiv c \pmod p$ is the reciprocal of 1/2, i.e. 2. For a discussion of other methods of taking square roots modulo a prime number, we refer the interested reader to [Pera86].

□ **The Four Solutions**

The final step in the decryption algorithm is of course to use the Chinese Remainder Theorem to combine each of the two solutions of $x^2 \equiv c \pmod p$ with each of the two solutions of $x^2 \equiv c \pmod q$.

Example 9.16 (Part 3)

We continue with the parameters of Example 9.16. So, $p = 9733$, $q = 10177$, $n = p \times q = 99052741$, and the solutions of

$$a \equiv 1 \pmod{9733} \quad \text{and} \quad a \equiv 0 \pmod{10177},$$

$$b \equiv 0 \pmod{9733} \quad \text{and} \quad b \equiv 1 \pmod{10177}.$$

are given by $a = 45287650$ and $b = 53765092$.

```
p = 9733; q = 10177; n = p * q;
a = 45287650; b = 53765092;
Mod[p, 8]
Mod[q, 8]
```

```
5
```

```
1
```

Let c = 9513124 be a ciphertext. Since p ≡ 5 (mod 8) and q ≡ 1 (mod 8), we follow Case 2 to find the square root of c modulo p and Case 3 to find the square root of c modulo q.

$\sqrt{9513124}$ modulo p by Case 2

We calculate $c^{(p-1)/4} \equiv 1 \pmod{p}$ with the Mathematica functions PowerMod and Mod

```
c = 9513124;
u = PowerMod[c, (p - 1) / 4, p]
```

```
1
```

and find 1. The square root of c modulo p is thus given by $\pm c^{(p+3)/8}$:

```
f = PowerMod[c, (p + 3) / 8, p]
```

```
868
```

$\sqrt{9513124}$ modulo q by Case 3

We want to find the zeros of $x^2 - 9513124$ modulo q. We take a random u in Z_q and compute $gcd((x - u)^2 - 9513124, x(x^{(q-1)/2} - 1))$ and hope to find a linear factor. We use the Mathematica functions <u>PowerMod</u>, <u>PolynomialGCD</u> *and*

```
u = 11; x =.;
PolynomialGCD[(x - u)^2 - c, x (x^(q-1)/2 - 1), Modulus -> q]
```

```
2492 + 10155 x + x^2
```

We try again

```
u = 111; x =.;
PolynomialGCD[(x - u)^2 - c, x (x^{(q-1)/2} - 1), Modulus -> q]
```

```
1438 + x
```

It follows that one of the square roots is given by $x - 111 - g \equiv x + 1438 \pmod{q}$. So, by

```
g = Mod[-111 - 1438, q]
```

```
8628
```

It follows from the Chinese Remainder Theorem (Thm. A.19) that the four square roots of $x^2 \equiv 9513124 \pmod{99052741}$ are given by

```
Mod[a * f + b * g, n]
Mod[a * f - b * g, n]
Mod[-a * f + b * g, n]
Mod[-a * f - b * g, n]
```

```
6969696
```

```
63567091
```

```
35485650
```

```
92083045
```

9.5.3 How to Distinguish Between the Solutions

Let f be one of the two solutions of $x^2 \equiv c \pmod{p_U}$ and let g be one of the two solutions of $x^2 \equiv c \pmod{q_U}$. Further, let a and b be the solutions of the linear congruence relations (9.30) and (9.31).

Then, by the Chinese Remainder Theorem (Thm. A.19), the four solutions of (9.29) are given by

$$\pm f.a \pm g.b \pmod{n_U}.$$

One would like the sender and receiver to be able to distinguish between the four solutions in such a way that they can agree on one of them. In some cases this can be done quite easily. Indeed, if

p_U and q_U are both congruent to 3 mod 4, one has by Corollary A.24 that -1 is a NQR both modulo p_U and modulo q_U. Hence, exactly one of f and $-f$ is QR and the same is true for g and $-g$. Replacing f by $-f$ and/or g by $-g$, if necessary, one has without loss of generality that

$$\begin{array}{ll}
+f.a + g.b \text{ is QR mod } p_U, & +f.a + g.b \text{ is QR mod } q_U, \\
+f.a - g.b \text{ is QR mod } p_U, & +f.a - g.b \text{ is QR mod } q_U, \\
-f.a + g.b \text{ is QR mod } p_U, & -f.a + g.b \text{ is QR mod } q_U, \\
-f.a - g.b \text{ is QR mod } p_U, & -f.a - g.b \text{ is QR mod } q_U.
\end{array}$$

By Definition A.11 and the second statement in Theorem A.26 we have that $(f.a + g.b/n_U) = (-f.a - g.b/n_U) = 1$, while $(f.a - g.b/n_U) = (-f.a + g.b/n_U) = -1$. Of the two solutions with Jacobi value $+1$, one will lie in between 1 and $(n_U - 1)/2$, the other will lie between $(n_U + 1)/2$ and $n_U - 1$ (or both are equal to 0).

We conclude that there is a unique solution m satisfying $0 \le m \le (n_U - 1)/2$ and $(m/n_U) = 1$. So, sender and receiver can agree to use only messages of this form.

Example 9.20 (Part 1)

Let $n_B = 77$ and let $c = 53$ be a received message. Repeating the decryption process explained in the previous subsection, we get $f = 2$, $g = 8$, $a = 22$, and $b = 56$.

With the Mathematica functions Mod *and* JacobiSymbol, *we get the following four possible messages with their respective Jacobi symbol value.*

```
nB = 77;
f = 2; g = 8;
a = 22; b = 56;
m1 = Mod[a * f + b * g, nB];
m2 = Mod[a * f - b * g, nB];
m3 = Mod[-a * f + b * g, nB];
m4 = Mod[-a * f - b * g, nB];
Print[m1, "  ", JacobiSymbol[m1, nB]]
Print[m2, "  ", JacobiSymbol[m2, nB]]
Print[m3, "  ", JacobiSymbol[m3, nB]]
Print[m4, "  ", JacobiSymbol[m4, nB]]
```

```
30   -1

58   1

19   1

47   -1
```

We conclude that $m = 19$ is the unique solution with $(m / 77) = 1$ and $0 \le m \le 33$, so $m = 19$ was the message transmitted by the sender.

If p_U (or q_U) is congruent to 1 modulo 4, one can still agree to use only messages with $0 \le m \le (n_U - 1)/2$. To get $(m/n_U) = 1$ the sender and receiver could restrict themselves to shorter messages, say 20 digits shorter, and fill up the remaining 20 digits in such a way that the resulting message has Jacobi symbol 1 modulo n_U.

9.5.4   The Equivalence of Breaking Rabin's Scheme and Factoring n

We shall now show that breaking Rabin's variant of RSA is equivalent to factoring n_U. Of course, when the factorization of n_U is known to the cryptanalist, Rabin's system is in fact broken, because the cryptanalist can use the same methods to decrypt as the receiver can (see Subsection 9.5.2).

> **Theorem 9.18**
> Let $n = p \times q$, where p and q are prime. Let $\mathcal{A}$ denote an algorithm that for every c, which is the square of an integer, finds a solution of $x^2 \equiv c \pmod{n}$ with $F(n)$ operations. Then a probabilistic algorithm exists that factors n with an expected number of operations that is $2(F(n) + 2\log_2 n)$.

Proof: Select a random m, $0 < m < n$, compute $c \equiv m^2 \pmod{n}$ and solve $x^2 \equiv m \pmod{n}$ with algorithm $\mathcal{A}$ in $F(n)$ steps. Let k be the solution found by $\mathcal{A}$. The following four possibilities each have probability 1/4:

$$
\begin{array}{lll}
i) & k \equiv +m \pmod{p} & \text{and} \quad k \equiv +m \pmod{q}, \\
ii) & k \equiv +m \pmod{p} & \text{and} \quad k \equiv -m \pmod{q}, \\
iii) & k \equiv -m \pmod{p} & \text{and} \quad k \equiv +m \pmod{q}, \\
iv) & k \equiv -m \pmod{p} & \text{and} \quad k \equiv -m \pmod{q}.
\end{array}
$$

Indeed, there are four different messages that are mapped to c and they are all four equally likely.

In case ii), $\gcd(k - m, n) = p$ and in case iii) $\gcd(k - m, n) = q$. So, the calculation of $\gcd(k - m, n)$ will yield the factorization of n with probability 1/2. This computation involves less than $2\log_2 n$ calculations by Theorem A.9, therefore, each choice of m involves at most $F(n) + 2\log_2 n$ operations.

Since the probability of success is 1/2, one expects to need two tries.

□

Example 9.20 (Part 2)

Suppose that $n = 77$ and that the value of m that we have picked is 30. Then $c \equiv 30^2 \equiv 53 \pmod{77}$. Now assume that Algorithm $\mathcal{A}$ finds $k = 19$ as solution to $x^2 \equiv 53 \pmod{77}$ (see Example 9.20 for these parameters).

Then one of the factors of n will be found from $\gcd(k - m, n)$. This would also have happened if $\mathcal{A}$ had found $k = 58$, but not with 30 or 47.

All these calculations can easily be checked with the Mathematica function GCD.

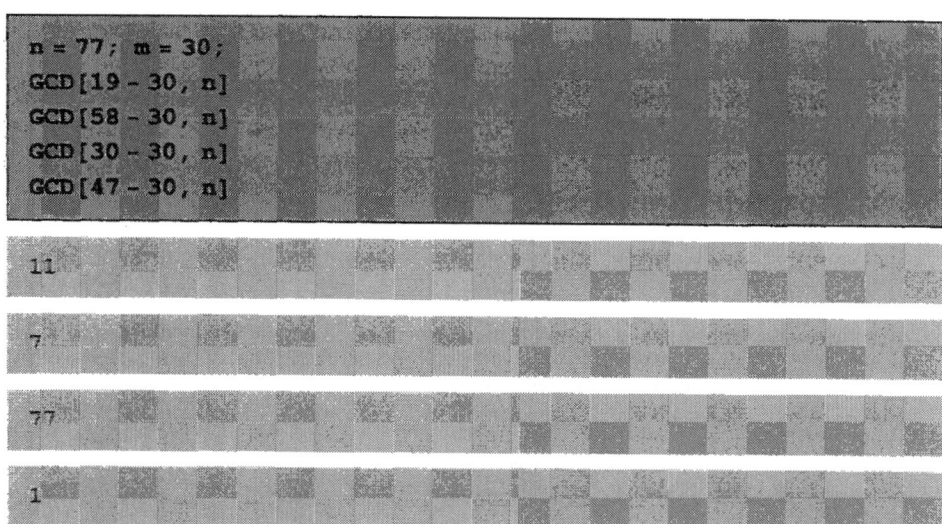

```
n = 77;  m = 30;
GCD[19 - 30, n]
GCD[58 - 30, n]
GCD[30 - 30, n]
GCD[47 - 30, n]

11

7

77

1
```

9.6   Problems

Problem 9.1
Consider the RSA system with $n = 383 \times 563$ (so $n = 215629$) and public key $e = 49$. So, a plaintext m will be encrypted into $c = E(m)$, where

$$E(m) = m^{49} \pmod{n}.$$

Prove that every ciphertext c satisfies $E^{10}(c) \equiv c \pmod{n}$. (Hint: use Fermat's Theorem and the Chinese Remainder Theorem.) The notation $E^{10}(c)$ stands for $\overbrace{E(E(\ldots E(c)))}^{10}$.
Give an easy way for a cryptanalist to recover plaintext m from ciphertext c.

Problem 9.2
Verify that the RSA secrecy system (or signature scheme) works correctly when a message m has a non-trivial factor in common with the modulus $n = p \times q$, i.e. show that

$$(m^e)^d \equiv m \pmod{n}$$

when $\gcd(m, n) = p$ or q (as always e and d denote the public resp. secret exponents).
(Hint : use Fermat's Theorem and the Chinese Remainder Theorem.)

Problem 9.3
Consider the RSA cryptosystem with modulus $n = p \times q$ and public exponent e.
a) Prove that the number of solutions of the equation $m^u \equiv 1 \pmod{p}$, when u divides $p - 1$, is exactly u (hint: use the multiplicative structure of $GF(p)$, Theorem B.21)
b) Show that each solution of $m^{e-1} \equiv 1 \pmod{p}$ is a solution of $m^{\gcd(e-1, p-1)} \equiv 1 \pmod{p}$ and vice versa (use Fermat's Theorem, use identity A.8 for the extended version of Euclid's algorithm).

c) Prove that the number of solutions of the equation $m^e \equiv m \pmod{p}$ is given by $1 + \gcd(e - 1, p - 1)$.

d) Prove that the number of plaintexts m satisfying

$$m^e \equiv m \pmod{n}$$

(in which case encryption does not conceal a message), is given by

$$\{1 + \gcd(e - 1, p - 1)\}.\{1 + \gcd(e - 1, q - 1)\}.$$

(Hint: use the Chinese Remainder Theorem.)

Problem 9.4
Demonstrate the principle of the Solovay and Strassen primality test on the number m = 33. The number m has been made small in this problem to keep the calculations simple. So, do not make use of numbers that "incidentally" have a factor in common with m.

Problem 9.5 ^{M}
Give a *Mathematica* implementation of Algorithm 9.14 and test it out for two values of m, $24 < m < 24^2$.

Problem 9.6 ^{M}
Give a complete factorization of $n = 110545695839248001$ by means of Pollard's ϱ Algorithm.

Problem 9.7 ^{M}
Complete Example 9.7. (Hint: extend the search to $(-105, 105)$.)

Problem 9.8 ^{M}
Apply the Wiener attack to $n = 122714980793$ and $e = 34587422599$.

Problem 9.9 ^{M}
Find a nontrivial strong liar for the composite number $m = 85$.

Problem 9.10 ^{M}
Suppose that Alice has sent the same secret message to B, C, D, E, and F by means of the RSA system. Let the public moduli of these people be given by $n_B = 324059$, $n_C = 324371$, $n_D = 326959$, $n_E = 324851$, and $n_F = 324899$. Assume that they all have the same public exponent $e = 5$.
Let the intercepted messages be given by $c_B = 68207$, $c_C = 96570$, $c_D = 251415$, $c_E = 273331$, resp. $c_F = 154351$.
Determine Alice's message (see Example 9.8).

Problem 9.11 ^{M}
Suppose that Alice has sent secret messages $m_1 = m$ and $m_2 = m^2 + 10\,m + 20$ to Bob by means of the RSA system. Let Bob's modulus be $n_B = 483047$ and $e_B = 3$. Suppose that you have intercepted the transmitted ciphertexts $c_1 = 346208$ resp. $c_2 = 230313$ and that you know the above relation between m_1 and m_2. Determine m_1 (see Example 9.10).

Problem 9.12
Consider the Rabin variant of the RSA system. So, only the number n is public.
Suppose that a message m, $1 < m < n$, has been sent that has a non-trivial factor in common with n.
How many possible plaintexts will the receiver find at the end of the decryption process?

Problem 9.13 [M]

The Rabin variant of the RSA system is used as cryptosystem with $n = 17419 \times 17431$. Demonstrate the decryption algorithm of this system for the ciphertext $c = 234279292$.

Which solution will come up if the method described in Subsection 9.5.3 is being followed? Why can this method be applied?

Problem 9.14

Let $p \equiv 5 \pmod 8$ and let c be a quadratic residue modulo p.

a) Show that $c^{(p-1)/4} \equiv \pm 1 \pmod p$.

b) Show that the solution of $x^2 \equiv c \pmod p$ is given by $\pm c^{(p+3)/8}$ if $c^{(p-1)/4} \equiv 1 \pmod p$.

c) Show that the solution of $x^2 \equiv c \pmod p$ is given by $\pm 2 c (4 c)^{(p-5)/8}$ if $c^{(p-1)/4} \equiv -1 \pmod p$. (Hint: use Theorem A.25 which implies that 2 is not a quadratic residue modulo p)

10   Elliptic Curves Based Systems

It will turn out in this chapter that discrete-logarithm-based cryptosystems can also be defined over elliptic curves. For RSA-based systems the same can be done, but there seems to be little reason to do so. For discrete-logarithm-like systems over elliptic curves, it may very well be that smaller parameters are possible with the same level of security as the regular systems over finite fields.

However, many questions regarding EC-systems are still open at this moment, making it unclear what the future of these systems will be.

10.1   Some Basic Facts of Elliptic Curves

Let $GF(q)$ be a finite field with q elements, where $q = p^m$. The number p is prime and is called the characteristic of $GF(q)$. If $m = 1$, we have $GF(q) = \mathbb{Z}_p$, the set of integers modulo p.

The so-called (affine) *Weierstrass equation* is given by

$$y^2 + u.x.y + v.y = x^3 + a.x^2 + b.x + c. \tag{10.1}$$

It is defined over any field (like $\mathbb{R}$ or $\mathbb{C}$), but for cryptographic purposes we shall always assume that the coefficients are in $GF(q)$.

If $p \neq 2$, one can simplify the Weierstrass equation by means of the transformation $y \rightarrow y - (u.x + v)/2$. One obtains (with new values for a, b, and c)

$$y^2 = x^3 + a.x^2 + b.x + c. \tag{10.2}$$

If also $p \neq 3$, one can apply $x \rightarrow x - a/3$ to further reduce this form to:

$$y^2 = x^3 + b.x + c. \tag{10.3}$$

If $p = 2$, two standard simplifications of (10.1) are possible. They are given by

$$y^2 + x.y = x^3 + a.x^2 + c. \tag{10.4}$$

$$y^2 + v.y = x^3 + b.x + c. \tag{10.5}$$

> **Definition 10.1**
> An *elliptic curve* $\mathcal{E}$ over $GF(q)$ is defined as the set of points (x, y) satisfying (10.1) together with a single element O, called the *point at infinity*.

To verify if a point (u, v) lies on a particular elliptic curve, say $y^2 = x^3 + 2x + 3$ over $\mathbb{Z}_5$, is quite easy.

```
p = 5;
a = 0; b = 2; c = 3;
EC[x_, y_] = y^2 - x^3 - a*x^2 - b*x - c;
{u, v} = {1, 4};
Mod[EC[u, v], p] == 0
```

```
True
```

To see if $\mathcal{E}$ contains a point with a given x-coordinate we can use the *Mathematica* function `Solve`. Since the Weierstrass equation is quadratic in y, there will be at most two values of y (see Theorem B.14).

```
p = 11;
Solve[ {y^2 == x^3 - 5 x + 3, x == 3, Modulus == p}, {y}]
```

```
{{Modulus → 11, x → 3, y → 2}, {Modulus → 11, x → 3, y → 9}}
```

So, $x = 3$ leads to the values $y = \pm 2$, i.e. to the points $(3, 2)$ and $(3, 9)$. The reader should try some other values of x.

The reader is referred to Subsection 9.5.2 to find a discussion on how the square root of a quadratic residue modulo a prime number can be determined by mathematical means.

It follows from the above that a point $P = (x, y)$ on an elliptic curve is completely characterized by its x-coordinate and the "sign" of y. This reduces the storage requirement of P by almost a factor 2. If $q = p$, $p > 2$, the "sign" of y can be defined as being plus one when $0 \leq y \leq (p-1)/2$ and as minus one otherwise.

If $q = p^m$, $p > 2$, one can use likewise the "sign" of the left-most nonzero coordinate in the p-ary representation of y.

For small values of p, one can find all points on $\mathcal{E}$ by trying out all possible value of x and check in each case if (10.1) has a solution. Below, we use the *Mathematica* functions `Flatten`, `Table`, and `Solve`.

```
Clear[x, y];
p = 11;
Flatten[
  Table[ Solve[ {y^2 == x^3 - 5 x + 3, x == u, Modulus == p}],
          {u, 0, p - 1}] , 1]
```

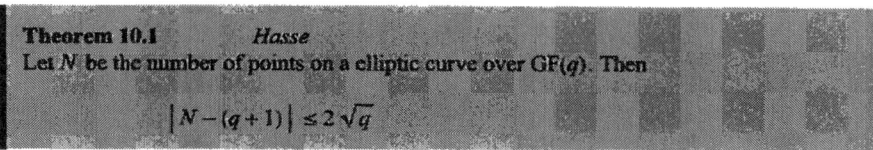

```
{{Modulus → 11, y → 5, x → 0}, {Modulus → 11, y → 6, x → 0},
 {Modulus → 11, y → 1, x → 2}, {Modulus → 11, y → 10, x → 2},
 {Modulus → 11, y → 2, x → 3}, {Modulus → 11, y → 9, x → 3},
 {Modulus → 11, y → 5, x → 4}, {Modulus → 11, y → 6, x → 4},
 {Modulus → 11, y → 2, x → 5}, {Modulus → 11, y → 9, x → 5},
 {Modulus → 11, y → 5, x → 7}, {Modulus → 11, y → 6, x → 7},
 {Modulus → 11, y → 4, x → 9}, {Modulus → 11, y → 7, x → 9}}
```

We see that for $p = 11$, there are 14 solutions (not counting O). There is a (imprecise) probabilistic argument to predict the number of points on $\mathcal{E}$: for each value of x, equation (10.1) will have two solutions with probability 1/2 and no solutions with probability 1/2, leading to about q solutions.

As supporting evidence of this statement, consider the right hand side in (10.2) and assume that $p > 2$. If, for a given value of x, the right hand side is a square in GF(p) (there are $(p-1)/2$ squares, namely all even powers of a primitive element in GF(p); or see Theorem A.20), there will be two solutions for y. If the right hand side is 0, there is only one solution, namely $y = 0$. There are no other solutions.

A famous theorem by Hasse [Silv86] states:

Theorem 10.1 *Hasse*
Let N be the number of points on a elliptic curve over GF(q). Then

$$|N - (q+1)| \le 2\sqrt{q}$$

Note that in the example above, we have indeed that $|15 - 12| \le 2\sqrt{11}$.

In general, it is very hard to find the precise number of points on an elliptic curve. There is however an algorithm by Schoof [Scho95] which computes this number (see also [Mene93] for a further discussion).

Although it is not necessary for the understanding of the rest of this chapter, we like to remind the reader of the possibilities in *Mathematica* to make calculations over fields GF(p^m) with $m > 1$.

Example 10.1

As an example of a curve over $GF(2^4) = GF(2)[\alpha]/(1 + \alpha^3 + \alpha^4)$ (see Table B.2), we can consider the equation $y^2 = x^3 + \alpha x + 1$. To test if (α^2, α^{14}) is on the curve we first load the Mathematica package `Algebra`FiniteFields`*.*

```
<< Algebra`FiniteFields`
```

```
f16 = GF[2, {1, 0, 0, 1, 1}];
al = f16[{0, 1, 0, 0}];
EC[x_, y_] = y^2 - x^3 - al * x - 1;
{u, v} = {al^2, al^14};
EC[u, v]
```

```
0
```

Indeed, $(\alpha^{14})^2 = (\alpha^2)^3 + \alpha(\alpha^2) + 1$, as can be checked with

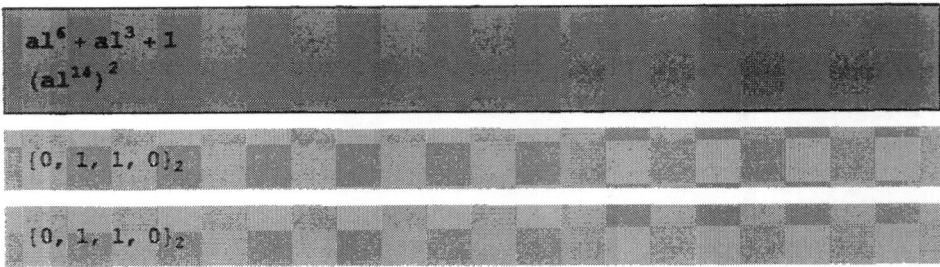

```
al^6 + al^3 + 1
(al^14)^2
```

```
{0, 1, 1, 0}_2
```

```
{0, 1, 1, 0}_2
```

10.2   The Geometry of Elliptic Curves

The reason that we are interested in elliptic curves is the addition operation that can be defined on them. This operation will have $O \in \mathcal{E}$ (the point at infinity) as its unit-element and will have the structure of an additive group.

To be able to define a suitable addition on $\mathcal{E}$, we shall make use of the property that any line intersecting $\mathcal{E}$ in at least two points, will intersect it in a third. Here, a tangent point should be counted twice. The point O at infinity is the intersection point of all vertical lines.

We shall first show a picture of an elliptic curve over the reals. We use the *Mathematica* function ImplicitPlot for which the package Graphics`ImplicitPlot` has to be loaded first.

```
<< Graphics`ImplicitPlot`
```

```
elliptic = ImplicitPlot[ y² == x³ - 5 x + 3, {x, -3, 3}]
```

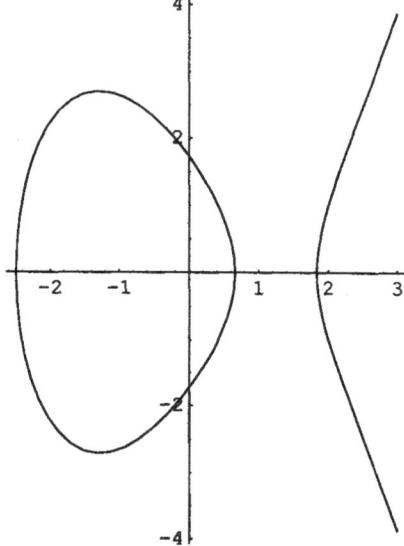

The reader is invited to change the coefficient of x in the function plotted above from -5 to -4 and -3 and observe how the graph changes.

To see how the line $y = x + 1$ intersects $y^2 = x^3 - 5x + 3$ we use the additional functions Epilog and Line.

```
ImplicitPlot[ y² == x³ - 5 x + 3, {x, -3, 4},
        PlotRange -> {-4, 4},
        Epilog -> Line[ {{-3, 4}, {4, -3}}]]
```

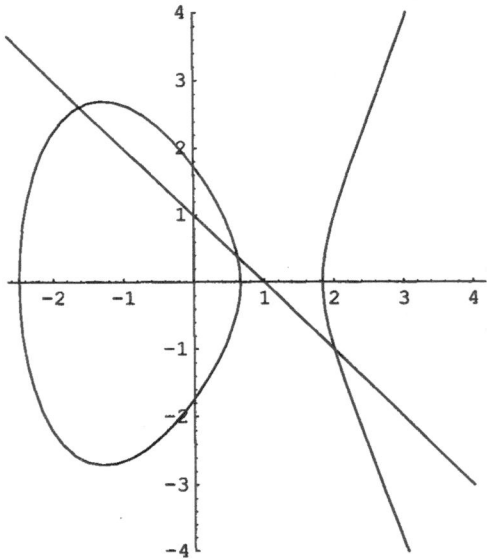

To find the intersection points numerically, one can use NSolve.

```
NSolve[ {y² == x³ - 5 x + 3, y == -x + 1}, {x, y}]
```

```
{{y → -1., x → 2.}, {y → 0.381966, x → 0.618034},
 {y → 2.61803, x → -1.61803}}
```

When the curve is defined over $\mathbf{Z}_p$ we can find the intersection points of a line with the curve by means of the Solve function as follows.

```
p = 11;
Solve[ {y² == x³ - 5 x + 3, y == x - 1, Modulus == p}, {x, y}]
```

```
{{Modulus → 11, y → 1, x → 2},
 {Modulus → 11, y → 2, x → 3}, {Modulus → 11, y → 6, x → 7}}
```

A different way to find the intersection points of a line $y = u.x + v$ with an elliptic curve is to substitute $y = u.x + v$ in (10.1), obtain a third degree equation in x and find its factorization.

Example 10.2

Suppose that we are working over $\mathbb{Z}_{11}$. To find the intersection points of $y = 4x + 1$ with $y^2 = x^3 - 5x + 1$, we factor $(4x + 1)^2 - (x^3 - 5x + 1)$ with the Mathematica function Factor.

```
p = 11;
Clear[x];
ec = x^3 - 5 x + 3;
il = 4 x + 1;
Factor[il^2 - ec, Modulus -> p]
```

```
10 (2 + x) (7 + x) (8 + x)
```

We get as x-values of the intersection points: -2, -7, and -8. From $y = 4x + 1$ we find the solutions $(9, 4)$, $(4, 6)$, and $(3, 2)$.

```
x = Mod[{-2, -7, -8}, p]
y = Mod[4 * x + 1, p]
```

```
{9, 4, 3}
```

```
{4, 6, 2}
```

□ A Line Through Two Distinct Points

Let $P_1 = (x_1, y_1)$ and $P_2 = (x_2, y_2)$ be two distinct points on an elliptic curve $\mathcal{E}$ (both not at infinity). Let $\mathcal{L}$ be the line through P_1 and P_2. How do we find the third point on the intersection of $\mathcal{L}$ with $\mathcal{E}$? If $x_1 = x_2$ and $y_1 = -y_2$ the point O will be defined as this third point.

So, let us consider the case that $x_1 \neq x_2$. The line $\mathcal{L}$ though P_1 and P_2 is given by:

$$y - y_1 = \lambda(x - x_1), \qquad \text{with } \lambda = \frac{y_2 - y_1}{x_2 - x_1}. \tag{10.6}$$

We discuss two cases.

$p \neq 2$

Assume that the elliptic curve is already in reduced form (see (10.2)). Substitution of (10.6) into this relation yields $(\lambda(x - x_1) + y_1)^2 = x^3 + a.x^2 + b.x + c$. Since we know two roots of this third degree equation, there must be a third one (to be called x_3). So, the same equation can also be written as $(x - x_1)(x - x_2)(x - x_3) = 0$. Comparing the coefficient of x^2 in both notations, we get

$$x_3 = \lambda^2 - a - x_1 - x_2, \tag{10.7}$$

and, by (10.6),

$$y_3 = \lambda(x_3 - x_1) + y_1.$$ (10.8)

Example 10.3

Consider the elliptic curve $y^2 = x^3 + 11x^2 + 17x + 25$ over $\mathbb{Z}_{31}$. The points $P_1 = (x_1, y_1) = (2, 7)$ and $P_2 = (x_2, y_2) = (23, 9)$ lie on $\mathcal{E}$ as can be verified with the $\underline{\text{Mod}}$ function as follows:

```
p = 31;
a = 11; b = 17; c = 25;
x1 = 2; y1 = 7; x2 = 23; y2 = 9;
F[x_, y_] := y^2 - (x^3 + a*x^2 + b*x + c);
Mod[F[x1, y1], p] == 0
Mod[F[x2, y2], p] == 0
```

```
True
```

```
True
```

The slope λ of the line $\mathcal{L}$ through P_1 and P_2 is given by (10.6): $\lambda = \frac{9-7}{23-2} = 2 \times 3 = 6$. Here we use the $\underline{\text{PowerMod}}$ function to get the multiplicative inverse of 21 modulo 31.

```
PowerMod[21, -1, p]
```

```
3
```

The coordinates (x_3, y_3) of the third intersection point of $\mathcal{L}$ with $\mathcal{E}$ are given by (10.7) and (10.8):

```
lam = 6;
x3 = Mod[lam^2 - a - x1 - x2, p]
y3 = Mod[lam (x3 - x1) + y1, p]
```

```
0
```

```
26
```

That the point $P_3 = (0, 26)$ indeed lies on $\mathcal{E}$ can be verified with the calculation

```
Mod[F[x3, y3], p] == 0
```

```
True
```

$p=2$

We now assume reduced form (10.4). As above, we substitute (10.6) into (10.4) and look at the coefficient of x^2. We get

$$x_3 = a - \lambda^2 - \lambda - x_1 - x_2, \tag{10.9}$$

$$y_3 = \lambda(x_3 - x_1) + y_1. \tag{10.10}$$

Note that all minus signs can be replaced by plus signs, when $p = 2$.

□ **A Tangent Line**

There is one more possibility that we want to discuss, namely that $P_1 = (x_1, y_1) = P_2$. Let $\mathcal{L}$ be the *tangent line* to $\mathcal{E}$ though P. This means that $\mathcal{L}$ meets $\mathcal{E}$ in $P = (x_1, y_1)$, and that the slope of $\mathcal{L}$ is the same as the derivative of $\mathcal{E}$ in P. One usually views P as point of intersection with multiplicity two.

Over $\mathbb{R}$ this situation looks like:

```
ImplicitPlot[ y^2 == x^3 - 5 x - 3, {x, -3, 4},
    PlotRange -> {-4, 4},
      Epilog -> Line[ {{-3, 3}, {4, -4}}]]
```

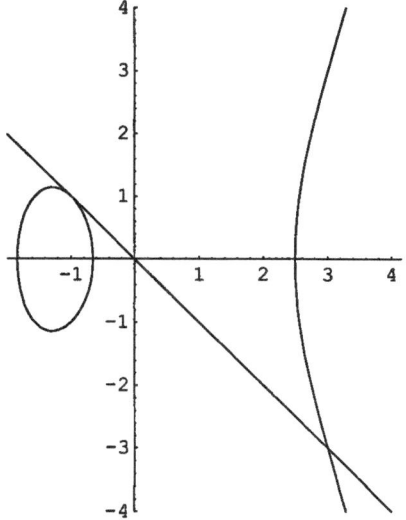

At this moment we exclude the possibility that $\mathcal{L}$ is a double tangent line to $\mathcal{E}$ (meaning that its multiplicity is 3). If it were, the tangent line already intersects $\mathcal{E}$ in a point with multiplicity 3.

In the sequel, when we speak of taking a *derivative* of a polynomial over a finite field we mean to take the formal derivative and then reduce the coefficients modulo the characteristic of the field.

For instance, in GF(3^m) the derivative of $x^4 + 2x^3 + x^2 + 1$ is given by $4x^3 + 6x^2 + 2x$, which reduces to $x^3 + 2x$.

$p \neq 2$

The slope of the tangent line through a point $P = (x_1, y_1)$ on the curve $y^2 = x^3 + a.x^2 + b.x + c$ (see (10.2)) is given by the value of y' determined through implicit differentiation, so $2y_1.y' = 3x_1^2 + 2a.x_1 + b$. We conclude that the tangent line through P is given by

$$y - y_1 = \lambda(x - x_1), \quad \text{with } \lambda = \frac{3x_1^2 + 2a.x_1 + b}{2y_1}. \tag{10.11}$$

To find the third point of the line $\mathcal{L}$ through $\mathcal{E}$ we can still use (10.7) and (10.8).

$p = 2$

The slope of the tangent line through a point $P = (x_1, y_1)$ on the curve $y^2 + x.y = x^3 + a.x^2 + c$ (see (10.4)) is given by the value of y' determined from $2y_1.y' + y_1 + x_1.y' = 3x_1^2 + 2a.x_1$, i.e. by $y_1 + x_1.y' = x_1^2$. Hence, the tangent line through P is given by

$$y - y_1 = \lambda(x - x_1), \quad \text{with } \lambda = \frac{x_1^2 + y_1}{x_1} = x_1 + \frac{y_1}{x_1}. \tag{10.12}$$

To find the third point on $\mathcal{L}$ through $\mathcal{E}$ we observe that (10.9) (take $x_2 = x_1$) reduces to

$$x_3 = a - \lambda^2 - \lambda = a + x_1^2 + \left(\frac{y_1}{x_1}\right)^2 + x_1 + \frac{y_1}{x_1} =$$

$$a + x_1^2 + x_1 + \frac{x_1^2 + x_1 y_1}{x_1^2} \stackrel{(10.4)}{=} a + x_1^2 + x_1 + \frac{x_1^3 + a.x_1^2 + c}{x_1^2},$$

i.e.

$$x_3 = x_1^2 + \frac{c}{x_1^2}, \tag{10.13}$$

and that (10.10) reduces to

$$y_3 = x_1^2 + \left(x_1 + \frac{y_1}{x_1}\right)x_3. \tag{10.14}$$

Example 10.4

Consider the elliptic curve $y^2 + x.y = x^3 + \alpha^9 x^2 + \alpha$ over GF(16), where $\alpha^4 = \alpha + 1$. The point (α^2, α^{12}) lies on this curve, as can be easily checked, once we have loaded the Mathematica package `Algebra`FiniteFields`.

```
<< Algebra`FiniteFields`
```

```
f16 = GF[2, {1, 1, 0, 0, 1}];
a = f16[{0, 1, 0, 0}];
EC[x_, y_] = y^2 + x * y - x^3 - a^9 * x^2 - a;
{x1, y1} = {a^2, a^12};
EC[x1, y1]
```

```
0
```

The tangent through (α^2, α^{12}) *has slope* λ *given by (10.12). So,*

```
lam = x1 + y1 / x1
```

```
{1, 1, 0, 0}_2
```

which is α^4. *To find the other point where the tangent intersects* $\mathcal{E}$, *we use (10.13) and (10.14).*

```
x3 = x1^2 + a / x1^2
y3 = x1^2 + (x1 + y1 / x1) x3
```

```
{0, 0, 1, 1}_2
```

```
{0, 0, 1, 0}_2
```

So, $(x_3, y_3) = (\alpha^6, \alpha^2)$. *This can all be checked easily.*

```
a^6
a^2
EC[x3, y3]
```

```
{0, 0, 1, 1}_2
```

```
{0, 0, 1, 0}_2
```

```
0
```

10.3   Addition of Points on Elliptic Curves

In the previous section, we have shown how the line through two points on an elliptic curve $\mathcal{E}$ intersects that curve in a third point and how that point can be computed efficiently. The same holds for a line that is tangent to $\mathcal{E}$, with the understanding that the tangent point is counted twice.

We are now ready to define an addition on $\mathcal{E}$. The geometric idea behind the formulas below is the following. First of all, if $P = (x, y)$ is a point on an elliptic curve $\mathcal{E}$ determined by (10.1), then

$$-P = (x, -y - u.x - v).$$

If $u = v = 0$, like in (10.2), this reduces to

$$-P = (x, -y).$$

Geometrically, this can be described as follows: compute the line $\mathcal{L}$ through O and P. It intersects $\mathcal{E}$ in a third point, namely $-P$. As noted before, the point O at infinity should be interpreted as the intersection point of all vertical lines.

To add points P_1 and P_2, both not at infinity, execute the following two steps:

1) Compute the line $\mathcal{L}$ through P_1 and P_2 (or tangent line though P_1, if $P_1 = P_2$) and find the third point of intersection with $\mathcal{E}$. Let this be Q.

2) The sum $P_1 + P_2$ is defined as $P_3 := -Q$.

The point O serves as unit element of this addition and is its own inverse.

Definition 10.2   *addition*

Let P be a point on an elliptic curve $\mathcal{E}$ (so, it defined by (10.1)), with O as point at infinity. Then we define the sums

$$P + O = O + P = P.$$

Further, let $P_1 = (x_1, y_1)$ and $P_2 = (x_2, y_2)$ be two points on $\mathcal{E}$, both not O. Then the sum $P_1 + P_2$ is defined by

i) $P_3 = -Q$ if $x_1 \neq x_2$.
                       Here, Q is the third point of intersection of $\mathcal{E}$ with of the line $\mathcal{L}$ through (x_1, y_1) and (x_2, y_2).

ii) $P_3 = -Q$ if $P_1 = P_2$ and the tangent line through P is a single tangent. Here, Q is the third point of intersection of $\mathcal{E}$ with the tangent $\mathcal{L}$ through P.

iii) $P_3 = -P_1$ if $P_1 = P_2$ and the tangent line through P is a double tangent.

iv) $P_3 = O$ if $P_1 = -P_2$.

Note that possibility iii) can be interpreted as a special case of ii).

We shall depict the two most typical cases, namely i) and ii), by means of elliptic curves over the reals. We need again package `Graphics`ImplicitPlot``.

```
<< Graphics`ImplicitPlot`
```

```
ImplicitPlot[ y² == x³ - 5 x + 3,  {x, -3, 4},
     Epilog -> {Line[{{-3, -2}, {4, 5}}],
             Line[{{3, -6}, {3, 6}}],
             Text["\!\(\*
StyleBox[\"P\",\nFontColor->RGBColor[0, 0, 1]]]\)\!\(\*
StyleBox[\"+\",\nFontColor->RGBColor[0, 0, 1]]]\)\!\(\*
StyleBox[\"Q\",\nFontColor->RGBColor[0, 0, 1]]]\)",
     {2.3, -4}],
             Text["\!\(\*
StyleBox[\"P\",\nFontColor->RGBColor[0, 0, 1]]]\)",
     {-2.4, -2.1}],
             Text["\!\(\*
StyleBox[\"Q\",\nFontColor->RGBColor[0, 0, 1]]]\)",
     {0.35, 1.9}], PointSize[0.03],
             Point[{-2.31, -1.4}],
             Point[{0.28, 1.26}],
             Point[{3.01, -3.9}]}];
```

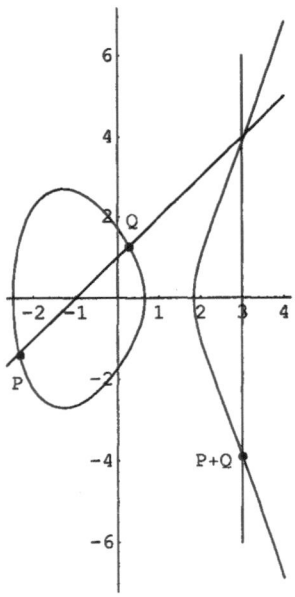

```
ImplicitPlot[ y² == x³ - 5 x - 3, {x, -3, 4},
    Epilog -> {Line[{{3, -6}, {3, 6}}],
            Line[{{-3, 3}, {4, -4}}],
            Text["\!\(\*
StyleBox[\"P\",\nFontColor->RGBColor[0, 0, 1]]\)",
    {-1.1, 1.65}],
            Text["\!\(\*
StyleBox[\"2\",\nFontColor->RGBColor[0, 0, 1]]\)\!\(\*
StyleBox[\"P\",\nFontColor->RGBColor[0, 0, 1]]\)",
    {2.8, 3.5}], PointSize[0.03],
            Point[{-1.1, 1.1}],
            Point[{3.01, +3}]}];
```

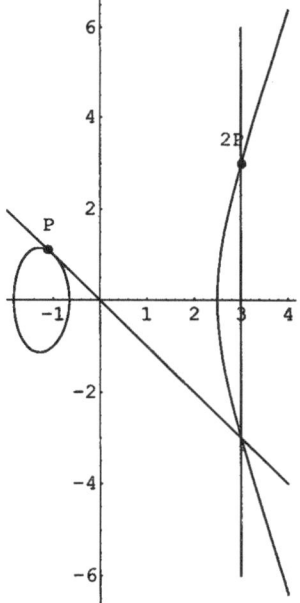

The points on an elliptic curve together with the addition defined above form an additive group. We shall not prove that here. The reader is referred to [Mene93] or [SilT92]. Note that the only non-trivial part to verify is the associativity of the addition.

Theorem 10.2
The points on an elliptic curve $\mathcal{E}$ together with the addition defined in Definition 10.2 form an additive group. The zero element is given by O.

With the following <u>Module</u> one can compute the sum of two points (the point O at infinity will be denoted by $\{O\}$) on an elliptic curve over $\mathbb{Z}_p$ with $p > 2$. We make use of formulas (10.6), (10.7), (10.8) and (10.11). and use the *Mathematica* function <u>Which</u> with the same order of cases as in Definition 10.2.

```
EllipticAdd[p_, a_, b_, c_, P_List, Q_List] :=
 Module[{lam, x3, y3, P3},
  Which[
   P == {O}, Q,
   Q == {O}, P,
   P[[1]] != Q[[1]],
       lam = Mod[
    (Q[[2]] - P[[2]]) PowerMod[Q[[1]] - P[[1]], p - 2, p], p];
        x3 = Mod[lam^2 - a - P[[1]] - Q[[1]], p];
        y3 = Mod[- (lam (x3 - P[[1]]) + P[[2]]), p];
        {x3, y3},
   (P == Q) ∧ (P[[2]] == 0), {O},
   (P == Q) ∧ (P != {O}),
       lam = Mod[ (3 * P[[1]]^2 + 2 a * P[[1]] + b)
    PowerMod[2 P[[2]], p - 2, p], p];
        x3 = Mod[lam^2 - a - P[[1]] - Q[[1]], p];
        y3 = Mod[- (lam (x3 - P[[1]]) + P[[2]]), p];
        {x3, y3},
   (P[[1]] == Q[[1]]) ∧ (P[[2]] != Q[[2]]), {O}]]
```

Below, we show the addition of points in a number of cases.

```
p = 11; a = 0; b = 6; c = 3;
EllipticAdd[p, a, b, c, {4, 6}, {9, 4}]
EllipticAdd[p, a, b, c, {9, 4}, {9, 4}]
EllipticAdd[p, a, b, c, {4, 6}, {4, 6}]
EllipticAdd[p, a, b, c, {4, 6}, {O}]
EllipticAdd[p, a, b, c, {4, 6}, {4, 5}]
EllipticAdd[p, a, b, c, {O}, {9, 4}]
```

```
{3, 9}
```

```
{7, 6}
```

```
{4, 5}

{4, 6}

{O}

{9, 4}
```

Observe that the tangent through (4, 6) is a double tangent, so by Definition 10.2, iii)

$$(4, 6) + (4, 6) = -(4, 6) = (4, 5).$$

As is common in additive groups, $2P$ will stand for $P + P$, similarly $3P$ stands for $P + P + P$, etc. Similarly, $0P$ stands for O and $-n.P$ stands for $-(n.P)$. These multiples of P are often called the *scalar multiples* of P.

The order of P is the smallest positive integer n with $n.P = O$. Since $\mathcal{E}$ is a finite group, this notion is well defined. The set $\{O, P, 2P, ..., (n-1)P\}$ is a cyclic subgroup of $\mathcal{E}$. It follows that n divides $|\mathcal{E}|$ (see Theorem B.5).

Now that we have the Module *EllipticAdd*, defined above, it is quite easy to compute $n.P$, $n \geq 1$, recursively as follows:

```
p = 11; a = 0; b = 6; c = 3; P = {9, 4};
f[1] = P;
f[n_] := f[n] = EllipticAdd[p, a, b, c, P, f[n-1]];
Table[f[n], {n, 1, 5}] // ColumnForm
```

```
{9, 4}
{7, 6}
{7, 5}
{9, 7}
{O}
```

So, on the curve $y^2 = x^3 + 6x + 3$ over $\mathbb{Z}_{11}$, the point $P = (9, 4)$ has order 5.

In the next section, it will be important to have points available on an elliptic curve $\mathcal{E}$ that have a very large order. If the cardinality of $\mathcal{E}$ is known and of a special form, for instance $|\mathcal{E}|$ is a small multiple of a large prime factor, then it is quite easy to find points on $\mathcal{E}$ with a known large order.

As an example, consider $|\mathcal{E}| = 3 \times 7919 = 23757$. Suppose that $3P \neq O$. Then P has order 7919 or 23757. If $7919P = 0$ then P has order 7919, otherwise $3P$ will have this order. To check these assertions, apply Lemma B.4 and Theorem B.5 (rewrite the multiplicative notation in the additive notation that we use here).

10.4   Cryptosystems Defined over Elliptic Curves

Most notions in this section can be viewed as direct translations of notions introduced in Chapter 8, but now using addition over an elliptic curve as principal operation instead of modular multiplication. Modular exponentiation will translate into scalar multiplication.

For the above reason, it will often suffice to just present the new formulations without copying all the proofs.

In [Demy94] one can find a RSA-like cryptosystem defined over elliptic curves. However, to break the system it is sufficient to factor its modulus. Since the original RSA system had the same security restriction and is faster in its calculations, there seems to be little reason to use this generalization of RSA to elliptic curves.

10.4.1 The Discrete Logarithm Problem over Elliptic Curves

We have seen in Section 10.3 how to add points on an elliptic curve $\mathcal{E}$. This is an operation with relatively low complexity. To compute scalar multiples of a point P, say $n.P$ for some integer n, we can use repeated addition, but it is much more efficient to copy the ideas of Subsection 8.1.1.

Example 10.5

Take $n=171$. Its binary expansion is 10101011, as follows from the Mathematica function `IntegerDigits`*.*

```
IntegerDigits[171, 2]
```

```
{1, 0, 1, 0, 1, 0, 1, 1}
```

So, to compute $171\,P$, it suffices to compute

$$2P = P + P,$$
$$4P = 2P + 2P,$$
$$8P = 4P + 4P$$
$$\vdots$$
$$128P = 64P + 64P$$

and add the suitable terms. This can be done on the fly as follows:

```
Clear[P];
2 (2 (2 (2 (2 (2 (2 P) + P)) + P)) + P) + P
```

> 171 P

Note that we only added partial results to themselves or to P. (The reader may want to look at Example 8.3 for the analogous modular arithmetic problem.)

Of course, addition chains may further reduce the complexity of these calculations.

The opposite problem of computing scalar multiples of a point is the following:

> **Definition 10.3**
> Let $\mathcal{E}$ be an elliptic curve. Let P be a point on $\mathcal{E}$ and let Q be a scalar multiple of P. The *discrete logarithm problem* over an elliptic curve is the problem of determining n for given P and Q from the relation

$$n.P = Q. \tag{10.15}$$

Although we shall see more efficient ways to solve (10.15) than by simply trying $n = 1, 2, \ldots$, all the methods have a complexity of the form n^{α}, $\alpha > 0$, and so they are exponentially slower than the (logarithmic) complexity of computing $n.P$ out of P.

10.4.2 The Discrete Logarithm System over Elliptic Curves

Now that we have formulated the discrete logarithm problem over elliptic curves, we can describe the analogue of the Diffie-Hellman key exchange protocol (see Subsection 8.1.2).

As system parameters one needs an elliptic curve $\mathcal{E}$ over a finite field GF(q) and a point P on the curve of high order, say the order n of P is 150-180 digits long.

Each user U of the system, selects a secret scalar m_U, computes the point $Q_U = m_U P$ and makes Q_U public. Alice and Bob can now agree on the common key $K_{A,B} = m_A m_B P$. Alice can find this common key by computing $m_A Q_B$ with her secret scalar m_A and Bob's public Q_B. Bob can do likewise.

This system is summarized in the following table.

system parameters	elliptic curve $\mathcal{E}$ over GF (q) point P on $\mathcal{E}$ of high order
secret key of U public key of U	m_U $Q_U = m_U\, P$
common key of A and B Ann computes Bob computes	$K_{A,B} = m_A\, m_B\, P$ $m_A\, Q_B$ $m_B\, Q_A$

The *Diffie-Hellman Key Exchange System* over Elliptic Curves

Table 10.1

Example 10.6

Consider the elliptic curve $\mathcal{E}$ over $\mathbb{Z}_{863}$ defined by $y^2 = x^3 + 100\,x^2 + 10\,x + 1$. The point $P = \{121,\ 517\}$ lies on it as can be checked with the Mathematica function Mod.

```
p = 863;
a = 100; b = 10; c = 1;
x = 121; y = 517;
Mod[y^2 - (x^3 + a*x^2 + b*x + c), p] == 0
```

```
True
```

The order of P is 432. To show this, we check that $432\,P = O$ and that $(432/p)\,P \neq O$ for the prime divisors of 432. We make use the binary expansion of these coefficients (to be found with the function IntegerDigits*). We also use of the EllipticAdd function defined in Section 10.3 and the* Do *function.*

```
FactorInteger[432]
IntegerDigits[432, 2]
IntegerDigits[432 / 2, 2]
IntegerDigits[432 / 3, 2]
```

```
{{2, 4}, {3, 3}}
```

```
{1, 1, 0, 1, 1, 0, 0, 0, 0}
```

```
{1, 1, 0, 1, 1, 0, 0, 0}
```

```
{1, 0, 0, 1, 0, 0, 0, 0}
```

```
p = 863; P = .;
a = 100; b = 10; c = 1;
P[0] = {121, 517};
P[i_] := P[i] = EllipticAdd[p, a, b, c, P[i - 1], P[i - 1]];
Q = EllipticAdd[p, a, b, c, EllipticAdd[p, a, b, c, P[8], P[7]],
        EllipticAdd[p, a, b, c, P[5], P[4]]]
EllipticAdd[p, a, b, c, EllipticAdd[p, a, b, c, P[7], P[6]],
        EllipticAdd[p, a, b, c, P[4], P[3]]]
EllipticAdd[p, a, b, c, P[7], P[4]]
```

```
{0}
```

```
{19, 0}
```

```
{341, 175}
```

Let Alice choose $m_A = 130$ and Bob $m_B = 288$. Then $Q_A = (162, 663)$ and $Q_B = (341, 688)$, as can be checked as follows (note that we have chosen very friendly secret scalars).

```
QAlice = EllipticAdd[p, a, b, c, P[7], P[1]]
QBob = EllipticAdd[p, a, b, c, P[8], P[5]]
```

```
{162, 663}
```

```
{341, 688}
```

Alice can compute the common key $K_{A,B}$ with the calculation $K_{A,B} = m_A Q_B$, where $m_A = 130$ is her secret key. She finds

```
QA[0] = {341, 688};
QA[i_] :=
  QA[i] = EllipticAdd[p, a, b, c, QA[i - 1], QA[i - 1]];
EllipticAdd[p, a, b, c, QA[7], QA[1]]
```

```
{341, 688}
```

Likewise, Bob can compute the common key $K_{A,B}$ with the calculation $K_{A,B} = m_B\,Q_A$, where $m_B = 288$ is his secret key. He also finds

```
QB[0] = {162, 663};
QB[i_] := QB[i] = EllipticAdd[p, a, b, c, QB[i - 1], QB[i - 1]];
EllipticAdd[p, a, b, c, QB[8], QB[5]]
```

```
{341, 688}
```

Now that the Diffie-Hellman key exchange system over elliptic curves has been described, it really is a straightforward exercise to show that the ElGamal protocol and the other systems, described in Section 8.2, can be rewritten in the language of elliptic curves.

10.4.3 The Security of Discrete Logarithm Based EC Systems

In Section 8.3, various methods are described to take the discrete logarithm over a finite field. The Pohlig-Hellman algorithm, the baby-step giant-step method, and the Pollard-ρ method can all be directly translated into elliptic curve terminology: just replace modular exponentiations by scalar multiplication on the elliptic curve.

At the time of this writing, the index-calculus method has defeated any attempt to transfer it efficiently to the elliptic curve setting (see [Mill86]). That is of great cryptographic significance, because the index-calculus method was the only one with a subexponential complexity. This means that in regular discrete-logarithm-like systems the index-calculus method is the governing factor in determining the size of its parameters (to keep the system computationally secure). Since the index-calculus method is no longer around in the elliptic curve setting, one can afford much smaller parameters to achieve the same level of security.

At the time of this writing, the *XEDNI method* has been proposed [Silv98] as an alternative to solve the elliptic curve discrete logarithm problem. Further analysis is needed to determine the implications of this method.

There are special attacks on discrete logarithm based elliptic curve cryptosystems. These attacks make it necessary to avoid special classes of elliptic curves. In particular, one should **not** use

> singular curves,
> supersingular curves,
> anomalous curves.

We shall not describe these attacks (see [MeOkV93], [SatA98], and [Smar98]. In each case the logarithm problem over an elliptic curve can be translated to the logarithm problem over a finite field (or an even simpler problem). We shall explain in one case that one can counter these attacks by simply avoiding these special curves.

Before we do so, we need to introduce a new notion. We *homogenize* the Weierstrass equation (10.1). This means that we multiply each term in it with the smallest power of z in such a way that all terms have the same degree:

$$F(x, y, z) = y^2 z + u.x.y.z + v.y.z^2 - x^3 - a.x^2 z - b.x.z - c.z^3 = 0. \qquad (10.16)$$

Note that if (x, y, z) satisfies (10.16), then so does $\lambda(x, y, z)$. For that reason, one often normalizes solutions to (10.16) by requiring the right-most non zero coordinate to be equal to 1.

Points (x, y) that satisfy (10.1) now lead to solutions $(x, y, 1)$ of (10.16). The (somewhat mysterious) point O at infinity can be represented by $(0, 1, 0)$.

A point on a curve $\mathcal{E}$ is a called *singular* if all partial derivatives $\partial F / \partial x$, $\partial F / \partial y$, and $\partial F / \partial z$ are zero. An elliptic curve can not contain two singular points. If a curve $\mathcal{E}$ contains a singular point then it is called a *singular curve*, otherwise it is called a *non-singular curve*.

With some effort one can show that (10.2) defines a non-singular curve if and only if the cubic expression on its right side has no multiple roots. For (10.3) with $c \neq 0$, this is equivalent to the condition $4 b^3 + 27 c^2 \not\equiv 0 \pmod{p}$.

When $p = 2$, (10.4) gives non-singular curves when $c \neq 0$ and (10.5) when $v \neq 0$.

The above means, that it is quite simple to test if a curve is non-singular or not.

We shall not give a definition of what *supersingular* means. Here it suffices to know that curves defined by (10.5) are supersingular and need to be avoided. Again, it is easy to avoid these curves.

Finally, *anomalous* curves are elliptic curves $\mathcal{E}$ over $\mathbf{Z}_p$ with the property that $|\mathcal{E}| = p$.

10.5   Problems

Problem 10.1 [M]
How many points lie on the elliptic curve defined in Example 10.1?

Problem 10.2
Find the intersection points over $\mathbb{Z}_{31}$ of the lines $y = 4x + 20$ and $y = 4x + 21$ with the elliptic curve $y^2 = x^3 + 25x + 10$.

Problem 10.3
Find the line that is tangent to the elliptic curve $y^2 = x^3 + 11x^2 + 17x + 25$ over $\mathbb{Z}_{31}$ in the point $(2, 7)$. Where else does this line intersect the curve?

Problem 10.4 [M]
Consider the elliptic curve $\mathcal{E}$ defined by $y^2 = x^3 + 11x^2 + 17x + 25$ over $\mathbb{Z}_{31}$.
Check that the points $P = \{12, 10\}$ and $Q = \{25, 14\}$ lie on $\mathcal{E}$. What is $-P$? Compute the sum of P and Q without using the *Mathematica* procedure presented in Subsection 10.3.

Problem 10.5
Consider an elliptic curve $\mathcal{E}$. Let P on $\mathcal{E}$ have order n. What is the order of $-P$?

Problem 10.6 [M]
Consider (again) the elliptic curve $\mathcal{E}$ defined by $y^2 = x^3 + 11x^2 + 17x + 25$ over $\mathbb{Z}_{31}$.
Determine the orders of $P = \{27, 10\}$ and $Q = \{24, 28\}$. What can you conclude about the cardinality of $\mathcal{E}$ (hint: use Theorem B.5)?
What is the cardinality of $\mathcal{E}$ (hint: use Theorem 10.1)?
Construct a point of maximal order from P and Q.

Problem 10.7 [M]
Duplicate Example 10.6 for the elliptic curve $\mathcal{E}$ over $\mathbb{Z}_{523}$ defined by the equation $y^2 = x^3 + 111x^2 + 11x + 1$. Use for P a point of order at least one hundred.

11   Coding Theory Based Systems

11.1   Introduction to Goppa codes

In this chapter it is assumed that the reader is familiar with algebraic coding theory. A reader without this background can freely skip this chapter and continue with Chapter 12. From [MacWS77] we recall the following facts about Goppa codes.

> **Theorem 11.1**
> Let $G(x)$ be any irreducible polynomial of degree t over GF (2^m). Then the set
>
> $$\Gamma(G(x), GF(2^m)) = \left\{ (c_\omega)_{\omega \in GF(2^m)} \in \{0, 1\}^n \;\middle|\; \sum_{\omega \in GF(2^m)} \frac{c_\omega}{x - \omega} \equiv 0 \,(\mathrm{mod}\, G(x)) \right\} \qquad (11.1)$$
>
> defines a binary *Goppa code* of length $n = 2^m$, *dimension* $k \geq n - t.m$ and *minimum distance* $d \geq 2t + 1$.
> A fast *decoding algorithm* with running time $n.t$, exists (see [Patt75]).

Note that we have used the elements in GF(2^m) as an index set for the coordinates of the vectors in $\{0, 1\}^n$. The notions used above mean that the elements in $\Gamma(G(x), GF(2^m))$ (which are called *codewords*) form a linear subspace in $\{0, 1\}^n$ of dimension at least $n - t.m$ and that different codewords differ in at least $2t + 1$ coordinates (one says that the *Hamming distance* $d_H (\underline{c}, \underline{c}')$ between different codewords is at least $2t + 1$).

A decoding algorithm will map any word in $\{0, 1\}^n$ that differs in at most t coordinates from a codeword $\underline{c}$ (which is unique by the triangle inequality) to that codeword. Hence, if a codeword $\underline{c}$ is transmitted and the received word $\underline{r}$ differs from $\underline{c}$ in no more than t coordinates ($d_H (\underline{c}, \underline{r}) \leq t$), the receiver is able to recover $\underline{c}$ from $\underline{r}$. For this reason, t is called the *error-correcting capability* of the code $\Gamma(G(x), GF(2^m))$.

Any $k \times n$ matrix of which the rows span a particular linear code is called a *generator matrix* of that code. It follows from this definition that the code can be described by

$$\{ \underline{m}.G \mid \underline{m} \in \{0, 1\}^k \}. \qquad (11.2)$$

Example 11.1 (Part 1)

Let α be the primitive element in GF(2^4) satisfying $\alpha^4 + \alpha^3 + 1 = 0$. After having loaded the Mathematica package Algebra`FiniteFields` *we can generate the log table of GF(2^4) with the functions* MatrixForm *and* PowerList.

```
<<Algebra`FiniteFields`
```

```
MatrixForm[PowerList[GF[2, {1, 0, 0, 1, 1}]]]
```

$$\begin{pmatrix} 1 & 0 & 0 & 0 \\ 0 & 1 & 0 & 0 \\ 0 & 0 & 1 & 0 \\ 0 & 0 & 0 & 1 \\ 1 & 0 & 0 & 1 \\ 1 & 1 & 0 & 1 \\ 1 & 1 & 1 & 1 \\ 1 & 1 & 1 & 0 \\ 0 & 1 & 1 & 1 \\ 1 & 0 & 1 & 0 \\ 0 & 1 & 0 & 1 \\ 1 & 0 & 1 & 1 \\ 1 & 1 & 0 & 0 \\ 0 & 1 & 1 & 0 \\ 0 & 0 & 1 & 1 \end{pmatrix}$$

Consider the binary Goppa code $\Gamma(G(x), GF(2^4))$ of length 16 defined by $G(x) = x^2 + x + \alpha$. That $G(x)$ is indeed an irreducible polynomial over $GF(2^4)$ can easily be checked with the Mathematica functions GF, Table, *and* TableForm *because it suffices to show that $G(x)$ has no linear factors.*

```
f16 = GF[2, {1, 0, 0, 1, 1}];
x = f16[{0, 1}];
a = f16[{0, 1}];
G[x_] := x^2 + x + a;
G[0]
Table[{i, G[x^i]}, {i, 0, 14}] // TableForm
```

```
{0, 1, 0, 0}₂
```

```
0       {0, 1, 0, 0}₂
1       {0, 0, 1, 0}₂
2       {1, 1, 1, 1}₂
3       {1, 0, 1, 0}₂
4       {1, 0, 1, 0}₂
5       {1, 1, 0, 0}₂
6       {0, 1, 1, 1}₂
7       {1, 0, 0, 1}₂
8       {0, 1, 1, 1}₂
9       {1, 1, 1, 1}₂
10      {1, 1, 0, 0}₂
11      {0, 0, 0, 1}₂
12      {0, 0, 1, 0}₂
13      {1, 0, 0, 1}₂
14      {0, 0, 0, 1}₂
```

To determine the inverses $1/(x - \omega)$ (mod $x^2 + x + \alpha$) in (11.1) we use the Mathematica package
Algebra`PolynomialExtendedGCD`

```
<<Algebra`PolynomialExtendedGCD`
```

and the Mathematica function PolynomialExtendedGCD. *For instance, $1/(x - \alpha^3)$ (mod $x^2 + x + \alpha$) can be found by*

```
x =.;
PolynomialExtendedGCD[x - a³, x² + x + a]
```

```
{1, {{0, 1, 0, 1}₂ + x {1, 1, 1, 1}₂, {1, 1, 1, 1}₂}}
```

With the logarithm table above we can rewrite these coefficients as follows:

$$0.1 + 1.\,\alpha + 0.\,\alpha + 1\,\alpha^3 = \alpha^{10},$$

$$1.1 + 1.\,\alpha + 1.\,\alpha^2 + 1.\,\alpha^3 = \alpha^6.$$

It follows from (A.8) that

$$(x - \alpha^3).(\alpha^{10} + \alpha^6\,x) + \alpha^6.G(x) = 1,$$

i.e. $1/(x - \alpha^3) = \alpha^{10} + \alpha^6\,x$. *This can be checked with the Mathematica function* PolynomialMod

```
Clear[x];
PolynomialMod[(x - a³) (a¹⁰ + a⁶ x), x² + x + a]
```

```
{1, 0, 0, 0}₂
```

We express all the inverses $1/(x-\omega)$, $\omega \in GF(2^4)$, in this way as polynomials $g_0^{(\omega)} + g_1^{(\omega)} x$, by means of

```
Clear[x];
PolynomialExtendedGCD[x, x² + x + a]
Do[Print[PolynomialExtendedGCD[x - aⁱ, x² + x + a]], {i, 0, 14}]
```

```
{1, {{0, 0, 1, 1}₂ + x {0, 0, 1, 1}₂, {0, 0, 1, 1}₂}}
```

$\quad$ {1, {x {0, 0, 1, 1}₂, {0, 0, 1, 1}₂}}

$\quad$ {1, {{0, 1, 0, 1}₂ + x {0, 1, 1, 0}₂, {0, 1, 1, 0}₂}}

$\quad$ {1, {{0, 0, 0, 1}₂ + x {1, 0, 1, 0}₂, {1, 0, 1, 0}₂}}

$\quad$ {1, {{0, 1, 0, 1}₂ + x {1, 1, 1, 1}₂, {1, 1, 1, 1}₂}}

$\quad$ {1, {{1, 0, 1, 0}₂ + x {1, 1, 1, 1}₂, {1, 1, 1, 1}₂}}

$\quad$ {1, {x {0, 0, 0, 1}₂ + {0, 1, 1, 0}₂, {0, 0, 0, 1}₂}}

$\quad$ {1, {{1, 0, 0, 0}₂ + x {1, 1, 1, 0}₂, {1, 1, 1, 0}₂}}

$\quad$ {1, {{1, 0, 1, 0}₂ + x {1, 0, 1, 1}₂, {1, 0, 1, 1}₂}}

$\quad$ {1, {{0, 1, 1, 0}₂ + x {1, 1, 1, 0}₂, {1, 1, 1, 0}₂}}

$\quad$ {1, {x {1, 0, 1, 0}₂ + {1, 0, 1, 1}₂, {1, 0, 1, 0}₂}}

$\quad$ {1, {x {0, 0, 0, 1}₂ + {0, 1, 1, 1}₂, {0, 0, 0, 1}₂}}

$\quad$ {1, {{1, 0, 1, 1}₂ + x {1, 1, 0, 0}₂, {1, 1, 0, 0}₂}}

$\quad$ {1, {{0, 0, 1, 1}₂ + x {0, 1, 1, 0}₂, {0, 1, 1, 0}₂}}

$\quad$ {1, {{0, 0, 0, 1}₂ + x {1, 0, 1, 1}₂, {1, 0, 1, 1}₂}}

$\quad$ {1, {{0, 1, 1, 1}₂ + x {1, 1, 0, 0}₂, {1, 1, 0, 0}₂}}

and put them as columns $\begin{pmatrix} g_0^{(\omega)} \\ g_1^{(\omega)} \end{pmatrix}$ in a 2×16 matrix H. Note that $1/(x-\alpha^3)$ appears as $\begin{pmatrix} \alpha^{10} \\ \alpha^6 \end{pmatrix}$ in column 5, because the first column corresponds to $\omega = 0$, the second column has index $\omega = 1$, etc.

$$H = \begin{pmatrix} \alpha^{14} & 0 & \alpha^{10} & \alpha^3 & \alpha^{10} & \alpha^9 & \alpha^{13} & 1 & \alpha^9 & \alpha^{13} & \alpha^{11} & \alpha^8 & \alpha^{11} & \alpha^{14} & \alpha^3 & \alpha^8 \\ \alpha^{14} & \alpha^{14} & \alpha^{13} & \alpha^9 & \alpha^6 & \alpha^6 & \alpha^3 & \alpha^7 & \alpha^{11} & \alpha^7 & \alpha^9 & \alpha^3 & \alpha^{12} & \alpha^{13} & \alpha^{11} & \alpha^{12} \end{pmatrix}$$

Here, we have made use of the log table of $GF(2^4)$, computed earlier.

The defining equation in (11.1) can be rewritten as

$$\sum_{\omega \in GF(2^4)} c_\omega (g_0^{(\omega)} + g_1^{(\omega)} x) \equiv 0 \, (mod \, x^2 + x + \alpha),$$

or, equivalently, as

$$\left(\sum\nolimits_{\omega \in GF(2^4)} c_\omega \, g_0^{(\omega)}\right) + \left(\sum\nolimits_{\omega \in GF(2^4)} c_\omega \, g_1^{(\omega)}\right) x \equiv 0 \, (mod \, x^2 + x + \alpha).$$

So, we have two linear equations for $\underline{c} = (c_\omega)_{\omega \in GF(2^4)}$:

$$\sum\nolimits_{\omega \in GF(2^4)} c_\omega \, g_0^{(\omega)} = 0 \qquad and \qquad \sum\nolimits_{\omega \in GF(2^4)} c_\omega \, g_1^{(\omega)} = 0.$$

These two equations can be efficiently denoted by

$$H.\underline{c}^T = \underline{0}^T.$$

Expressing each power of α *as binary linear combination of 1,* α, α^2, *and* α^3 *(or using the output of the PolynomialExtendedGCD-calculations directly) gives the 8×16 binary matrix* H':

$$H' = \begin{pmatrix} 0 & 0 & 0 & 0 & 0 & 1 & 0 & 1 & 1 & 0 & 1 & 0 & 1 & 0 & 0 & 0 \\ 0 & 0 & 1 & 0 & 1 & 0 & 1 & 0 & 0 & 1 & 0 & 1 & 0 & 0 & 0 & 1 \\ 1 & 0 & 0 & 0 & 0 & 1 & 1 & 0 & 1 & 1 & 1 & 1 & 1 & 1 & 0 & 1 \\ 1 & 0 & 1 & 1 & 1 & 0 & 0 & 0 & 0 & 0 & 1 & 1 & 1 & 1 & 1 & 1 \\ 0 & 0 & 0 & 1 & 1 & 1 & 0 & 1 & 1 & 1 & 1 & 0 & 1 & 0 & 1 & 1 \\ 0 & 0 & 1 & 0 & 1 & 1 & 0 & 1 & 0 & 1 & 0 & 0 & 1 & 1 & 0 & 1 \\ 1 & 1 & 1 & 1 & 1 & 1 & 0 & 1 & 1 & 1 & 1 & 0 & 0 & 1 & 1 & 0 \\ 1 & 1 & 0 & 0 & 1 & 1 & 1 & 0 & 1 & 0 & 0 & 1 & 0 & 0 & 1 & 0 \end{pmatrix}.$$

So, another way to describe $\Gamma(x^2 + x + \alpha, \, GF(2^4))$ *is*

$$C = \{\underline{c} \in \{0, 1\}^{16} \mid H'.\underline{c}^T = \underline{0}^T\}.$$

It is not difficult to check that C *is a binary, linear code of length 16, dimension 7 and minimum distance 5.*

We call a matrix H whose nullspace is a particular linear code C a *parity check matrix* of C. We write

$$C = \{\underline{c} \in \{0, 1\}^n \mid H.\underline{c}^T = \underline{0}^T\}. \tag{11.3}$$

The *syndrome* of a received vector $\underline{r}$ is defined by: $\underline{s}^T = H.\underline{r}^T$.

The number of irreducible polynomials of degree t over $GF(2^m)$ is about $2^{m.t}/t$ (see Corollary B.18). So, a randomly selected polynomial of degree t over $GF(2^m)$ will be irreducible with probability $1/t$. Since fast algorithms for testing irreducibility (see [Ber168], Ch. 6 or [Rabi80]) exist, one can find an irreducible polynomial of degree t over $GF(2^m)$, just like in Algorithm 9.3, by repeatedly guessing and testing.

11.2   The McEliece Cryptosystem

Based on the theory of error-correcting codes, McEliece [McE178] proposed the following secrecy system.

11.2.1 The System

□ **Setting Up the System**

1) Each user U chooses a suitable Goppa code of length $n_U = 2^{m_U}$ and with error-correcting capability t_U. To this end, user U selects a random, irreducible polynomial $p_U(x)$ of degree t_U over $GF(2^{m_U})$ and makes a generator matrix G_U of the corresponding Goppa code $\Gamma(p_U(x), GF(2^{m_U}))$. The size of G_U is $k_U \times n_U$.

2) User U chooses a random, dense $k_U \times k_U$ nonsingular matrix S_U and a random $n_U \times n_U$ permutation matrix P_U and computes

$$G_U^* = S_U \, G_U \, P_U. \qquad (11.4)$$

3) User U makes G_U^* and t_U public, but keeps G_U, S_U, and P_U secret.

□ **Encryption**

Suppose that user Alice wants to send a message to user Bob. She looks up Bob's publicly known parameters G_B^* (of size $k_B \times n_B$) and t_B represents her message as a binary string $\underline{m}$ of length k_B. Next Alice chooses a random vector $\underline{e}$ (error pattern) of length n_B with at most t_B coordinates are equal to 1. As encryption of $\underline{m}$ Alice sends to Bob

$$\underline{r} = \underline{m}.G_B^* + \underline{e}. \qquad (11.5)$$

(One usually says: the weight of $\underline{e}$ is at most t_B, denoted by $w_H(\underline{e}) \le t_B$, where the *weight* function w counts the number of non-zero coordinates in a vector.)

□ **Decryption**

Upon receiving $\underline{c}$, Bob computes with his secret permutation matrix P_B

$$\underline{r}.P_B^{-1} \overset{(11.5)}{=} \underline{m}.G_B^*(P_B)^{-1} + \underline{e}(P_B)^{-1} \overset{(11.4)}{=} \underline{m}.S_B\, G_B\, P_B\, P_B^{-1} + \underline{e}' = (\underline{m}.S_B)\, G_B + \underline{e}'.$$

where $\underline{e}' = \underline{e}.P_B^{-1}$ is a permutation of $\underline{e}$, so it also has weight $(\le t)_B$. With the decoding algorithm of the Goppa code $\Gamma(p_U(x), GF(2^{m_U}))$ Bob can efficiently decode $\underline{r}.P_B^{-1}$. He will find $\underline{e}'$ as error pattern and can retrieve $\underline{m}.S_B$. Multiplication of this expression on the right with S_B^{-1} (known to Bob) yields the originally transmitted message $\underline{m} \in \{0, 1\}^{k_B}$.

11.2.2 Discussion

□ **Summary and Proposed Parameters**

The McEliece cryptosystem introduced in the previous section can be summarized as follows.

Public	G_U^* and t_U of all users U
	G_U^* has size $k_U \times n_U$
Secret	$p_U(x)$, S_U, and P_U by each user U
Property	$S_U^{-1}\,G_U^*\,P_U$ is the generator
	matrix of the Goppa code
	defined by $p_U(x)$ of degree t_U
Format of message of Ann to Bob	$\underline{m} \in \{0,\,1\}^{k_B}$
Encryption	$\underline{c} = \underline{m}.G_B^* + \underline{e}$,
	weight of $\underline{e}$ is $\le t_B$
Decryption	compute $\underline{c}' = \underline{c}.P_B^{-1}$
	decode $\underline{c}'$ to find $\underline{m}' = \underline{m}.S_B$
	compute $\underline{m}'.S_B^{-1} = \underline{m}$

The McEliece cryptosystem

Table 11.1

The reason that an error pattern e is introduced in (11.5), is of course to make it impossible for the cryptanalist to retrieve $\underline{m}$ from $\underline{c}$ by a straightforward Gaussian elimination process.

McEliece suggests in his original proposal [McEl78] to take $m_B = 10$ (so $n_B = 1024$) and $t_B = 50$ (so $k_B \approx 1025 - 50 \times 10 = 524$).

□ **Heuristics of the Scheme**

The heuristics behind this scheme are not difficult to guess. Take a sufficiently long, binary, linear block code, that can correct a large number, say t, of errors and for which an efficient decoding algorithm exists. The code should belong to a large class of codes, making it impossible to guess which particular code has been selected. Let n be the length of the code and k its dimension. Manipulate the generator matrix to such an extent, that the resulting matrix looks like a random $k \times n$ matrix of full rank. The decoding complexity of a randomly generated code with these parameters should be infeasible. In the next section the complexity of several decoding methods will be discussed.

In [BerMT77] it is shown that the general decoding problem of linear codes, i.e. how to find the closest codeword to any word of length n, is *NP-complete*. We shall not explain what this notion means exactly. We refer the interested reader to [GarJ79].

Here, it suffices to know that this characterization implies that no known algorithm can decode an arbitrary word to its closest codeword neighbor in a running time that depends in a polynomial way on the size of the input.

Moreover, if one were to find such an algorithm, it could be adapted to solve a large class of equally hard problems.

□ **Not a Signature Scheme**

The encryption function of the McEliece cryptosystem maps binary k-tuples to binary n-tuples. This mapping is not surjective. Indeed, for the proposed parameter set the number of vectors of length 1024 at distance ≤ 50 to a codeword is

$$2^k \sum_{i=0}^{50} \binom{n}{i} \approx 2^{524} \sum_{i=0}^{50} \binom{1024}{i} \approx 2^{808.4}.$$

which is an ignorable fraction of the total number of 1024-length words. So, the (secret) function S_U mentioned in Property PK4 (in Subsection 7.1.1) is not defined for most words in $\{0, 1\}^n$. Consequently, the McEliece system can not be turned into a signature scheme. See, also Table 7.2.

11.2.3 Security Aspects

We shall now discuss the security of the McEliece cryptosystem by analyzing four possible attacks on the specific parameters that McEliece suggests. (The most powerful attack at this moment seems to be [CanS98].)

```
n = 1024; k = 524; t = 50;
```

□ **Guessing S_B and P_B**

As a cryptanalist, one may try to guess S_B and P_B to calculate G_B from G_B^* by means of (11.4). Once G_B has been recovered, it is not so difficult for the cryptanalist to find the defining Goppa polynomial $p_U(x)$ of the Goppa code $\Gamma(p_U(x), GF(2^{m_U}))$ that has G_B as generator polynomial. One can now follow the decryption algorithm of Bob to find the transmitted message $\underline{m}$.

However the number of invertible matrices S_B and permutation matrices P_B is so astronomical ($\prod_{i=0}^{k-1} (2^k - 2^i)$ resp. $n!$), that the probability of success of this attack is smaller than the probability of correctly guessing vector $\underline{m}$ directly.

□ **Exhaustive Codewords Comparison**

The cryptanalist can compare the received vector $\underline{r}$ with all 2^k codewords in the code generated by G_B^*. Let $\underline{c}$ be the closest codeword. It is at distance $\leq t$ from $\underline{r}$ (by the encryption rule (11.5)) and is unique because the minimum distance of the code is at least $2t + 1$. It also follows from (11.5) that $\underline{c} = \underline{m}.G_B^*$. With a simple Gaussian elimination process one can now retrieve the transmitted message $\underline{m}$ from $\underline{c}$.

This approach involves the following number of comparisons!

```
N[2^k, 5]
```

```
5.4918×10^157
```

Example 11.2 (Part 1)

Consider the binary code of length n = 7 and dimension k = 4, generated by

$$G = \begin{pmatrix} 1 & 0 & 0 & 0 & 1 & 1 & 0 \\ 0 & 1 & 0 & 0 & 1 & 0 & 1 \\ 0 & 0 & 1 & 0 & 0 & 1 & 1 \\ 0 & 0 & 0 & 1 & 1 & 1 & 1 \end{pmatrix};$$

and suppose that $\underline{r} = (1, 1, 0, 0, 1, 0, 1)$ is a intercepted ciphertext which is a codeword $\underline{c}$ plus an error vector of weight at most 1 (so t = 1).

We shall compare $\underline{r}$ with two codewords (instead of $2^k = 16$) and use again the <u>Mod</u> function:

```
r = {1, 1, 1, 0, 1, 0, 1};
i1 = {1, 1, 1, 1};
c = Mod[i1.G, 2]
Mod[r - c, 2]
```

```
{1, 1, 1, 1, 1, 1, 1}
```

```
{0, 0, 0, 1, 0, 1, 0}
```

So, $\underline{c} = \underline{i_1}.G$ lies at distance ≥ 2 from $\underline{r}$, which is too much.

```
i2 = {1, 0, 1, 0};
c = Mod[i2.G, 2]
Mod[r - c, 2]
```

```
{1, 0, 1, 0, 1, 0, 1}
```

```
{0, 1, 0, 0, 0, 0, 0}
```

Now $\underline{c} = \underline{i}_2.G$ lies at distance 1 from $\underline{r}$ and we conclude that (1, 0, 1, 0) was the transmitted information.

□ **Syndrome Decoding**

The cryptanalist may compute the parity check matrix H_B^* corresponding to G_B^* from the equation $H_B^*.G_B^* = O$ (see (11.3)). It has rank $n - k$. Next, generate all error vectors $\underline{e}$ of weight at most t, compute the syndrome $H_B^* \underline{e}^T$ for each of them, and put these in a table.

For the intercepted vector $\underline{r}$ one first computes the syndrome $\underline{s}^T = H.\underline{r}^T$. From the table one can find the corresponding error vector $\underline{e}$. Subtracting $\underline{e}$ from $\underline{r}$ one gets the codeword $\underline{c} = \underline{m}.G_B^*$ (see (11.5)). With a simple Gaussian elimination process one can now retrieve the transmitted message $\underline{m}$ from this vector $\underline{c}$.

The work load of this attack is $\sum_{i=0}^{50} \binom{n}{i}$:

```
N[∑_{i=0}^{50} Binomial[n, i], 5]
```

```
3.3623 × 10^85
```

Example 11.2 (Part 2)

The parity check matrix of the code introduced in Example 11.2 is given by

```
H = ( 1 1 0 1 1 0 0 )
    ( 1 0 1 1 0 1 0 );
    ( 0 1 1 1 0 0 1 )

MatrixForm[H]
```

$$\begin{pmatrix} 1 & 1 & 0 & 1 & 1 & 0 & 0 \\ 1 & 0 & 1 & 1 & 0 & 1 & 0 \\ 0 & 1 & 1 & 1 & 0 & 0 & 1 \end{pmatrix}$$

as can be checked with the Mathematica function Transpose *(and* MatrixForm*) as follows*

```
U = Mod[G.Transpose[H], 2];
MatrixForm[U]
```

$$\begin{pmatrix} 0 & 0 & 0 \\ 0 & 0 & 0 \\ 0 & 0 & 0 \\ 0 & 0 & 0 \end{pmatrix}$$

Next, we generate all error vectors $\underline{e}$ *of weight* ≤ 1 *and compute their syndrome* $H_B^* \, \underline{e}^T$. *We put these in a table. Apart from the Mathematica functions* Mod, Do, *and* Print, *we also make use of* ReplacePart, *which replaces the i-th coordinate of* $\underline{e}$ *by the specified value (here its compliment).*

```
e = {0, 0, 0, 0, 0, 0, 0};
Print[e, "     ", Mod[H.e, 2]];
Do[ {er = ReplacePart[e, Mod[e[[i]] + 1, 2], i],
         Print[er, "     ", Mod[H.er, 2]]},
      {i, 1, 7}]
```

{0, 0, 0, 0, 0, 0, 0}	{0, 0, 0}
{1, 0, 0, 0, 0, 0, 0}	{1, 1, 0}
{0, 1, 0, 0, 0, 0, 0}	{1, 0, 1}
{0, 0, 1, 0, 0, 0, 0}	{0, 1, 1}
{0, 0, 0, 1, 0, 0, 0}	{1, 1, 1}
{0, 0, 0, 0, 1, 0, 0}	{1, 0, 0}
{0, 0, 0, 0, 0, 1, 0}	{0, 1, 0}
{0, 0, 0, 0, 0, 0, 1}	{0, 0, 1}

With this table it is now easy to find a codeword at distance ≤ 1 *from* $\underline{r}$.

```
r = {1, 1, 1, 0, 1, 0, 1};
Mod[H.r, 2]
```

```
{1, 0, 1}
```

This is the syndrome corresponding to $\underline{e} = (0, 1, 0, 0, 0, 0, 0)$, so the closest codeword is given by

```
e = {0, 1, 0, 0, 0, 0, 0};
Mod[r - e, 2]
```

```
{1, 0, 1, 0, 1, 0, 1}
```

Since the generator matrix G in this example has the form $(I_4 \mid P)$, we can recover the transmitted information $\underline{m}$ from the first four coordinates in $\underline{c}$:

$$\underline{m} = (1, 0, 1, 0).$$

□ Guessing k Correct and Independent Coordinates

The cryptanalist selects k random positions and hopes that they are not in error, i.e. he hopes that $\underline{e}$ is zero on these k positions. If the restriction of matrix G_B^* to these k positions still has rank k, one can find a candidate $\underline{m}'$ for the information vector $\underline{m}$ with a Gaussian elimination process.

If the rank is less than k it will very likely still be close to k (see Problem 11.2). So, the Gaussian elimination process will either lead to only a few possibilities for $\underline{m}'$ or to no solution at all.

For each possible candidate $\underline{m}'$ compute $\underline{m}'.G_B^*$ and check if it lies at distance $\leq t$ from the intercepted vector $\underline{r}$. If so, one has found the correct $\underline{m}$.

The probability that the k positions are correct is about $(1 - t/n)^k$. The Gaussian elimination process involves k^3 steps. So, the expected workload of this method is

```
N[k^3 (1 - t/n)^-k, 5]
```

```
3.5504×10^19
```

Although this attack is the most efficient thus far, it is still not a feasible attack.

Example 11.2 (Part 3)

Guessing that coordinates 2, 4, 5, and 7 are error-free in Example 11.2 we use the Mathematica functions `Transpose` *and* `MatrixForm` *to get the restriction G' of the generator matrix G to this guess and the restriction $\underline{r}'$ of the intercepted vector $\underline{r}$ of Example 11.2 to this guess.*

```
     ( 1  0  0  0  1  1  0 )
     | 0  1  0  0  1  0  1 |
G =  | 0  0  1  0  0  1  1 | ;
     ( 0  0  0  1  1  1  1 )
Guess = {2, 4, 5, 7}
RestrG = Transpose[G][[Guess]];
MatrixForm[Transpose[RestrG]]
```

```
{2, 4, 5, 7}
```

```
( 0  0  1  0 )
| 1  0  1  1 |
| 0  0  0  1 |
( 0  1  1  1 )
```

```
r = {1, 1, 1, 0, 1, 0, 1};
rRestr = r[[Guess]]
```

```
{1, 0, 1, 1}
```

We use the Mathematica functions LinearSolve, NullSpace, *and* Transpose *to see if the equation*

```
LinearSolve[RestrG, rRestr, Modulus -> 2]
NullSpace[RestrG, Modulus -> 2]
```

```
{0, 1, 0, 0}
```

```
{}
```

has a solution.

Apparently the restriction of G to the four coordinates has full rank. The solution $(0, 1, 0, 0)$ gives rise to a codeword that has distance ≥ 2 to $\underline{r}$.

```
m1 = {0, 1, 0, 0};
Mod[r - m1.G, 2]
```

```
{1, 0, 1, 0, 0, 0, 0}
```

Let us now try another guess.

```
Guess = {1, 3, 6, 7};
RestrG = Transpose[G] [[ Guess]] ;
MatrixForm[Transpose[RestrG]]
```

$$\begin{pmatrix} 1 & 0 & 1 & 0 \\ 0 & 0 & 0 & 1 \\ 0 & 1 & 1 & 1 \\ 0 & 0 & 1 & 1 \end{pmatrix}$$

```
r = {1, 1, 1, 0, 1, 0, 1};
rRestr = r[[Guess]]
```

```
{1, 1, 0, 1}
```

```
LinearSolve[RestrG, rRestr, Modulus -> 2]
NullSpace[RestrG, Modulus -> 2]
```

```
{1, 0, 1, 0}
```

```
{}
```

The solution (1, 0, 1, 0) now turns out to generate a codeword at distance ≤ 1 to r.

```
m = {1, 0, 1, 0};
Mod[r - m.G, 2]
```

```
{0, 1, 0, 0, 0, 0, 0}
```

We conclude that (1, 0, 1, 0) was the transmitted information.

To let *Mathematica* make guesses one first has to load the package
`DiscreteMath`Combinatorica``

```
<<DiscreteMath`Combinatorica`
```

and one can then use the *Mathematica* function RandomKSubset.

```
RandomKSubset[{1, 2, 3, 4, 5, 6, 7}, 4]
```

```
{2, 3, 4, 6}
```

□ **Multiple Encryptions of the Same Message**

It is not safe to encrypt the same message several times with the same encryption matrix G_B. To see this, let us consider two different encryptions of the same message $\underline{m}$, say $\underline{r} = \underline{m}.G_B^* + \underline{e}$ and $\underline{r}' = \underline{m}.G_B^* + \underline{e}'$ (see (11.5)). On the coordinates where $\underline{r}$ and $\underline{r}'$ disagree, we know for sure that either $\underline{e}$ or $\underline{e}'$ has a 1. On the coordinates where $\underline{r}$ and $\underline{r}'$ agree, we know almost for sure that both $\underline{r}$ and $\underline{r}'$ are error-free.

To be more precise, if the error vectors $\underline{e}$ and $\underline{e}'$ are truly randomly chosen, as they should be, one expects the following values

(e_i, e_i')	# coordinates
$(0, 0)$	$(n - t)^2 / n$
$(0, 1)$ or $(1, 0)$	$2 t (n - t) / n$
$(1, 1)$	t^2 / n

For instance, when the parameters are $n = 1024$ and $t = 50$, one expects $e_i = e_i' = 1$ on roughly $50^2 / 1024 \approx 2.44$ coordinates.

Also, one expects

```
n = 1024; t = 50;
N[(n - t)^2 / n, 3]
```

```
926.
```

coordinates where $\underline{r}$ and $\underline{r}'$ agree. At most three of these coordinates are likely to be corrupted.

By removing in every possible way t^2 / n coordinates from the coordinate set where $\underline{r}$ and $\underline{r}'$ agree, one almost surely finds a coordinate set that is error free and on which the matrix G_B^* still has full rank (see Problem 11.2). With a simple Gaussian elimination process one recover $\underline{m}$ from $\underline{r}$.

When the same message has been encrypted more than two times, it is correspondingly easier to break the system.

11.2.4 A Small Example of the McEliece System

Example 11.1 (Part 2)

The Goppa code $\Gamma(x^2 + x + \alpha,\ GF(2^4))$ of Example 11.1 has a generator matrix G that can be computed from the parity check matrix H by means of the Mathematica function `Nullspace`.

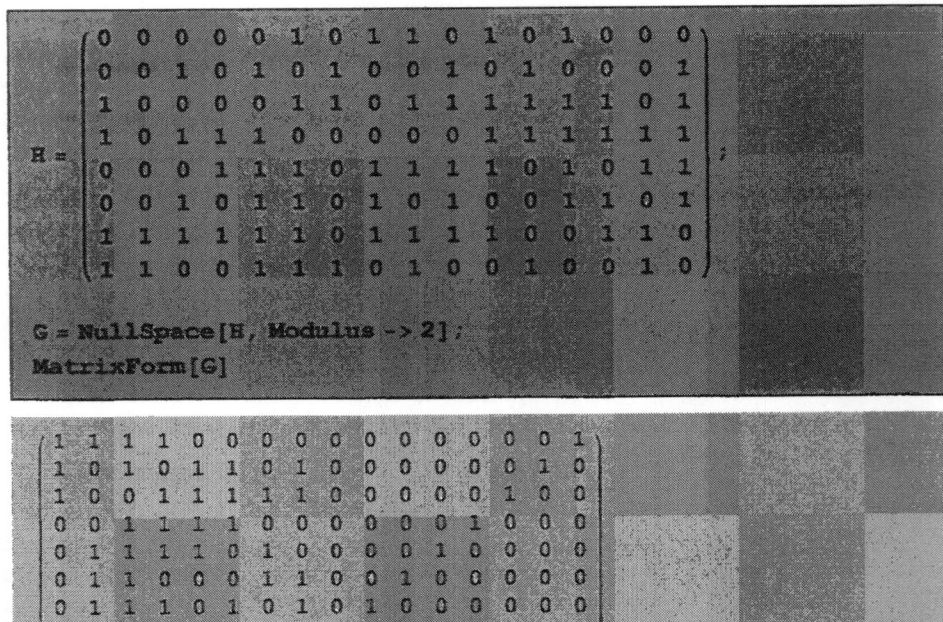

The generator matrix G of $\Gamma(x^2 + x + \alpha,\ GF(2^4))$ will be transformed into $G^* = S.G.P$, where S is an invertible matrix and P a permutation matrix, as follows:

```
      ( 1  1  1  1  1  1  1  1 )
      | 0  1  1  1  1  1  1  1 |
      | 0  0  1  1  1  1  1  1 |
      | 0  0  0  1  1  1  1  1 |
  S = | 0  0  0  0  1  1  1  1 | ;
      | 0  0  0  0  0  1  1  1 |
      | 0  0  0  0  0  0  1  1 |
      ( 0  0  0  0  0  0  0  1 )


      ( 1  0  0  0  0  0  0  0  0  0  0  0  0  0  0  0 )
      | 0  0  1  0  0  0  0  0  0  0  0  0  0  0  0  0 |
      | 0  1  0  0  0  0  0  0  0  0  0  0  0  0  0  0 |
      | 0  0  0  0  1  0  0  0  0  0  0  0  0  0  0  0 |
      | 0  0  0  0  0  1  0  0  0  0  0  0  0  0  0  0 |
      | 0  0  0  1  0  0  0  0  0  0  0  0  0  0  0  0 |
      | 0  0  0  0  0  0  0  1  0  0  0  0  0  0  0  0 |
  P = | 0  0  0  0  0  0  0  0  1  0  0  0  0  0  0  0 | ;
      | 0  0  0  0  0  0  0  0  0  1  0  0  0  0  0  0 |
      | 0  0  0  0  0  0  1  0  0  0  0  0  0  0  0  0 |
      | 0  0  0  0  0  0  0  0  0  0  0  1  0  0  0  0 |
      | 0  0  0  0  0  0  0  0  0  0  0  0  1  0  0  0 |
      | 0  0  0  0  0  0  0  0  0  0  0  0  0  1  0  0 |
      | 0  0  0  0  0  0  0  0  0  0  0  0  0  0  1  0 |
      | 0  0  0  0  0  0  0  0  0  0  0  0  0  0  0  1 |
      ( 0  0  0  0  0  0  0  0  0  0  1  0  0  0  0  0 )

Gstar = Mod[S.G.P, 2];
MatrixForm[Gstar]
```

```
( 0  1  0  1  1  0  1  0  0  1  1  1  1  1  1  1 )
| 1  0  1  1  0  0  1  0  0  1  0  1  1  1  1  1 |
| 0  1  1  0  0  1  1  0  1  1  0  1  1  1  1  0 |
| 1  1  1  1  1  0  1  1  0  1  0  1  1  1  0  0 |
| 1  0  1  0  0  1  1  1  0  1  0  1  1  0  0  0 |
| 1  1  0  0  1  0  1  0  0  1  0  1  0  0  0  0 |
| 1  0  1  0  1  0  1  1  1  1  0  0  0  0  0  0 |
( 1  1  0  1  0  0  0  1  0  1  0  0  0  0  0  0 )
```

A possible encoding of the information sequence (1, 1, 0, 0, 1, 0, 0, 1) is given by

```
m = {1, 1, 0, 0, 1, 0, 0, 1};
err = {0, 0, 0, 0, 1, 0, 0, 0, 1, 0, 0, 0, 0, 0, 0, 0};
cw = Mod[m.Gstar + err, 2]
```

```
{1, 0, 0, 1, 0, 1, 1, 0, 1, 0, 1, 1, 1, 0, 0, 0}
```

Note that errors have been introduced at coordinates 5 and 9.

An eavesdropper has no efficient algorithm to find the information vector $\underline{m}$ from the word $\underline{cw}$.

The legitimate receiver will first compute $\underline{cd} = \underline{cw}.P^{-1}$ with the Mathematica function Inverse.

```
PInv = Inverse[P, Modulus -> 2];
cd = cw.PInv
```

```
{1, 0, 0, 0, 1, 1, 0, 1, 0, 1, 1, 1, 0, 0, 0, 1}
```

Next, this vector has to be decoded with a decoding algorithm of the Goppa code $\Gamma(x^2 + x + \alpha, GF(2^4))$. Such a method has not been discussed here. The outcome turns out to be the vector $\underline{m'} = \{1, 0, 0, 0, 1, 1, 1, 0\}$. This can be checked by computing $m'.G$ and compare that with $\underline{cd}$. The difference is an error vector $\underline{err'}$ of weight 2 which is exactly $\underline{err}.P^{-1}$.

```
mpr = {1, 0, 0, 0, 1, 1, 1, 0};
errpr = Mod[mpr.G - cd, 2]
err.PInv
```

```
{0, 0, 0, 1, 0, 0, 0, 1, 0, 0, 0, 0, 0, 0, 0, 0}
```

```
{0, 0, 0, 1, 0, 0, 0, 1, 0, 0, 0, 0, 0, 0, 0, 0}
```

To find $\underline{m}$, the legitimate receiver computes $\underline{m'}.S^{-1}$.

```
mpr = {1, 0, 0, 0, 1, 1, 1, 0};
InvS = Inverse[S, Modulus -> 2];
Mod[mpr.InvS, 2]
```

```
{1, 1, 0, 0, 1, 0, 0, 1}
```

This is indeed the original message.

11.3 Another Technique to Decode Linear Codes

A large research effort has been made in the past to find decoding algorithms for general linear codes. The McEliece cryptosystem has only intensified this quest. Most of these algorithms are of the type that was discussed before: find k coordinates where the generator matrix has full rank and where the received vector is error free. Such a technique is called *information set decoding*.

Here we describe a technique introduced by Van Tilburg [vTbu88] (see also [LeeB88]).

Algorithm 11.2   *Bit Swapping Technique*

Let G be the generator matrix of a binary code C of length n, dimension k, and minimum distance d.
Let $\underline{r} = \underline{c} + \underline{e}$ be a received vector, where $\underline{c} \in C$ (say $\underline{c} = \underline{m}.G$) and $\underline{e}$ has weight at most t, with $2t + 1 \leq d$.

Step 1: Apply suitable elementary row operations and a column permutation to G to bring G in so-called standard form i.e. $S.G.P = (I_k \mid A)$.
Put $\underline{r}' = \underline{r}.P$ and write $\underline{r}' = (\underline{r_1}', \underline{r_2}')$, where $\underline{r_1}'$ has length k.
Note that $\underline{r} = \underline{m}.G.P + \underline{e}.P = \underline{m}.S^{-1}(I_k \mid A) + \underline{e}'$, where $\underline{e}$ and $\underline{e}'$ have the same weight.

Step 2: Put $\underline{c}' = \underline{r_1}'.(I_k \mid A)$. The first k coordinates of $\underline{c}'$ and $\underline{r}'$ are identical.

Step 3: If $\underline{c}'$ and $\underline{r}'$ differ in at most t coordinates, conclude that the first k coordinates are error-free. Compute $\underline{m}$ from $\underline{r}' = \underline{m}.S^{-1}$ with Gaussian elimination.
Let the algorithm terminate.

Step 4: If $\underline{c}'$ and $\underline{r}'$ differ in more than t coordinates, pick a random row index i, $1 \leq i \leq k$, and column index j, $1 \leq j \leq n-k$, with $A_{i,j} \neq 0$. Construct a new matrix G from $(I_k \mid A)$ by interchanging the i-th and the $(k+j)$-th column of G (the i-th column of I_k is swapped with the j-th column of A).
Return to Step 1, but use there only elementary row operations with the i-th row to bring the matrix in standard form again.

Let us demonstrate one cycle of the above algorithm. We continue with Example 11.2.

Example 11.2 (Part 4)

$$
G = \begin{pmatrix}
1 & 0 & 0 & 0 & 1 & 1 & 0 \\
0 & 1 & 0 & 0 & 1 & 0 & 1 \\
0 & 0 & 1 & 0 & 0 & 1 & 1 \\
0 & 0 & 0 & 1 & 1 & 1 & 1
\end{pmatrix};
$$

```
MatrixForm[G]
r = {1, 1, 1, 0, 1, 0, 1}
```

$$\begin{pmatrix} 1 & 0 & 0 & 0 & 1 & 1 & 0 \\ 0 & 1 & 0 & 0 & 1 & 0 & 1 \\ 0 & 0 & 1 & 0 & 0 & 1 & 1 \\ 0 & 0 & 0 & 1 & 1 & 1 & 1 \end{pmatrix}$$

```
{1, 1, 1, 0, 1, 0, 1}
```

The matrix G is already in standard form. We also see that the first four coordinates of $\underline{r}$ lead to a codeword $\underline{c}'$ that has distance 2 to $\underline{r}$.

```
r1 = Take[r, 4]
cc = Mod[r1.G, 2]
Mod[r - cc, 2]
```

```
{1, 1, 1, 0}
```

```
{1, 1, 1, 0, 0, 0, 0}
```

```
{0, 0, 0, 0, 1, 0, 1}
```

To make a swap we pick $G_{2,5}$ as non-zero entry from columns 5-7 in G. We perform a swap of the 2-nd and 5-th column of G, by using the function:

```
ColumnSwap[B_, i_, j_] := Module[{U, V}, U = Transpose[B];
  V = U[[i]]; U[[i]] = U[[j]]; U[[j]] = V; Transpose[U]]
```

```
G1=ColumnSwap[G,2,5];
MatrixForm[G1]
```

$$\begin{pmatrix} 1 & 1 & 0 & 0 & 0 & 1 & 0 \\ 0 & 1 & 0 & 0 & 1 & 0 & 1 \\ 0 & 0 & 1 & 0 & 0 & 1 & 1 \\ 0 & 1 & 0 & 1 & 0 & 1 & 1 \end{pmatrix}$$

To bring this in systematic form we use the Mathematica function RowReduce.

```
G2 = RowReduce[G1, Modulus -> 2];
MatrixForm[G2]
```

$$\begin{pmatrix} 1 & 0 & 0 & 0 & 1 & 1 & 1 \\ 0 & 1 & 0 & 0 & 1 & 0 & 1 \\ 0 & 0 & 1 & 0 & 0 & 1 & 1 \\ 0 & 0 & 0 & 1 & 1 & 1 & 0 \end{pmatrix}$$

In order to analyze the complexity of the bit-swapping algorithm, we let $\Pr(l + u \mid l)$ denote the conditional probability that exactly $l + u$ of the first k positions of $\underline{e}$ are in error after a swap given that precisely l were in error before the swap ($u = -1, 0, 1$).

Let $a = \min\{t, k\}$. Then the following straightforward relations hold:

$$\Pr(l - 1 \mid l) = \frac{l}{k} \times \frac{n-k-l+l}{n-k}, \qquad \text{if } 1 \leq l \leq a, \tag{11.6}$$

$$\Pr(l + 1 \mid l) = \frac{k-l}{k} \times \frac{t-l}{n-k}, \qquad \text{if } 1 \leq l \leq a-1, \tag{11.7}$$

$$\Pr(l \mid l) = \begin{cases} 1 - \Pr(l-1 \mid l) - \Pr(l+1 \mid l), & \text{if } 1 \leq l \leq a-1, \\ 1 - \Pr(l-1 \mid l), & \text{if } l = a. \end{cases} \tag{11.8}$$

Example 11.3 (Part 1)

Consider a (binary) code with parameters $n = 23$, $k = 12$, and $t = 3$. Then $a = \min\{k, t\} = 3$. The values of $\Pr(l - 1 \mid l)$ and $\Pr(l + 1 \mid l)$ can be computed (and printed) from (11.6) and (11.7) with the Mathematica functions Min, Do, *and* Print.

```
n = 23; k = 12; t = 3;
a = Min[k, t];
PrDown[l_] := l * (n - k - t + 1) / (k * (n - k));
PrUp[l_] := (k - l) * (t - 1) / (k * (n - k));
Do[
    Print["Pr(", i - 1, "|", i, ")=", PrDown[i]], {i, a, 1, -1}];
Print["and"];
Do[Print["Pr(", i + 1, "|", i, ")=", PrUp[i]],
    {i, a - 1, 1, -1}]
```

$$\Pr(2 \mid 3) = \frac{1}{4}$$

$$\Pr(1 \mid 2) = \frac{5}{33}$$

$$\Pr(0 \mid 1) = \frac{3}{44}$$

and

$$\Pr(3 \mid 2) = \frac{5}{66}$$

$$\Pr(2 \mid 1) = \frac{1}{6}$$

Note that the probability of a successful swap gets smaller for smaller values of l.

Lemma 11.3

Let N_l, $1 \le l \le a$, denote the expected number of swaps needed to pass from a state with l errors to a state with $l - 1$ errors.

Then, the N_l's can be computed recursively by

$$N_a = \frac{1}{\Pr(a-1|a)} = \frac{1}{1 - \Pr(a|a)}, \qquad (11.9)$$

$$N_{l-1} = \frac{1 + \Pr(l|l-1) N_l}{\Pr(l-2|l-1)}. \qquad (11.10)$$

Proof:

The first equality in equation (11.9) follows directly from the definition of $\Pr(a-1|a)$. The second equality follows from (11.8).

To show (11.10), we note that from state $l-1$ there are three possible directions for the algorithm to follow:

i) with probability $\Pr(l-2|l-1)$ it goes to state $l-2$ in one step.

ii) with probability $\Pr(l-1|l-1)$ it stays in state $l-1$ and so one can expect the algorithm to reach state $l-2$ in $1 + N_{l-1}$ steps.

iii) With probability $\Pr(l|l-1)$ it goes back to state l and so one expects it to reach state $l-2$ in $1 + N_l + N_{l-1}$ steps.

The above proves the following recurrence relation

$$N_{l-1} = \Pr(l-2|l-1).1 + \Pr(l-1|l-1).\{1 + N_{l-1}\} + \Pr(l|l-1).\{1 + N_l + N_{l-1}\},$$

which reduces to (11.10) because $\Pr(l-2|l-1) = 1 - \Pr(l-1|l-1) - \Pr(l|l-1)$.

□

Note that in the calculations of N_l only probabilities of the form $\Pr(i-1|i)$ play a role.

Example 11.3 (Part 2)

Continuing with Example 11.3, we see that the values of N_l can be computed recursively with (11.9) and (11.10).

```
Numb[a] = 1 / PrDown[a];
Do[Numb[i - 1] = (1 + PrUp[i - 1] * Numb[i]) / PrDown[i - 1],
  {i, a, 2, -1}]
Do[Print["Numb(", i, ")=", Numb[i]], {i, a, 1, -1}]
```

Numb$(3) = 4$

Numb$(2) = \dfrac{43}{5}$

$$\text{Numb}(1) = \frac{1606}{45}$$

Theorem 11.4

The expected number of swaps for the bit swapping algorithm to find k error-free coordinates is given by

$$\sum_{j=1}^{a} \frac{\binom{k}{j}\binom{n-k}{t-j}}{\binom{n}{t}} \sum_{i=1}^{j} N_i. \tag{11.11}$$

Proof:

The expected number of steps to reach state 0 when one starts in state j, $1 \le j \le a$, is given by the expected number of steps to reach state $j-1$ from state j, plus the expected number of steps to reach state $j-2$ from state $j-1$, etc. This explains the inner sum in (11.11):

$$N_j + N_{j-1} + \dots + N_1.$$

The probability of starting in state j is equal to the probability that a randomly selected k tuple contains j errors. This probability is equal to the fraction of the number of t-tuples out of n that have intersection j with a given k-tuple (and intersection $t - j$ with the other $n - k$ positions). So, this probability is given by

$$\frac{\binom{k}{j}\binom{n-k}{t-j}}{\binom{n}{t}}.$$

Now, take the product of the two factors above and sum it over all values of j.

□

Example 11.3 (Part 3)

It follows from Theorem 11.4 that the expected number of swaps that are needed in a code with $n = 23$, $k = 12$, and $t = 3$ (as introduced in Example 11.3) to get 12 error-free coordinates is given by:

```
NS = Sum[
       j=1
        a
      (Binomial[k, j] * Binomial[n - k, t - j] / Binomial[n, t])
        
       Sum[Numb[l];
       l=1
        j
      ]

N[NS, 5]
```

The above bit swapping algorithms gives a significant improvement (also asymptotically) over the methods explained in Subsection 11.2.3. For the strongest result in this area we refer the reader to [BaKT99].

11.4   The Niederreiter Scheme

The Niederreiter scheme [Nied86] is a variation of the McEliece cryptosystem. It applies the very same idea to the parity check matrix of a linear code. The scheme is summarized in the Table 11.2 below.

So, again we have a Goppa code $\Gamma(p_U(x), GF(2^m))$, (see (11.1)) defined by user's U Goppa polynomial $p_U(x)$ over $GF(2^m)$ of degree t_U. Let H_U be a parity check matrix of this code. It has size $(n_U - k_U) \times n_U$, where k_U is the dimension of the code.

The code $\Gamma(p_U(x), GF(2^m))$ is t_u-error correcting which implies that every vector $\underline{y}$ of weight $(\leq t)_U$ has a unique syndrome $H_U.\underline{y}$. Existing decoding algorithms for Goppa codes find $\underline{y}$ efficiently from its syndrome.

Just like in the McEliece system, the structure of the Goppa code has to be hidden from the matrix H_U. This is done by computing

$$H_U^* = S_U H_U P_U, \tag{11.12}$$

where S_U is a $(n_U - k_U) \times (n_U - k_U)$ invertible matrix and P_U a permutation matrix of size n_U (see (11.4)).

The matrix H_U^* has to be made public, together with the value t_U.

If Alice wants to send a message to Bob, she looks up Bob's public parameters H_B^* and t_B. She represents her message by means of a (column) vector $\underline{m}$ of weight $\leq t_B$. She computes $\underline{y} = H_B^*.\underline{m}$ and sends that as her ciphertext to Bob.

Bob first multiplies $\underline{y}$ on the left with S_B^{-1}. He obtains $\underline{y}' = S_B^{-1} \underline{m} \equiv H_B (P_B \underline{m})$ by (11.12). Since $P_B \underline{m}$ is a permutation of $\underline{m}$, and thus also of weight $(\leq t)_B$, the decoding algorithm of Bob's Goppa code will find $\underline{m}' = P_B \underline{m}$ efficiently. The message $\underline{m}$ can now be recovered by multiplying $\underline{m}'$ on the left with P_B.

Public	H_U^* and t_U of all users U
	H_U^* has size $(n_U - k_U) \times n_U$
Secret	$p_U(x)$, S_U, and P_U by each user U
Property	$S_U^{-1} H_U^* P_U$ is the parity check
	matrix of the Goppa code
	defined by $p_U(x)$ of degree t_U
Format of message	$\underline{m} \in \{0, 1\}^{n_B}$
of Ann to Bob	weight $(\underline{m}) \le t_B$
Encryption	$\underline{v} = H_B^* . \underline{m}$
Decryption	compute $\underline{v}' = S_B^{-1} . \underline{v}$
	use decoding algorithm to
	find $\underline{m}'$ with $H_B^* . \underline{m}' = \underline{v}'$
	compute $\underline{m}' . P_B^{-1} = \underline{m}$

The Niederreiter cryptosystem

Table 11.2

11.5   Problems

Problem 11.1
What is the probability that k columns in a random $k \times n$ binary matrix have rank k? How about the probability that $k + 1$ columns in this matrix have rank?
Compute these two probabilities for $n = 16$ and $k = 5$.

Problem 11.2
Let C be a linear code of length $n = 23$ and dimension $k = 12$. Assume that at most three errors have occurred. What is the complexity of the various attacks described in Subsection 11.2.3.

Problem 11.3^M
Let C be a linear code of length 11 and dimension 6. Suppose that two errors have occurred. How many swaps are expected to get 6 error-free coordinates if one follows Algorithm 11.2?

12   Knapsack Based Systems

12.1   The Knapsack System

12.1.1 The Knapsack Problem

In [MerH78], Merkle and Hellman propose a public key cryptosystem that is based on the difficulty of solving the knapsack problem. Since then, other knapsack related cryptosystems have been suggested, most of which turned out to be insecure. An exception, up to now, is the Chor-Rivest scheme proposed in [ChoR85], but in [Vaud98] it is shown that the suggested parameters in [ChoR85] are also insecure.

> **Definition 12.1**
> Let $a_1, a_2, ..., a_n$ be a sequence of n positive integers. Let also S be an integer. The question if the equation
>
> $$x_1 a_1 + x_2 a_2 + ... + x_n a_n = S \qquad\qquad (12.1)$$
>
> has a solution with each x_i in $\{0, 1\}$ is called the *knapsack problem*.

Note that we do not ask for a solution of (12.1), the question is only if there exists a solution. Finding a $\{0, 1\}$-solution to (12.1) is of course at least as difficult as just finding out whether a solution exists.

For large n the knapsack problem is intractable to solve. In fact it has been shown that the knapsack problem is NP-complete (see [GarJ79] or a very short discussion in Subsection 11.2.2).

For some sequences $\{a_i\}_{i=1}^n$ it is not difficult to find a $\{0, 1\}$-solution to (12.1), resp. to show that no such solution exists. For example, with the sequence $a_i = 2^{i-1}, 1 \le i \le n$, equation (12.1) will have a solution if and only if $0 \le S \le 2^n - 1$. Finding the solution is very easy in this case.

A much more general class of sequences $\{a_i\}_{i=1}^n$ exists, for which (12.1) is easily solvable. This is the class of so-called super-increasing sequences.

A sequence $\{a_i\}_{i=1}^n$ is called *super-increasing*, if for all $1 \le k \le n$,

$$\sum_{i=1}^{k-1} a_i < a_k. \qquad\qquad (12.2)$$

Algorithm 12.1 solves the knapsack problem for super-increasing sequences. It actually finds the solution $\{x_i\}_{i=1}^n$ for each right hand side S for which (12.1) is solvable. The idea is very simple: since $\sum_{i=1}^{n-1} a_i < a_n$, it follows that in a solution

$$x_n = 1 \iff S \geq a_n.$$

Now, subtract $x_n a_n$ from S and determine x_{n-1} in the same way. So, recursively for $k = n-1, n-2, \ldots, 1$

$$x_k = 1 \iff (S - \sum_{i=k+1}^n x_i.a_i) \geq a_k.$$

If at the end $S - \sum_{i=1}^n x_i.a_i = 0$ one has found the solution to (12.1), otherwise one may conclude that (12.1) does not admit a solution.

> **Algorithm 12.1** Solving the knapsack problem for a super-increasing sequence.
> **input** $\{x_k\}_{k=1}^n$ a super-increasing sequence of positive integers,
> S integer
> **initialize** $k = n$
> **while** $k \geq 1$ **do   begin**
>   **if** $S \geq a_k$ **then** $x_k = 1$ **else** $x_k = 0$.
>   **put** $S = S - x_k.a_k$.
>   **put** $k = k - 1$
>   **end**
> **if** $S = 0$ **then print** $\{x_k\}_{k=1}^n$ **else print** "no solution"

Example 12.1 (Part 1)

Consider the super-increasing sequence $\{a_i\}_{i=1}^6 = \{22, 89, 345, 987, 4567, 45678\}$ and the right hand side $S = 5665$. To see if (12.1) has a solution we apply Algorithm 12.1.

Because $S < a_6$, we get $x_6 = 0$. Next, we see that $S \geq a_5$, so we have $x_5 = 1$. We subtract a_5 from S and get 1098. We see that this new value of S satisfies $S \geq a_4$, so $x_4 = 1$, etc. The final solution is $\{1, 1, 0, 1, 1, 0\}$.

Below the same process is written in Mathematica. We make use of the functions Length, While, If, and Join. The solution $\{x_i\}_{i=1}^6$ is formed by prepending each newly found value x_i to $\{x_{i+1}, \ldots, x_6\}$, $i = 6, 5, \ldots, 1$.

```
KnapsackForSuperIncreasingSequence[a_List, S_] :=
    Module[{n, x, X, T},
        n = Length[a]; X = {}; T = S;
        While[n ≥ 1,
                If[T ≥ a[[n]], x = 1, x = 0];
                T = T - x * a[[n]];
                X = Join[{x}, X]; n = n - 1];
        If[T != 0, Print["No solution"], X]]
```

```
a = {22, 89, 345, 987, 4567, 45678}; S = 5665;
X = KnapsackForSuperIncreasingSequence[a, S]
```

```
{1, 1, 0, 1, 1, 0}
```

Indeed

```
X.a
```

```
5665
```

12.1.2 The Knapsack System

□ **Setting Up the Knapsack System**

The knapsack cryptosystem, as proposed in [MerH78] is based on the apparent difficulty of solving the knapsack problem and the ease of solving this problem for super-increasing sequences.

Each user U makes a super-increasing sequence $\{u_i\}_{i=1}^{n} U$ of length n_U. Next, U selects integers W_U and N_U such that

$$N_U > \sum_{i=1}^{n} U u_i \qquad (12.3)$$

and

$$\gcd(W_U, N_U) = 1. \qquad (12.4)$$

User U computes the numbers

$$u_i' = (W_U.u_i \bmod N_U), \qquad 1 \le i \le n, \qquad (12.5)$$

and makes the sequence $\{u_i'\}_{i=1}^{n_U}$ known as his public key.

As a precalculation for the decryption, user U also computes $W_U^{-1} \bmod N_U$.

The number $W_U^{-1} \bmod N_U$ can be computed with the extended version of Euclid's Algorithm (Alg. A.8). Indeed, since $\gcd(W_U, N_U) = 1$, this algorithm will give X and Y such that $1 = X.W_U + Y.N_U$. It follows that $X.W_U \equiv 1 \pmod{N_U}$, i.e. $X = W_U^{-1}$.

Each user keeps the super-increasing sequence $\{u_i\}_{i=1}^{n} U$ and the numbers W_U, $(W_U)^{-1}$, and N_U secret.

Example 12.1 (Part 2)

We continue with the parameters of Example 12.1. So, Bob chooses $\{b_i\}_{i=1}^{6} = \{22, 89, 345, 987, 4567, 45678\}$ as his super-increasing sequence. Further, he selects

$N_B = 56789$, which satisfies $N_B > \sum_{i=1}^{6} b_i$ and $W_B = 12345$ which is coprime with N_B.

Next, he calculates $b_i' = (W_B\, b_i \bmod N_B)$. Here, we do this with the Mathematica function <u>Mod</u>. To check the conditions above we need the <u>GCD</u> function.

```
b = {22, 89, 345, 987, 4567, 45678};
WB = 12345; NB = 56789;
6
∑ b[[i]] < NB
i=1
GCD[WB, NB] == 1
bb = Mod[WB * b, NB]
```

```
True
```

```
True
```

```
{44434, 19714, 56639, 31669, 44927, 36929}
```

So, $\{b_i'\}_{i=1}^{6} = \{44434, 19714, 56639, 31669, 44927, 36929\}$ is the public key.

For this small value of n_B it already takes some effort to solve the knapsack problem (try 101077).

The number $W_B^{-1} \bmod N_B$ can be found with the <u>ExtendedGCD</u> and <u>Mod</u> functions.

```
WB = 12345; NB = 56789;
Mod[ExtendedGCD[WB, NB], NB]
```

```
{1, {39750, 3704}}
```

It follows that $W_B^{-1} = 39750$. Indeed

```
WBinverse = 39750;
Mod[WB * WBinverse, NB]
```

```
1
```

□ **Encryption**

Suppose that Alice wants to send a message to Bob. She looks up the public encryption key $\{b_i{}'\}_{i=1}^{n_B}$ of Bob. Next, she represents her message by a binary vector $(m_1, m_2, ..., m_{n_B})$ of length m_B (or by more vectors of this length if the messages is too long).

Alice will send to Bob the ciphertext

$$C = \sum_{i=1_B}^{n} m_i.b_i'.\qquad(12.6)$$

Example 12.1 (Part 3)

We continue with the parameters of Example 12.1.So, Bob's public key is given by $\{b_i{}'\}_{i=1}^{6} = \{44434, 19714, 56639, 31669, 44927, 36929\}$.

Let Alice's message be $\{m_i\}_{i=1}^{6} = \{1, 1, 0, 0, 0, 1\}$. Then the ciphertext that she will send will be $\sum_{i=1}^{6} m_i.b_i' = 101077$.

```
bb = {44434, 19714, 56639, 31669, 44927, 36929};
m = {1, 1, 0, 0, 0, 1};
CipherText = m.bb
```

```
101077
```

□ **Decryption**

When Bob receives a ciphertext C he will first multiply it with W_B^{-1} and reduce the answer modulo N_B (both are his secret parameters). It follows that

$$W_B^{-1}.C \overset{(12.6)}{\equiv} W_B^{-1}.\sum_{i=1_B}^{n} m_i.b_i' \overset{(12.5)}{\equiv} \sum_{i=1_B}^{n} m_i.b_i \pmod{N_B}.$$

Inequality (12.3) implies that $\sum_{i=1_B}^{n} m_i.b_i < N_B$. So, we can rewrite the above equation as follows:

$$\sum_{i=1_B}^{n} m_i.b_i = (W_B^{-1}.C \bmod N_B).\qquad(12.7)$$

Since the sequence $\{b_i\}_{i=1_B}^{n}$ is super-increasing, Bob can now apply Algorithm 12.1 with $(W_B^{-1}.C \bmod N_B)$ as right hand side to recover the message $\{m_i\}_{i=1_B}^{n}$

Example 12.1 (Part 4)

We continue with the parameters of Example 12.1.

Assume that Bob has received $C = 101077$. First Bob computes $(W_B^{-1}.C \bmod N_B)$ with $W_B^{-1} = 39750$ and $N_B = 56789$.

```
CipherText = 101077;
S = Mod[WBinverse * CipherText, NB]
```

```
45789
```

He gets 45789. To solve (12.1) $\sum_{i=1}^{6} m_i.b_i = S$, he can use the KnapsackForSuperIncreasingFunction defined earlier.

```
b = {22, 89, 345, 987, 4567, 45678}; S = 5665;
X = KnapsackForSuperIncreasingSequence[b, S]
```

```
{1, 1, 0, 1, 1, 0}
```

□ **A Further Discussion**

The knapsack system is summarized in the table below.

Public	$\{u_i{}'\}_{i=1}^{n_U}$ of all users
Secret to U	$\{u_i\}_{i=1}^{n_U}$, W_U^{-1}, N_U
Properties	$u_i{}' \equiv W_U.u_i \pmod{N_U}$,
	$\{u_i{}'\}_{i=1}^{n_U}$ super-increasing,
	$\gcd(W_U, N_U) = 1$
Message for B	$\{m_i\}_{i=1}^{n_B}$
Encryption	$C = \sum_{i=1}^{n_B m_i} .b_i'$
Decryption by B	Apply Algorithm 12.1 to
	$\{u_i{}'\}_{i=1}^{n_U}$ and $W_B^{-1}.C \bmod N_B$

The Knapsack Cryptosystem

Table 12.1

Even though the knapsack cryptosystem does not have the signature property, for a short while it gained an enormous popularity. The main reason is the low complexity of its implementation. In applications, both encryption and decryption can take place at very high data rates.

The authors [MerH78] recommend the users to take length $n_U = 100$, a sequence $\{u_i\}_{i=1}^{n_U}$ satisfying

$$(2^i - 1).2^{100} < u_i < 2^i.2^{100}, \quad 1 \le i \le 100,$$

(it will automatically be super-increasing), and a modulus N_U such that

$$2^{101} + 1 < N_U < 2^{202}.$$

Note that also (12.3) is satisfied.

It is further recommended that user U makes a permuted version $\{u_i'\}_{i=1U}^n$ public instead of $\{u_i'\}_{i=1U}^n$ itself to disguise the order of the original super-increasing sequence. In this way, a cryptanalist has no information about which element u_i' in the public knapsack came from (the smallest knapsack element) u_1, for instance.

The idea of multiplying a super-increasing sequence with a constant W_U modulo N_U is of course to obtain a knapsack that looks random. To increase this effect and thus to increase the security of the knapsack cryptosystem, [MerH78] advises to iterate this multiplication.

Hence, each user U also selects $N_U' > \sum_{i=1U}^n u_i'$ and $1 < W_U' < N_U'$ with $\gcd(W_U', N_U')=1$, computes $u_i'' \equiv W_U' . u_i' \pmod{N_U'}$, $1 \leq i \leq n_U$, and makes $\{u_i''\}_{i=1U}^n$ public instead of $\{u_i'\}_{i=1U}^n$.

It makes sense to iterate this process of modulo-multiplication, as is illustrated in the following example.

Example 12.2

Let $n = 3$ and consider $\{u_i\}_{i=1}^3 = \{5, 10, 20\}$. Multiplying this sequence with 17 modulo 47 gives $\{u_i'\}_{i=1}^3 = \{38, 29, 11\}$. Multiplying this sequence with 3 modulo 89 gives $\{u_i''\}_{i=1}^3 = \{25, 87, 33\}$.

These calculations can be verified with the Mod function.

```
u = {5, 10, 20}
uu = Mod[17 u, 47]
uuu = Mod[3 uu, 89]
```

```
{5, 10, 20}
```

```
{38, 29, 11}
```

```
{25, 87, 33}
```

It is impossible to find integers W and N that map $\{u_i\}_{i=1}^3$ directly into $\{u_i''\}_{i=1}^3$. Indeed the congruence relations

$$5W \equiv 25 \pmod{N},$$
$$10W \equiv 87 \pmod{N}$$

imply that N divides $87 - 2 \times 25 = 37$. Since 37 is a prime, it follows that $N = 37$. It also follows that $W = 5$. These values of W and N however violate the third congruence relation

$$20W \equiv 33 \pmod{N}.$$

This shows that an iteration of modulo-multiplications can not always be replaced by a single modulo-multiplication.

The above example also demonstrates something else. Note that the second iteration mapped the not-super-increasing knapsack {38, 29, 11} into {25, 87, 33}, which after a reordering is a super-increasing sequence.

This also makes it clear that cryptanalist Eve does not have to guess the original integers W_U and N_U (and also W_U' and N_U' in the iterated case) to convert the public key back into a super-increasing sequence. Eve can also decrypt the ciphertext, if she is able to obtain another super-increasing sequence from $\{u_i'\}_{i=1U}^n$ (resp. $\{u_i''\}_{i=1U}^n$).

These observations demonstrate two important things:

> 1) Iteration does not necessarily increase the security of the system.

> 2) It may be easier for a cryptanalist to map the public knapsack into a super-increasing sequence other than the original.

Some critics of the knapsack cryptosystem did not trust the linearity of the system. Their intuition/experience told them that the knapsack cryptosystem was bound to be broken.

The reader should remember that the general knapsack problem is NP-complete. This implies in particular that no known algorithm solves it in polynomial time. However, the property of NP-completeness has never been proved for the restriction of the knapsack problem to the subclass of knapsacks, obtained by a single modulo-multiplication of a super-increasing sequence. In 1982, Shamir [Sham82] showed that the single iteration version of the knapsack system can be broken with very high probability in polynomial time. This attack was later generalized by others (see [Adle83] and [Bric85])

In Section 12.2, an outline of the much more general attack by Lagarias and Odlyzko [LagO83] will be given.

12.2   The L^3-Attack

12.2.1 Introduction

In the original knapsack cryptosystem it is assumed that the secret sequence $\{u_i\}_{i=1}^{n_U}$ is super-increasing. However, this is not crucial for a knapsack-based cryptosytem. It only makes the decryption easy, because of Algorithm 12.1. The only essential requirement is that the plaintext-to-ciphertext mapping $\{m_i\}_{i=1}^{n_U} \longrightarrow C$ in (12.6) is one-to-one.

Since the general knapsack problem is NP-complete, no known algorithm solves it in polynomial time. Still, it is quite possible that polynomial-time algorithms do exist, which solve with some positive probability any knapsack problem in a large subclass of knapsack problems. Such an algorithm would make the knapsack system unsuitable for cryptographic purposes.

In this section, we shall often use the vector notation $\underline{u} = (u_1, u_2, \ldots, u_n)$ for a knapsack $\{u_i\}_{i=1}^n$. Before we give an outline of the *Lagarias and Odlyzko attack* (also called the L^3-attack) [LagO83], we have to define a few new notions.

> **Definition 12.2**
> The *density* $d(\underline{u})$ of a knapsack $\underline{u} = (u_1, u_2, \ldots, u_n)$ is defined by
> $$d(\underline{u}) = \frac{n}{\max_{1 \le i \le n} \log_2 u_i}.$$

Example 12.3

For instance, the density of the knapsack $\{22, 89, 345, 987, 4567, 45678\}$ *is* $6 / \log_2 45678 \approx 0.39$, *as can be checked with the Mathematica functions* Max, Log, Length, *and* N.

```
a = {22, 89, 345, 987, 4567, 45678};
N[Length[a] / Log[2, Max[a]], 2]
```

```
0.39
```

The density $d(\underline{u})$ serves as measure for the information rate of a knapsack system. Indeed, the numerator is the number of message bits that are stored in the sum C of the knapsack (see (12.6)). The denominator is a good approximation of the average number of bits needed for the binary representation of C. For instance, with $u_i = 2^{i-1}$, $1 \le i \le n$, the density is $n / (n-1) \approx 1$ as it should be.

We shall show further on that the Lagarias and Odlyzko attack is more likely to break the knapsack system if its density is smaller.

This may sound like a heavy restriction, but one should realize that nobody likes to use a cryptosystem that has a non-trivial positive chance to be broken.

12.2.2 Lattices

> **Definition 12.3**
> Let $\{\underline{v}_1, \underline{v}_2, \ldots, \underline{v}_n\}$ be a set of vectors in $\mathbf{Z}^n$ that are linearly independent over $\mathbf{R}$. Then the set of all integer linear combinations of $\{\underline{v}_1, \underline{v}_2, \ldots, \underline{v}_n\}$ is called an *integer lattice*. In formula:
> $$\Lambda = \{\textstyle\sum_{i=1}^n \alpha_i \underline{v}_i \mid \alpha_i \in \mathbf{Z}, 1 \le i \le n\}$$
> or
> $$\Lambda = \mathbf{Z}\underline{v}_1 + \mathbf{Z}\underline{v}_2 + \ldots + \mathbf{Z}\underline{v}_n.$$

We say that the n independent vectors $\underline{v}_1, \underline{v}_2, \ldots, \underline{v}_n$ form a *basis* for the lattice Λ. Note that the basis of a lattice is certainly not unique. Normally, the order of the basis vectors does not matter, but in the sequel such an order will matter. We shall use the notation $[\underline{v}_1, \underline{v}_2, ..., \underline{v}_n]$ to indicate a particular ordering.

Example 12.4

Consider the lattice Λ in $\mathbf{Z}^2$ with basis $\underline{u} = (3, 1)$ and $\underline{v} = (1, 2)$. It consists of all points of the form $\alpha.(3, 1) + \beta.(1, 2)$, with $\alpha, \beta \in \mathbf{Z}$. Below part of this lattice is depicted.

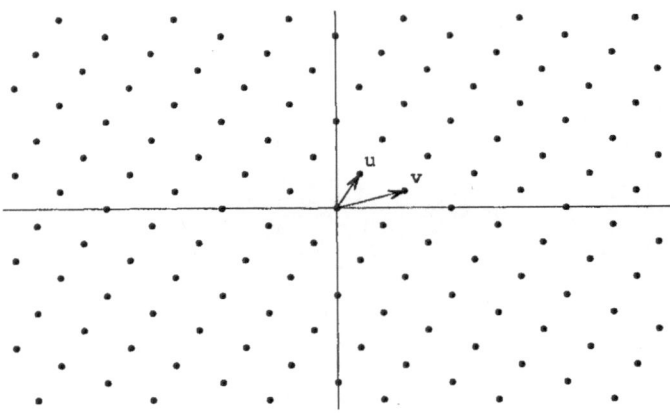

Lattice in $\mathbf{R}^2$ with basis (3, 1) and (2, 1)

Figure 12.1

For the L^3-attack that we shall describe later on, it is of great importance to find a vector in Λ of short length, or even better to find a complete basis of short vectors for Λ. For this reason, we need to study basis transformations more carefully.

The *Gram-Schmidt* process is a well known algorithm from linear algebra to transfer a basis $\{\underline{v}_1, \underline{v}_2, ..., \underline{v}_n\}$ of a linear (sub)space into an orthogonal basis, i.e. in a basis $\{\underline{u}_1, \underline{u}_2, ..., \underline{u}_n\}$ with the property that all vectors $\underline{u}_i$ are orthogonal to each other, i.e. $(\underline{u}_i, \underline{u}_j) = 0$, for $i \neq j$. It goes as follows:

$$\underline{u}_1 = \underline{v}_1,$$
$$\underline{u}_2 = \underline{v}_2 - \mu_{1,2}\,\underline{u}_1,$$
$$\underline{u}_3 = \underline{v}_3 - \mu_{1,3}\,\underline{u}_1 - \mu_{2,3}\,\underline{u}_2,$$

$$\vdots$$

$$\underline{u}_n = \underline{v}_n - \mu_{1,n}\,\underline{u}_1 - \mu_{2,n}\,\underline{u}_2 - ... - \mu_{n-1,n}\,\underline{u}_{n-1}.$$

where

$$\mu_{i,j} = \frac{(\underline{v}_j, \underline{u}_i)}{(\underline{u}_i, \underline{u}_i)}, \quad 1 \leq i \leq j \leq n.$$

Example 12.5

To demonstrate the Gram-Schmidt process we take $\underline{v}_1 = (3, 4, 2)$, $v_2 = (2, 5, 2)$, and $\underline{v}_3 = (1, 2, 6)$ in $\mathbb{R}^3$.

```
v1={3,4,2};v2={2,5,2};v3={1,2,6};
u1=v1
u2=v2-((u1.v2)/(u1.u1))u1
u3=v3-((u1.v3)/(u1.u1))u1-((u2.v3)/(u2.u2))u2
```

```
{3, 4, 2}
```

$$\left\{ -\frac{32}{29}, \frac{25}{29}, -\frac{2}{29} \right\}$$

$$\left\{ -\frac{24}{19}, -\frac{24}{19}, \frac{84}{19} \right\}$$

This can also be done in Mathematica. We first load the Mathematica package `LinearAlgebra`Orthogonalization`` *and then run GramSchmidt. The result will be orthonormal basis, i.e. we obtain a set of n orthogonal vectors $\underline{u}_i$ that have been further divided by their length to give them unit-length.*

```
<<LinearAlgebra`Orthogonalization`
```

```
v1={3,4,2};v2={2,5,2};v3={1,2,6};
{u1,u2,u3}=GramSchmidt[{v1,v2,v3}]
```

$$\left\{ \left\{ \frac{3}{\sqrt{29}}, \frac{4}{\sqrt{29}}, \frac{2}{\sqrt{29}} \right\}, \right.$$
$$\left. \left\{ -\frac{32}{\sqrt{1653}}, \frac{25}{\sqrt{1653}}, -\frac{2}{\sqrt{1653}} \right\}, \left\{ -\frac{2}{\sqrt{57}}, -\frac{2}{\sqrt{57}}, \frac{7}{\sqrt{57}} \right\} \right\}$$

As we can see in the example above, the vectors $\underline{u}_i$, $1 \le i \le n$, will, in general, no longer have integer coordinates. In the context of integer lattices that is an undesirable situation.

In the next subsection we shall discuss an (integer-valued) basis for lattice Λ, that is not completely orthonormal, but has two other attractive properties.

12.2.3 A Reduced Basis

Let $\|\underline{u}\|$ denote the standard Euclidean norm or length of a vector $\underline{u}$. So, $\|\underline{u}\| = (\underline{u}, \underline{u})^{1/2} = \sum_{i=1}^{n} (u_i)^2$.

Definition 12.4

A basis $\{\underline{v}_1, \underline{v}_2, ..., \underline{v}_n\}$ of an integer lattice Λ is called y-*reduced*, where $1/4 < y < 1$, if the orthogonal basis $\{\underline{u}_1, \underline{u}_2, ..., \underline{u}_n\}$ obtained from $\{\underline{v}_1, \underline{v}_2, ..., \underline{v}_n\}$ through the Gram-Schmidt process satisfies

$$\|\underline{u}_i + \mu_{i,i-1} \underline{u}_{i-1}\|^2 \geq y \cdot \|\underline{u}_{i-1}\|^2, \quad 2 \leq i \leq n,$$
$$(|\mu|)_{i,j} \leq 1/2, \quad 1 \leq i \leq j \leq n.$$

An alternative definition of a y-reduced basis can be given as follows. Let V_k be the k-dimensional linear subspace of $\mathbb{R}^n$, spanned by $\{\underline{v}_1, \underline{v}_2, ..., \underline{v}_k\}$ or, equivalently, by $\{\underline{u}_1, \underline{u}_2, ..., \underline{u}_k\}$. Let $V_k^{\perp}$ be the orthogonal complement of V_k. Define $\underline{v}_j^{(k)}$, $k + 1 \leq j \leq n$, as the projection of $\underline{v}_j$ onto $V_k^{\perp}$. In particular, $\underline{v}_{k+1}^{(k)} = \underline{u}_{k+1}$. Then it can be shown (see [LagO83]) that the two conditions in Definition 12.4 are equivalent to

$$\|\underline{v}_i^{(i-2)}\|^2 \geq y \cdot \|\underline{v}_{i-1}^{(i-2)}\|^2 = y \cdot \|\underline{u}_{i-1}\|^2, \quad 2 \leq i \leq n, \tag{12.8}$$

resp.

$$\|\underline{v}_j^{(i)} - \underline{v}_j^{(i-1)}\| \leq \tfrac{1}{2} \|\underline{v}_i^{(i-1)}\|, \quad 1 \leq i \leq j \leq n. \tag{12.9}$$

Note that (12.8) implies that the projection of $\underline{v}_i$ onto $V_{i-2}^{\perp}$ should not be too small in size (when compared with the length of $\underline{u}_{i-1}$). The inequality in (12.9) says that the projection of $\underline{v}_j$ onto $\underline{u}_i$ is relatively small.

These two statements can be interpreted by saying that the vectors in a y-reduced basis are of comparable size and all point in different directions.

In the sequel, y will always be 3/4. The L^3 – Algorithm (see [LenLL82]) is a very effective way to find a y-reduced basis for a lattice Λ. It will not be described in full detail here (see however Subsection 12.2.5). We quote the following facts from [LenLL82].

Theorem 12.2

Let $\{\underline{v}_1, \underline{v}_2, ..., \underline{v}_n\}$ be a basis of an integer lattice Λ in $\mathbb{Z}^n$ and let $B = \max_{1 \leq i \leq n} \|\underline{v}_i\|$. Then the L^3-lattice basis reduction algorithm produces a reduced basis $\{\underline{w}_1, \underline{w}_2, ..., \underline{w}_n\}$ for Λ in about $n^6 (\log B)^3$ bit operations.

Theorem 12.3

Let $\{\underline{w}_1, \underline{w}_2, ..., \underline{w}_n\}$ be a reduced basis for an integer lattice Λ. Then

$$\|\underline{w}_1\|^2 \leq 2^{n-1} \cdot \min\{\|\underline{x}\|^2 \mid \underline{x} \in \Lambda \setminus \{\underline{0}\}\}.$$

In fact, Prop.1.12 in [LenLL82] shows that no vector in a reduced basis can be very long.

12.2.4 The L^3-Attack

We can now present the idea behind the L^3-attack. We want to find a solution to the knapsack problem $\sum_{i=1}^n x_i a_i = C$ (see (12.1)).

The idea of the attack will be to convert the parameters of the knapsack problem into a basis of some integer lattice Λ. Then we find a short vector in this lattice with the L^3-lattice basis reduction algorithm. The hope will be that this short vector can be transformed back into the solution $\{x_i\}_{i=1}^n$ of (12.1).

L^3-**attack** on $\sum_{i=1}^n a_i x_i = S$.

Step 1:

Define the vectors

$$
\begin{aligned}
\underline{v}_1 &= (1, & 0, & ..., & 0, & -a_1) \\
\underline{v}_2 &= (0, & 1, & ..., & 0, & -a_2) \\
&\vdots \\
\underline{v}_n &= (0, & 0, & ..., & 1, & -a_n) \\
v_{n+1} &= (0, & 0, & ..., & 0, & S)
\end{aligned}
\tag{12.10}
$$

Together they form a basis for a $(n+1)$-dimensional lattice Λ in $\mathbb{Z}^{n+1}$.

Note that for the solution $\{x_i\}_{i=1}^n$ one has

$$\sum_{i=1}^n x_i \underline{v}_i + \underline{v}_{n+1} = (x_1, x_2, ..., x_n, 0).$$

So, this vector has length $\sqrt{n}$, which is relatively very short, e.g., if the knapsack has length $n = 100$, we have $\| \sum_{i=1}^n x_i \underline{v}_i + \underline{v}_{n+1} \| \leq 10$.

Step 2:

Find a reduced basis $\{\underline{w}_1, \underline{w}_2, ..., \underline{w}_n\}$ for Λ with the L^3-algorithm ([LenLL82]).

Step 3:

Check if one of the $n+1$ "short" vectors $\underline{w}_i$, $1 \leq i \leq n+1$, has the property that $(\underline{w}_i)_{n+1} = 0$ and that each of the first n coordinates is either 0 or α, for some constant α.

If so, check if the vector $\frac{1}{\alpha}((\underline{w}_i)_1, (\underline{w}_i)_2, ..., (\underline{w}_i)_n)$ is a solution of (12.1). If it does, STOP, otherwise continue with Step 4.

Step 4:

Repeat Steps 1, 2 and 3 with S replaced by $\sum_{i=1}^n a_i - S$. If these steps result in a solution $\{x_i'\}_{i=1}^n$ for this new knapsack problem then $\{x_i\}_{i=1}^n$, defined by $x_i = 1 - x_i'$, $1 \leq i \leq n$, will be the solution of the original knapsack.

Example 12.6

Consider the knapsack problem with $\{a_i\}_{i=1}^{10} = \{541,400,259,1059,895,590,498,973,41,649\}$ and $S = 4517$.

Let us first make the vectors $\underline{v}_i$, $1 \le i \le 10$, as indicated by (12.10). We use the Mathematica functions `Transpose`, `Append`, `IdentityMatrix`, `Do`, `Table`, *and* `MatrixForm`.

```
a = {541, 400, 259, 1059, 895, 590, 498, 973, 41, 649};
s = 4517;
aux = Transpose[ Append[IdentityMatrix[10], -a]];
Do[v[i] = aux[[i]], {i, 1, 10}];
v[11] = Append[Table[0, {10}], s];
Table[v[i], {i, 1, 11}] // MatrixForm
```

$$
\begin{pmatrix}
1 & 0 & 0 & 0 & 0 & 0 & 0 & 0 & 0 & 0 & -541 \\
0 & 1 & 0 & 0 & 0 & 0 & 0 & 0 & 0 & 0 & -400 \\
0 & 0 & 1 & 0 & 0 & 0 & 0 & 0 & 0 & 0 & -259 \\
0 & 0 & 0 & 1 & 0 & 0 & 0 & 0 & 0 & 0 & -1059 \\
0 & 0 & 0 & 0 & 1 & 0 & 0 & 0 & 0 & 0 & -895 \\
0 & 0 & 0 & 0 & 0 & 1 & 0 & 0 & 0 & 0 & -590 \\
0 & 0 & 0 & 0 & 0 & 0 & 1 & 0 & 0 & 0 & -498 \\
0 & 0 & 0 & 0 & 0 & 0 & 0 & 1 & 0 & 0 & -973 \\
0 & 0 & 0 & 0 & 0 & 0 & 0 & 0 & 1 & 0 & -41 \\
0 & 0 & 0 & 0 & 0 & 0 & 0 & 0 & 0 & 1 & -649 \\
0 & 0 & 0 & 0 & 0 & 0 & 0 & 0 & 0 & 0 & 4517
\end{pmatrix}
$$

The vectors $\{\underline{v}_1, \underline{v}_2, ..., \underline{v}_{10}\}$ form the basis of a lattice Λ.

Next we use the Mathematica function `LatticeReduce` *to find a reduced basis.*

```
LatticeReduce[ Table[v[i], {i, 1, 11}]]
```

```
{{1, -2, 1, 0, 0, 0, 0, 0, 0, 0, 0},
 {-1, 0, -2, 1, 0, 0, 0, 0, 0, 0, 0}, {0, 1, -1, 1, -2, 1, 0, 0, 0, 0, 0},
 {1, -1, -1, 0, -1, 0, 0, 1, 1, 0, -1},
 {1, 1, -2, 0, 0, 1, 0, -1, -1, 0, 1},
 {1, 1, -1, 0, 0, -2, 1, 0, 0, 0, 0}, {1, -1, 0, 0, 1, 0, -2, 0, -1, 0, 1}
 {0, 1, 0, -1, 0, 1, -1, 0, -2, 1, 0},
 {0, 0, -1, -1, -1, 1, 0, 1, 0, 1, 1},
 {1, -1, 0, 0, 0, 1, 0, 0, -2, -1, 0}, {1, 1, 0, 1, 1, 0, 0, 1, 0, 1, 0}}
```

We see that only the last output is a two-valued vector on its first 10 coordinates. One of the values is indeed 0, the other value is $\alpha=1$. Trying out $\{a_i\}_{i=1}^{10} = \{1, 1, 0, 1, 1, 0, 0, 1, 0, 1\}$ gives indeed $\sum_{i=1}^{n} a_i x_i = S$.

```
x = {1, 1, 0, 1, 1, 0, 0, 1, 0, 1};
a.x == S
```

```
True
```

The computing time of Steps 1 and 3 in the L^3-attack is ignorable. Therefore, the running time of this algorithm is essentially (twice) the running time of the L^3-algorithm, as given by in Theorem 12.2. There is in no guarantee that the L^3-algorithm will find a solution of the knapsack problem. However the authors of [LagO83] give the following analysis of the L^3-algorithm.

Theorem 12.4
Let $B \geq 2^{(1+\beta)n^2}$ for some constant $\beta > 0$ and knapsack length n. Let $K(n, B)$ denote the number of knapsacks $\{a_i\}_{i=1}^n$ satisfying
1) $1 \leq a_i \leq B$ for all $1 \leq i \leq n$,
2) the L^3-attack will find a $\{0,1\}$-solution $\{x_i\}_{i=1}^n$ for (12.1) for each right hand side S for which there exists such a solution.
Then
$$K(n, B) = B^n(1 - \epsilon(B)),$$

where
$$0 < \epsilon(B) < \frac{C_1}{B^{C_2 - 3(\ln n)/n}}$$

for some constant C_1 and where $C_2 = 1 - (1 + \beta)^{-1} > 0$.

Theorem 12.4 states that for any $\beta > 0$ and n sufficiently large one can solve the knapsack problem for almost all knapsacks $\{a_i\}_{i=1}^n$ with density

$$d(\{a_i\}_{i=1}^n) \leq \frac{n}{\log_2 B} < \frac{1}{(1+\beta)n}.$$

With some additional work [LagO83], the inequality above can be weakened to

$$d(\{a_i\}_{i=1}^n) < (1 - \epsilon) \frac{1}{n.\log_2 4/3}.$$

for any fixed $\epsilon > 0$ and n. This inequality is probably not the best possible one.

12.2.5 The L^3-Lattice Basis Reduction Algorithm

Recall that the L^3-algorithm must find a basis $\{\underline{v}_1, \underline{v}_2, ..., \underline{v}_n\}$ for an integer lattice that meets the requirement given in Definition 12.3:

$$\| \underline{u}_i + \mu_{i,i-1} \underline{u}_{i-1} \|^2 \geq y. \| \underline{u}_{i-1} \|^2, \quad 2 \leq i \leq n,$$

$$|(\mu)_{i,j}| \leq 1/2, \quad 1 \leq i \leq j \leq n,$$

where $\mu_{i,j} = \frac{(v_j, u_i)}{(u_i, u_i)}$.

The L^3-algorithm makes use of the following procedure:

```
Procedure reduce[k, l]
        Input           1 ≤ l < k
        Compute μ_{l,k}
        If |(μ)_{l,k}| > 1/2 then begin
                        r = ⌊0.5 + μ_{l,k}⌋
                        v_k := v_k − r.v_l
                        end
```

The L^3-algorithm now runs as follows:

```
L^3-Algorithm
        Input           {v_1, v_2, ..., v_n}, basis of integer lattice
        Initialize      k=2
        While k ≤ n     do
                        begin
                        reduce(k, k − 1)
                        compute ‖u_k‖, ‖u_{k−1}‖ and μ_{k−1,k}
                        if ‖u_k‖² < (y − μ²_{k−1,k})·‖u_{k−1}‖²
                                then begin      exchange v_k and v_{k−1}
                                        k := max {2, k − 1}
                                end
                                else begin   reduce(k, l) for l = k − 1, ..., 2, 1
                                        k = k + 1
                                end
                        end
```

For further reading see [LenLL82]. Notice that only the basis $\{v_1, v_2, ..., v_n\}$ is adjusted in this algorithm. No vector u_i enters the reduced basis, they are only used in the calculations.

12.3   The Chor-Rivest Variant

The Chor-Rivest scheme [ChoR85] is a knapsack based cryptosystem that does not convert a secret knapsack, for which the knapsack problem is easy to solve, into the public knapsack, for which the knapsack problem should be intractable. It does make use of the standard conversion of integers to binary sequences of fixed length. Further, it employes a fixed constant, a fixed choice of an irreducible polynomial, a fixed choice of a primitive element, a fixed permutation, and an exponentiation in a finite field for which the logarithm problem is tractable.

In [Vaud98], it is shown that the parameters suggested in [ChoR85] are not secure. The author gives suggestions to repair the original proposal. Here we shall only explain the original idea of the Chor-Rivest scheme.

□ **Setting Up the System**

1)   Each user U selects a finite field GF (q) for which the logarithm problem is feasible (also by the cryptanalist). For instance, in view of the Pohlig-Hellman Algorithm explained in Subsection 8.3.1, this can be achieved by letting $q - 1$ have only small prime factors. Further, the characteristic p of GF(q), so $q = p^k$ for some k, should satisfy $p > k$.

To represent GF(q), U uses a   random irreducible polynomial $f(x)$ of degree k over $\mathbf{Z}_p$ The elements of GF(q) can be represented by p-ary polynomials of degree $< k$ (see Theorem B.15).

Note that, for reasons of clarity, we have omitted the subscript U in the above choices by U).

2)   User U selects a random primitive element α in GF(q). Primitive means that each non-zero element in GF(q) can be written as some power α^i of α, where $i < q - 1$. Note that α, being an element in GF(q), is also a p-ary polynomial of degree less than k.

3)   For each $i \in \mathbf{Z}_p$, user U determines the discrete logarithm of the field elements $x + i$ with respect to the primitive element α. In other words, one needs to find exponents U_i, $i \in \mathbf{Z}_p$, satisfying

$$\alpha^{U_i} \equiv x + i \,(\mathrm{mod}\, f(x)). \tag{12.11}$$

This is feasible by our assumption in 1).

4)   Finally, user U has to select a random permutation π_U of $\{0, 1, ..., p - 1\}$ and a random element D_U, $0 \leq D < q - 1$. He computes the numbers

$$u_i \equiv U_{\pi(i)} + D_U \,(\mathrm{mod}\, q - 1). \tag{12.12}$$

and makes these numbers $u_0, u_1, ..., u_{p-1}$ public together with the value $q = p^k$.

(The reader should recall that $q - 1$ is the order of the multiplicative group of GF(q), see Theorem B.20).

Example 12.7 (Part 1)

Bob selects the finite field GF(7^3), so p = 7 and k = 3. An irreducible, binary polynomial f(x) of degree 3 over $\mathbb{Z}_7$ can be found with the Mathematica function IrreduciblePolynomial, *once the package* Algebra`FiniteFields` *has been loaded.*

```
<<Algebra`FiniteFields`
```

```
p = 7; k = 3; q = p^k;
f = IrreduciblePolynomial[x, p, k]
```

```
4 + x + 2 x^2 + x^3
```

So, f(x) = $x^3 + 2x^2 + x + 4$. It turns out that $\omega = x$ is a primitive element in GF(7^3). This can be checked as follows. From q − 1 = 7^3 − 1 = 11 × 31, we see that the order of any element is either 1, 11, 31, or 342 (see Theorem B.5). But $\omega = x$ does not have order 11 or 31, as can be checked with the following calculations. (We use the GF-function. Note that f342 represents GF(7^3) = $\mathbb{Z}_7[x]/(f(x))$.)

```
f341 = GF[7, {4, 1, 2, 1}];
om = f341[{0, 1}];
om^11
om^31
```

```
{6, 1, 3}_7
```

```
{3, 3, 6}_7
```

To get a random primitive element α in GF(7^3), Bob raises ω to the power i with gcd(i, q − 1) = 1 (see Lemma B.4). We use the functions Random, GCD, *and* While.

```
i = q - 1;
While[GCD[i, q - 1] != 1, i = Random[Integer, {1, q - 2}]];
  i
```

```
239
```

We find i=239. The random primitive element will be $\alpha = \omega^i$, which is $3 + 4x + 5x^2$ by

```
a = om^i
```

```
{3, 4, 5}_7
```

It follows from $83 \times 239 \equiv 1 \pmod{q - 1}$ that $\omega = \alpha^{83}$.

To determine the numbers B_i satisfying $\alpha^{B_i} \equiv x + i \pmod{f(x)}$ we use

```
B = Table[Mod[83 * FieldInd[om + i], q - 1], {i, 0, p - 1}]
```

```
{83, 101, 175, 90, 170, 321, 213}
```

We conclude that $B_0 = 83$, $B_1 = 101$, $B_2 = 175$, $B_3 = 90$, $B_4 = 170$, $B_5 = 321$, $B_6 = 213$.

This can be checked with:

```
B = {83, 101, 175, 90, 170, 321, 213};
a^B
```

```
{{0, 1, 0}_7, {1, 1, 0}_7, {2, 1, 0}_7,
 {3, 1, 0}_7, {4, 1, 0}_7, {5, 1, 0}_7, {6, 1, 0}_7}
```

A few more things need to be done by Bob. He has to select a random number D, $0 \le D < q - 1$, and a random permutation π of $\{0, 1, ..., 6\}$. We load the Mathematica package `DiscreteMath`Combinatorica`` *and use the function* <u>RandomPermutation</u>.

```
<<DiscreteMath`Combinatorica`
```

```
RD = Random[Integer, {0, q - 2}]
pi = RandomPermutation[7]
```

```
244
```

```
{6, 3, 7, 4, 5, 2, 1}
```

So, $D = 244$ and $\pi = \{6, 3, 7, 4, 5, 2, 1\}$, meaning that $\pi(1) = 6$, $\pi(2) = 3$, ..., $\pi(7) = 1$.

(The reader should watch out here. Mathematica labels the entries in a list starting with 1, while we start with 0.)

The public key is given by the sequence (12.12): $b_i = B_{\pi(i)} + D$. *We use the functions* <u>Table</u> *and* <u>Mod</u>.

```
BPerm = Table[B[[pi[[i]]]], {i, 1, 7}]
b = Mod[BPerm + RD, q - 1]
```

```
{321, 175, 213, 90, 170, 101, 83}
```

```
{223, 77, 115, 334, 72, 3, 327}
```

Bob makes $\{b_i\}_{i=0}^6 = \{223, 77, 115, 334, 72, 3, 327\}$ *public and also* $k = 3$.

□ **Encryption**

Now suppose that Alice wants to send a secret message to Bob. She looks up the public parameters b_0, b_1, ..., b_{p-1} and k of Bob. She calculates $q_B = p^k$. Alice's message is a number M in between 1 and $\binom{p}{k}$.

Alice represents her message (in a manner that is shown below) as a binary string m_0, m_1, ..., m_{p-1} of length p and weight k (exactly k of the m_i's are equal to 1), so

$$\sum_{i=1}^{p-1} m_i = k. \tag{12.13}$$

Alice will send

$$c = \left(\sum_{i=1}^{p-1} m_i b_i \bmod q_B\right). \tag{12.14}$$

Example 12.7 (Part 2)

Suppose that Alice wants to send a message to Bob. She looks up Bob's public parameters $k = 3$ *and* $\{b_i\}_{i=0}^6 = \{223, 77, 115, 334, 72, 3, 327\}$ *(see Example 12.7). So, she knows that* $p = 7$ *(and* $q = 7^3 = 341$).

Let Alice's message be $M = 19$ *(which indeed lies in between 1 and* $\binom{7}{3}$).

This can be represented by the binary sequence $\{m_i\}_{i=0}^6 = \{0, 1, 1, 0, 1, 0, 0\}$, *as shown below.*

The ciphertext c *that Alice will send will be* $\sum_{i=0}^6 m_i b_i$, *which is 264 in this case.*

```
m = {0, 1, 1, 0, 1, 0, 0};
ct = m.b
```

```
264
```

There is a recursive way to map a number M, $1 \le M \le \binom{p}{k}$, into a binary string $m_0, m_1, ..., m_{p-1}$ of length p and weight k. It makes use of the well-known identity:

$$\binom{p}{k} = \binom{p-1}{k} + \binom{p-1}{k-1}.$$

If $M > \binom{p-1}{k}$, we put $m_{p-1} = 1$ and decrease M by $\binom{p-1}{k}$. This new value will be in between 1 and $\binom{p-1}{k-1}$ and can be described by a string $m_0, m_1, ..., m_{p-2}$ of length $p-1$ and weight $k-1$.

On the other hand, if $M \le \binom{p-1}{k}$, put $m_{p-1} = 0$ and describe M by a string $m_0, m_1, ..., m_{p-2}$ of length $p-1$ and weight k.

Algorithm 12.5 Conversion from M **to** $m_0, m_1, ..., m_{p-1}$ **of weight** k

Input	M, $1 \le M \le \binom{p}{k}$.
Initialize	$l = k$
For $i = 1$ to p do	if $M > \binom{p-i}{l}$
	then begin $\quad m_{p-i} = 1$
	$M := M - \binom{p-i}{l}$
	$l := l - 1$
	end
	else $\quad m_{p-i} = 0$

Example 12.8

Let $p = 7$ and $k = 3$. Then $\binom{7}{3} = 35$.

To find out into which binary sequence of length 7 and weight 3 the integer $M = 19$ will be mapped, we follow the algorithm below, which makes use of the Mathematica functions Table, If, Do, *and* Binomial.

```
p = 7; k = 3;
Me = 19;
l = k;
m = Table[0, {i, 1, p}];
Do[If[Me > Binomial[p - i, l],
                    {m[[i]] = 1,
     Me = Me - Binomial[p - i, l], l = l - 1}],
                {i, 1, p}];
m
```

```
{0, 1, 1, 0, 1, 0, 0}
```

□ **Decryption**

Bob receives c, which is in fact $c = \left(\sum_{i=1}^{p-1} m_i\, b_i \bmod q_B\right)$ by (12.14). He computes $C = c - k.D_B$ with his secret D_B (see (12.12)).

Next, Bob computes α^C. Now note that in $\mathrm{GF}(q)$:

$$\alpha^C = \alpha^{c-k.D_B} = \alpha^{\left(\sum_{i=1}^{p-1} m_i\, b_i\right)-k.D_B} \overset{(12.12)}{=} \alpha^{\left(\sum_{i=1}^{p-1} m_i(B_{\pi(i)}+D_B)\right)-k.D_B}$$

$$\overset{(12.13)}{=} \alpha^{\sum_{i=1}^{p-1} m_i\, B_{\pi(i)}} = \prod_{i=1}^{p-1}\left(\alpha^{B_{\pi(i)}}\right)^{m_i} \overset{(12.11)}{=} \prod_{i=1}^{p-1}\left(x+\pi(i)\right)^{m_i}.$$

This means that

$$\alpha^C \equiv \prod_{i=1}^{p-1}\left(x+\pi(i)\right)^{m_i}\ (\bmod f(x)).$$

Next, we add a suitable multiple of $f(x)$ to α^C to make its polynomial representation monic. So, for some $\beta \in \mathrm{GF}(q)$: $a(x) = \alpha^C + \beta.f(x)$ is monic.

Since also $\prod_{i=1}^{p-1}\left(x+\pi(i)\right)^{m_i}$ is monic, the above in fact implies that

$$a(x) = \prod_{i=1}^{p-1}\left(x+\pi(i)\right)^{m_i}.$$

It follows that $m_i = 1$, $0 \le i \le p-1$, if and only if $-\pi(i)$ is a zero of $a(x)$.

We summarize the decryption process in the following algorithm.

Algorithm 12.6	Decryption of Chor-Rivest Cryptosystem by Bob
Input	ciphertext c
Bob's Secret	$D_B, k, \alpha, f(x), \pi$.
Compute	$C = c - k.D_B$ with secret D_B (see (12.12)).
Compute	α^C, where α is Bob's primitive element
Add	multiple of $f(x)$ to α^C to get monic $a(x)$
Put	$m_i = 1$ if and only if $-\pi(i)$ is a zero of $a(x)$.

Example 12.7 (Part 3)

We continue with Example 12.7. Assume that Bob receives the ciphertext c = 264.

Bob's secret parameters are $k = 3$, $D = 244$, $\pi = \{6, 3, 7, 4, 5, 2, 1\}$, $f(x) = 4 + x + 2x^2 + x^3$ and $\alpha = 3 + 4x + 5x^2$.

Bob subtracts k.D from c,

```
CT = Mod[ct - RD * k, q - 1]
```

```
216
```

Next he raises α to the power C. To write this as a polynomial we use the function `ElementToPolynomial`.

```
aCT
u = ElementToPolynomial[aCT, x]
```

```
{2, 1, 3}₇
```

```
2 + x + 3 x²
```

Next, Bob has to add $f(x)$ to get the monic polynomial $a(x)$. We use the function PolynomialMod.

```
AX = PolynomialMod[u + f, 7]
```

```
6 + 2 x + 5 x² + x³
```

We factor this by means of the function Factor.

```
Factor[AX, {Modulus -> 7}]
```

```
{(2 + x) (4 + x) (6 + x)}
```

The inverse permutation of π can be computed with InversePermutation *(in the package* DiscreteMath`Combinatorica` *that we have already loaded)*

```
InversePermutation[pi]
```

```
{7, 6, 2, 4, 5, 1, 3}
```

We subtract 1 from these elements because π acts on $\{0, 1, ..., 6\}$ instead of $\{1, 2, ..., 7\}$. We get

```
InversePermutation[pi] - 1
```

```
{6, 5, 1, 3, 4, 0, 2}
```

From this we see that the numbers 2, 4, and 6 are mapped to 1, 4, and 2 under π^{-1}. In other words, π maps 1, 2, and 4 to 2, 4, resp. 6.

We conclude that the message vector has ones on the coordinates 1, 2, and 4 (and thus zeros on the coordinates 0, 3, 5, and 6), i.e. the message vector is given by $\{m_i\}_{i=0}^{6} = \{0, 1, 1, 0, 1, 0, 0\}$.

This is indeed equal to the value that was chosen during encryption.

12.4   Problems

Problem 12.1
Solve the knapsack problem if the elements are given by 333, 41, 4, 172, 19, 3, 80, and 11 and if the total size of the knapsack equals 227.

Problem 12.2
Solve the knapsack problem if the elements are given by 31, 32, 46, 51 63, 72 and 87 and if the total size of the knapsack equals 227.

Problem 12.3 ^{M}
A knapsack cryptosystem has the numbers 381, 424, 2313, 2527, 2535, 3832, 3879, and 4169 as public key. They are obtained by multiplying the elements of a super-increasing sequence by $W = 4673$ and reducing the result modulo 5011.
Decrypt message 11678.

Problem 12.4
Let $p_1, p_2, ..., p_n$ be a sequence of different prime numbers and let P be their product. The numbers a_i, $1 \leq i \leq n$, are defined by $a_i = P / p_i$.
Let $S = \sum_{i=1}^n x_i.a_i$, where each element x_i is either 0 or 1.
Give a simple algorithm to recover the numbers x_i, $1 \leq i \leq n$, from S.

Problem 12.5 ^{M}
Let $C = 5738$ be the ciphertext obtained through a knapsack encryption with $\{u_i\}_{i=1}^n$ $= \{437, 1654, 1311, 625, 1250, 1720, 663, 1420, 63, 319\}$ as public knapsack.
Apply the L^3-attack to find the plaintext.

Problem 12.6
Which integer will be mapped to the binary vector (1, 1, 0, 1, 1, 0, 1, 0, 1, 1) by Algorithm 12.5?

Problem 12.7 ^{M}
Work out a complete Chor-Rivest cryptosystem example (including encryption and decryption) for the parameters $p = 11$, $k = 2$.

13   Hash Codes & Authentication Techniques

13.1   Introduction

In Section 1.1 we mentioned confidentiality (privacy) as the first reason why people use cryptosystem. Of course, this goal is very important and it does lead to interesting mathematical issues, but for the vast majority of data secrecy is not the user's prime concern.

Authentication and integrity on the other hand are almost always essential. Think, for instance, of receivers of data files, E-mail messages, fax, etc. Violation of the confidentiality does (in general) little harm, but significant damage may be done if somebody else is able to tamper with data files.

When studying authentication schemes one needs to distinguish between the following goals:

i)   Does one want unconditional security or just computational security?

ii)   Do the various parties trust each other or not?

iii)   Is there a mutually trusted third party?

iv)   Are the data files typically very long or just short?

v)   Is confidentiality also an issue?

vi)   Is the system intended for multiple use or just for single use?

The first two distinctions especially, have lead to completely different research areas. The main topic of Section 13.3 will be authentication schemes with *unconditional security*. This means that even with unlimited computing power the opponent can not break the system.These schemes are usually called *authentication codes* and a particular subclass of them is called *A-codes*.

Computationally secure systems are based on mathematical assumptions like the infeasibility of factoring large numbers or of taking discrete logarithms. These methods are called *digital signature schemes* and have already been discussed in Sections 8.1.2, 8.2.1, 8.2.2, and 9.1.4.

If a file is very long and confidentiality is not an issue a very common technique to add proof of authenticity and/or integrity to it, is to send it just like it is and then add a relatively short sequence of bits (e.g. 100-200) that depend in an intricate way on all the bits in the original message. This tail should be proof that the message indeed came from the assumed sender and that its contents have not been changed.

The standard way to realize this is to *hash* the file in a cryptographically secure way into a short sequence and compute a signature on this hash value. It is the signature of the hash value that is appended to the original file. If an authentication scheme is slow in its implementations (as is the case with digital signature schemes), this two-step approach may make them very practical.

In many applications, the hash function also makes use of a secret key that sender and receiver share. These systems, which are called Message Authentication Codes (MAC's) are not unconditionally secure, because somebody with unlimited computing power can, in principle, try out all keys.

Hash functions and MAC's are the topic of Section 13.2.

13.2 Hash Functions and MAC's

We do not intend to give a formal description of various types of hash codes. For our purposes, a global understanding of these codes and their properties suffices.

A *hash function* (or *hash code*) is a mapping h from $\mathcal{A}^*$, the set of all sequences of symbols from an alphabet $\mathcal{A}$, to $\mathcal{A}^m$, where m is some fixed positive integer. So, each sequence over $\mathcal{A}$ (of arbitrary length) will be mapped to a sequence over $\mathcal{A}$ of length m. In typical applications $\mathcal{A} = \{0, 1\}$ and the value of m ranges somewhere between 64 and 256.

Since one normally wants very fast implementations of hash functions h, we also require that it is easy to evaluate the hash value for any sequence over $\mathcal{A}$.

To make a hash function cryptographically secure, one often requires one or more of the following properties to hold.

H1: The hash function h is a one-way function (see Section 7.1.2), i.e. for almost all outputs b it is computationally infeasible to find an input $a \in \mathcal{A}^*$ such that $b = h(a)$.

H2: The hash function h is *weak collision resistant*. This means that for a given value of a it is computationally infeasible to find a second value $a' \in \mathcal{A}^*$, $a \neq a'$, such that $h(a) = h(a')$.

H3: The hash function h is *strong collision resistant* This means that it is computationally infeasible to find a pair of values $a, a' \in \mathcal{A}^*$, $a \neq a'$, such that $h(a) = h(a')$.

The implications of these requirements may be clear to the reader. For instance, H2 implies that if the hash values $h(a)$ of a file a is protected by a digital signature, one can not replace it by another file a' with the same hash value, simply because it is infeasible to find such an a'.

Property H3 is even much stronger and makes it possible to convince a judge that the system has been compromised.

Example 13.1

Consider $m = 1$ and $\mathcal{A} = Z_n$. To hash $a = (a_0, a_1, \ldots, a_l)$ one simply takes $b = (\sum_{i=0}^{l} a_i \bmod n)$. This hash value depends on all symbols in a and is easy to compute, but it does not meet any of the requirements H1-H3.

Example 13.2

Consider again $m = 1$ *and* $\mathcal{A} = \mathbb{Z}_n$. *To hash* $a = (a_0, a_1, \ldots, a_l)$ *one computes* $b = \left(\left(\sum_{i=0}^{l} a_i\right)^2 \bmod n\right)$. *If* n *is a large composite number, property H1 will hold, because taking square roots modulo such an integer* n *is considered to be infeasible (see Theorem 9.18).*

With the Mathematica functions Mod *and* Length *this hash function can be easily evaluated.*

```
h[inputfile_List, nn_Integer] :=
                Length[inputfile]         2
    Mod[(        ∑         inputfile[[i]])  , nn]
               i=1
n = 989;
in = {189, 632, 900, 722, 349};
h[in, n]
```

```
955
```

Properties H2 and H3 are not met, because $-a$ *will have the same hash value as* a. *Also, when one coordinate is increased and the next one decreased by the same amount, the hash value remains the same.*

```
alternative = Mod[-in, n]
h[alternative, n]
```

```
{800, 357, 89, 267, 640}
```

```
955
```

Even if a hash function meets properties H1-H3, it is still possible to intercept a transmission $(a, h(a))$ and replace it with another file $(a', h(a'))$. For this reason, one sometimes wants to introduce a secret key, shared by sender and receiver. The hash function h will now be called a *message authentication code* (MAC) and is a function of $\mathcal{A}^* \times \mathcal{K}$ to $\mathcal{A}^m$, where $\mathcal{K}$ is the key space, just as in conventional cryptosystems.

Example 13.3

Let $m = 64$ *and* $\mathcal{A} = \mathbb{Z}_2$. *With* $DES_k(u)$ *we denote a DES encryption of a block* u *of length under key* k . *Assume that* k *is the key that Alice and Bob share.*

Now, consider a binary file $\{a_1, a_2, \ldots, a_l\}$ *of length* l *that Alice is going to send to Bob. Alice first pads it with sufficient zeros to make the length a multiple of 64. Let L be this new length. To compute the hash value on* $\{a_1, a_2, \ldots, a_L\}$ *Alice follows the following algorithm:*

Algorithm 13.1 Using DES as Message Authentication Code

input   binary string $\{a_1, a_2, ..., a_L\}$, padded to make $64 \mid L$.

initialize $h = \overline{\{0, 0, ..., 0\}}^{64}$

for $i = 0$ to $(L/64) - 1$ **do** $h = DES_k(h \oplus \{a_{64\,i+1}, a_{64\,i+2}, ..., a_{64\,i+64}\})$

output   hash value h

The receiver duplicates the above calculations to verify that the file has not been changed and was indeed sent by Alice.

Of course, we could have used any other block cipher instead of DES in this example.

It is also possible to use a block cipher as a keyless hash function. To this end one also makes the key a public parameter.

The implicit assumption when using a block cipher for authentication purposes is that for a fixed key it behaves as a random permutation on the input set. Also, one hopes that the block cipher is cryptographically secure. In the next section, authentication codes will be discussed that are not based on any mathematical assumption.

There are many different standards for hash functions. The reader is referred to [MeOoV97] and [Schn96].

13.3   Unconditionally Secure Authentication Codes

13.3.1 Notions and Bounds

No authentication scheme can give an absolute guarantee that an accepted message comes from a particular user, say Alice. For instance, there is always a small probability that a (randomly or otherwise) generated sequence could have been made by Alice, but in fact was not. It will then be accepted by others as a genuine document from Alice.

It follows that it is necessary to define and compute the probability of a successful fraud. However, in such computations there is an essential difference between assuming the computational security of certain problems (as we do in public key cryptosystems), or not making any further assumptions at all (unconditional security). This last situation will be the topic of this section.

We shall assume that Alice and Bob trust each other and have agreed upon a secret key. This assumption is not really necessary, but then the notion of a trusted third party (like an arbitrator) must be introduced.

Let us start with a simple example.

Example 13.4

Alice wants to send a single bit of information (a yes or a no) to Bob by means of a word of length 2. Alice and Bob have 4 possible keys available. Alice and Bob make use of the following matrix:

key\sent	00	01	10	11
1	0	1	–	–
2	1	–	0	–
3	–	0	–	1
4	–	–	1	0

Authentication Code for two messages.

Table 13.1

So, message 1 will be sent as word 11 under the third key.

The probability that somebody else can successfully impersonate Alice is 1/2, because only two of the four words in {00, 01, 10, 11} are possible as transmitted word under the joint secret key of Alice and Bob.

An opponent Eve who tries to replace a transmitted message by another one will know that only two keys can possibly have been used, but she does not know which one. So, the probability of a successful substitution is also 1/2. For instance, if Eve intercepts 01, she knows that either message 1 was sent (under key 1) or message 0 was sent (under key 3). In the first case, she needs to transmit 00 and in the second case it should be 11, therefore, she succeeds with probability 1/2.

The above scheme even gives secrecy, because every transmitted word can come from message 0 or from message 1 (both with probability 1/2).

The general definition of an authentication code (we deviate here from the standard notation in the theory of authentication codes in order to avoid confusion with the standard notation in the theory of error-correcting codes) is as follows:

Definition 13.1
An *authentication code* is a triple $(M, \mathcal{K}, C)$ and a mapping $f : M \times \mathcal{K} \to C$ such that for all $m, m' \in M$ and for all $k \in \mathcal{K}$

$$f_k(m) = f_k(m') \implies m = m'. \tag{13.1}$$

The set M is called the *message set*, $\mathcal{K}$ the *key set*, and C the *codeword set*.

An authentication code can be depicted by a table U with the rows indexed by the keys k in $\mathcal{K}$, the columns indexed by the codewords c in C and entry (k, c) in U given by m if an $m \in M$

exists such that $f_k(m) = c$ (such an m is unique by (13.1)) and by a hyphen if such an m does not exist. We shall call this table the *authentication matrix* of the code.

In Example 13.4 above, $M = \{0, 1\}$, $\mathcal{K} = \{1, 2, 3, 4\}$ and $C = \{00, 01, 10, 11\}$. The authentication matrix of this code is given by Table 13.1.

Condition (13.1) implies that f_k is an injective mapping for each possible key.

When Bob receives codeword $c \in C$ from Alice, he will accept it as a signed version of message $m \in M$, where m is uniquely determined by $f_k(m) = c$. Here k is the key that Alice and Bob have agreed upon. To make the system practical, f_k should be easily invertible for each key. To this end, f_k (and C) will often have a much simpler structure.

> **Definition 13.2**
> An *A-code* is a triple $(M, \mathcal{K}, \mathcal{T})$ and a mapping $g : M \times \mathcal{K} \rightarrow \mathcal{T}$.
> Given key $k \in \mathcal{K}$, message $m \in M$ will be transmitted as (m, t), where $t = g_k(m)$ is called the *authenticator* of m.

By taking $f_k(m) = (m, g_k(m))$ and $C = M \times \mathcal{T}$ we see that an A-code is a special case of an authentication code.

A good authentication code is designed in such a way that fraudulent words $\hat{c}$ are spread evenly over C, while the subset of words that the legitimate receiver expects, knowing the common key $k \in \mathcal{K}$, is only a fraction of this set.

Thus the aim of an authentication code is that not only Bob, but also an arbitrator, can check the authenticity of a properly made c (in the case of an A-code by verifying that $g_k(m) = t$, in the case of a general authentication code by checking that c is in the image space of f_k), but an impersonator who does not know the key has only a small probability of getting a word $\hat{c}$ accepted. An attack by an impersonator is called an *impersonation attack*.

The same should be true if the enemy wants to replace a genuine codeword c (made with the proper key) by another one, say $\hat{c}$, that represents a different message. This kind of attack is called a *substitution attack*. Note that in this case, some information on the key is available to the opponent. We shall not discuss systems in which the same key can be safely used more than once by the legitimate users.

In the following definitions we shall assume that keys will be chosen from $\mathcal{K}$ with a uniform distribution and that messages will be chosen from M with a uniform distribution.

Let us assume that a general authentication code is being used by Alice and Bob. To maximize the probability of a successful impersonation, the opponent can do no better than select and send a codeword $c \in C$ that will have the highest probability of being accepted by the legitimate receiver. This is the case if for the maximum number of keys $k \in \mathcal{K}$ the codeword c will be in the image space of f_k.

Another way of saying this is that one looks for the column in the authentication matrix that has the maximum number of non-hyphen entries. The column index c of that column will be sent.

> **Definition 13.3**
> The probability P_I is the maximum probability of a successful impersonation attack, i.e.
>
> $$P_I = \max_{c \in C} \frac{|\{k \in \mathcal{K} | c \in f_k(\mathcal{M})\}|}{|\mathcal{K}|} \tag{13.2}$$

In Example 13.4, each codeword is the image of a message under exactly two of the four keys (each column counts two non-hyphens). So, $P_I = 2/4 = 1/2$.

In case of a substitution attack one has intercepted a codeword $c \in C$. This restricts the possible keys that may have been used by sender and receiver to $\{k \in \mathcal{K} | c \in f_k(\mathcal{M})\}$. The best attack for the opponent is to search among those codewords that are possible with these keys for the one that occurs the most often.

A different way of saying this is that in the authentication matrix of the code one looks at the column under the intercepted c and removes all rows from the matrix that have a hyphen in that column (these rows are indexed by a key that can not have been used). Also delete the column indexed by c. Among the remaining columns one looks for the one with the largest number of non-hyphen entries. The column index c' of that column will be substituted for c.

> **Definition 13.4**
> The probability P_S is the maximum probability of a successful substitution attack, i.e.
>
> $$P_S = \max_{c, c' \in C, c \neq c'} \frac{|\{k \in \mathcal{K} | c \in f_k(\mathcal{M}) \ \& \ c' \in f_k(\mathcal{M})\}|}{|\{k \in \mathcal{K} | c \in f_k(\mathcal{M})\}|} \tag{13.3}$$

In Example 13.4, each codeword is the image of a message under exactly two of the four keys. For each of these two keys, the other possible message will be mapped to a distinct codeword. So, $P_S = 1/2$.

The maximum of the two probabilities in (13.2) and (13.3) is often called the probability of successful *deception*. In formula

$$P_D = \max\{P_I, P_S\}. \tag{13.4}$$

Since an authentication function f_k is injective for each $k \in \mathcal{K}$, it follows that exactly $|\mathcal{M}|$ codewords must be authentic for any given key. In other words, it follows that each row of the authentication matrix U of an authentication code has exactly $|\mathcal{M}|$ non-hyphen entries. Since U has $|\mathcal{K}|$ rows and $|C|$ columns it follows that the average number of non-hyphen entries over the columns of U is $|\mathcal{K}| \times |\mathcal{M}| / |C|$. So, the maximum fraction of non-hyphen entries per column is at least $|\mathcal{M}| / |C|$. This proves the following theorem.

Theorem 13.2
The maximum probability P_I of a successful impersonation in an authentication scheme for $(\mathcal{M}, \mathcal{K}, C)$ satisfies

$$P_I \geq \frac{|\mathcal{M}|}{|C|}.$$

Similarly, in the case of the substitution attack the restriction of the authentication matrix U to the rows where an intercepted codeword c has non-hyphen entries consists of $|\{k \in \mathcal{K} \mid c \in f_k(\mathcal{M})\}|$ rows, each with $|\mathcal{M}| - 1$ non-hyphen entries. After deleting the column indexed by c, this restriction has $|C| - 1$ columns. So, the average value of the relative frequency of non-hyphen entries in this restriction of U is $(|\mathcal{M}| - 1)/(|C| - 1)$. This proves the following bound.

Theorem 13.3
The maximum probability P_S of a successful substitution in an authentication scheme for $(\mathcal{M}, \mathcal{K}, C)$ satisfies

$$P_S \geq \frac{|\mathcal{M}| - 1}{|C| - 1}.$$

If the messages and keys are not uniformly distributed over the message space and key space, it is still possible to derive lowerbounds on P_I, P_S, and P_D. In these lowerbounds, functions appear that we have discussed in Chapter 5. For the proofs of the next two theorems, we refer the interested reader to [Joha94b].

Theorem 13.4
Let M, K, and C denote random variables defined on $\mathcal{M}$, $\mathcal{K}$, and C, related by a function $f : \mathcal{M} \times \mathcal{K} \to C$, satisfying (13.1). Further, let $H(X \mid Y)$ and $I(X; Y)$ denote the conditional entropy function resp. the mutual information function. Then

$$P_I \geq 2^{-I(C;K)}. \tag{13.5}$$

$$P_S \geq 2^{-H(K \mid C)}. \tag{13.6}$$

$$P_D \geq \frac{1}{\sqrt{|K|}}. \tag{13.7}$$

The bound in (13.7) is called the *square root bound*. Authentication codes meeting this bound are called *perfect*.

Theorem 13.5
A necessary condition for an authentication code to be perfect is that

$$|\mathcal{M}| \leq \sqrt{|\mathcal{K}|} + 1.$$

For further reading on authentication codes, we refer the reader to [GilMW74], [MeOoV97], [Schn96], and [Simm92].

13.3.2 The Projective Plane Construction

In [GilMW74] one can find a nice description of a perfect authentication scheme. We first need to describe what a projective plane is, before we can explain this construction

□ A Finite Projective Plane

A projective plane is a kind of geometric object that differs somewhat from planes in regular Euclidean geometry. It is defined in a formal way by a set of axioms, that among others does not allow for parallel lines! After the definition we shall give a construction of these projective planes that will explain the name "projective".

We start with a finite set $\mathcal{P}$, whose elements are called *points*. Further, $\mathcal{L}$ is a collection of subsets $\ell \subset \mathcal{P}$, called *lines*. We shall say that a point P "lies" on a line ℓ, if $P \in \ell$. Also, two lines may "intersect" in a point, etc., so, we adopt all the regular terminology from geometry. To avoid trivialities, we shall assume that all lines contain at least two points $((\ell \in \mathcal{L}) \Longrightarrow (|\ell| \geq 2))$.

> **Definition 13.5**
> The pair $(\mathcal{P}, \mathcal{L})$ is called a finite *projective plane* if the following axioms hold:
>
> **PP-1:** There are at least four points, no three of which lie on the same line.
> **PP-2:** For every pair of points there is a unique line going through them.
> **PP-3:** Every pair of lines intersect in a unique point.

Property PP-1 is there to avoid the following object in our considerations. All lines have cardinality two and go through the same point (depicted below) except for one line which goes through the remaining points

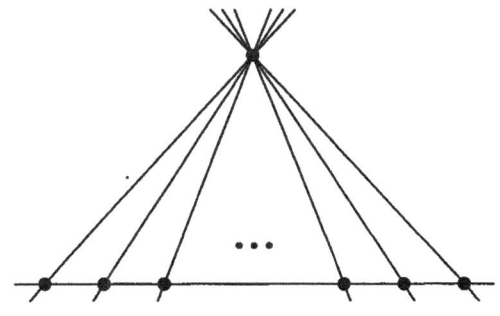

Theorem 13.6

Let $(\mathcal{P}, \mathcal{L})$ be a projective plane. Then there exists a constant n, called the *order* of the plane, such that:

PP-4   Every line contains exactly $n + 1$ points.
PP-5   Every point lies on exactly $n + 1$ lines.
PP-6 $|\mathcal{P}| = |\mathcal{L}| = n^2 + n + 1$.

Proof:

<u>Proof of PP-4</u>: Every line contains exactly $n + 1$ points.

Our first step is to show the claim that each point in $\mathcal{P}$ lies on at least three different lines. Let us start with four points P, Q, R, and S no three of which are colinear (see PP-1). For each of these points, any of the other three defines a unique line through them by PP-2. For a point T not on any of the lines going through two of the points P, Q, R, and S, the claim is also trivial (each of these four points defines a unique line through T). We leave it as an exercise to the reader to prove the claim for a point that is on one of the six lines going through two of the points P, Q, R, and S.

Now, consider an arbitrary point P. We know that at least three lines go through it. Let Q be a point on one of these lines, say on line ℓ. We shall show that all the other lines through P have the same cardinality. To this end, let $A_0 = P$, A_1, A_2, ..., A_m be the points on line m through P (where $m \neq \ell$) and let $B_0 = P$, B_1, B_2, ..., B_n be the points on line n through P (where $n \neq \ell$, $n \neq m$). We need to show that $m = n$.

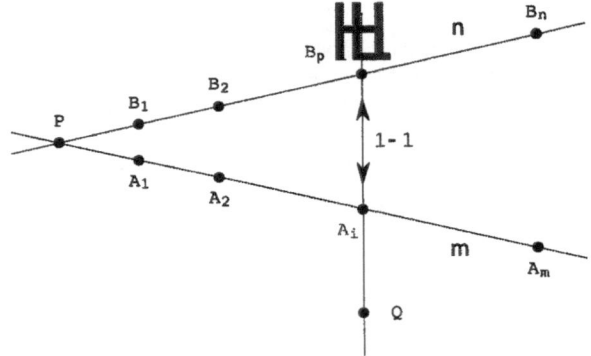

For each $0 \leq i \leq m$ there is a unique line through Q and A_i by PP-2. By PP-3 this line will intersect n in a point, say $B_{\pi(i)}$. This is a one-to-one mapping, because a line through Q and $B_{\pi(i)}$ can not intersect m in two points (by PP-3). We conclude that $m \geq n$. By interchanging the role of m and n we may conclude that $m = n$.

So, all the lines through P, except possibly for the line that also meets Q, have the same cardinality $n + 1$. By putting Q on one of the other lines through P, say n, and repeating the above argument, we may conclude that all lines through P have cardinality $n + 1$.

Let U be another point. For exactly the same reason as above, all the lines through U have the same cardinality, say $u + 1$. However one of these lines also goes through P by PP-2. It follows that $u = n$.

Proof of PP-5: Every point lies on exactly $n + 1$ lines.

Consider a point P and a line m not through P. Let the points on m be numbered $M_1, M_2, \ldots, M_{n+1}$. Each point M_i on m together with P defines a unique line passing through them (property PP-2). These lines are all different by the uniqueness property in PP-2. On the other hand, every line through P must intersect m in a unique point. We conclude that $n + 1$ lines pass through P.

Proof of PP-6: $|\mathcal{P}| = |\mathcal{L}| = n^2 + n + 1$.

Consider a point P. There are $n + 1$ lines through P, each containing n other points. This gives rise to $1 + (n + 1)n$ points. There are no other points in $\mathcal{P}$ by PP-2.

Similarly, consider a line ℓ. There are $n + 1$ points on it, each being on n other lines. This gives rise to $1 + (n + 1)n$ lines. There are no other lines in $\mathcal{L}$ by PP-3. (Notice the symmetry between points and lines in Definition 13.5.)

$\square$

Example 13.5

Take $n = 2$. Then $|\mathcal{P}| = |\mathcal{L}| = 7$. Each line contains three points and each point lies on three lines. This projective plane is depicted in the following figure.

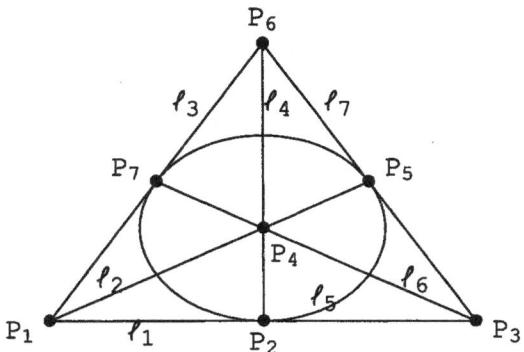

The 7 lines in this figure are the three outer edges, the three bisectors and the circle in the middle. So, $\mathcal{L}$ consists of the following seven lines:

$$\ell_1 = \{P_1, P_2, P_3\}, \quad \ell_2 = \{P_1, P_4, P_5\},$$
$$\ell_3 = \{P_1, P_6, P_7\}, \quad \ell_4 = \{P_2, P_4, P_6\},$$
$$\ell_5 = \{P_2, P_5, P_7\}, \quad \ell_6 = \{P_3, P_4, P_7\},$$
$$\ell_7 = \{P_3, P_5, P_6\}.$$

The projective plane of order 2 is unique and is called the Fano plane.

A projective plane is often described by means of its *incidence matrix*. This the matrix A of which the rows are indexed by the lines $\ell \in \mathcal{L}$, the columns by the points $P \in \mathcal{P}$ and where

$$A_{P,\ell} = \begin{cases} 1 & \text{if } P \text{ on } \ell, \\ 0 & \text{otherwise.} \end{cases}$$

The incidence matrix of the Fano plane (with the labeling given in the figure above) is

$$A = \begin{pmatrix} 1 & 1 & 1 & 0 & 0 & 0 & 0 \\ 1 & 0 & 0 & 1 & 1 & 0 & 0 \\ 1 & 0 & 0 & 0 & 0 & 1 & 1 \\ 0 & 1 & 0 & 1 & 0 & 1 & 0 \\ 0 & 1 & 0 & 0 & 1 & 0 & 1 \\ 0 & 0 & 1 & 1 & 0 & 0 & 1 \\ 0 & 0 & 1 & 0 & 1 & 1 & 0 \end{pmatrix};$$

The properties in Definition 13.5 and Theorem 13.6 can be directly translated into the following matrix requirements.

PP-2   Every two different columns of A have inner product 1.
PP-3   Every two different rows of A have inner product 1.
PP-4   Every row of A has $n + 1$ ones.
PP-5   Every column of A has $n + 1$ ones.
PP-6   Matrix A has $n^2 + n + 1$ rows and columns.

These properties can be summarized in the formula

$$A.A^T = A^T.A = n.I + J. \tag{13.8}$$

where J is the all-one matrix of size $(n^2 + n + 1) \times (n^2 + n + 1)$ and I the identity matrix (of the same size).

For the example above we can check this with the *Mathematica* functions Transpose and MatrixForm.

```
MatrixForm[A.Transpose[A]]
MatrixForm[Transpose[A].A]
```

$$\begin{pmatrix} 3 & 1 & 1 & 1 & 1 & 1 & 1 \\ 1 & 3 & 1 & 1 & 1 & 1 & 1 \\ 1 & 1 & 3 & 1 & 1 & 1 & 1 \\ 1 & 1 & 1 & 3 & 1 & 1 & 1 \\ 1 & 1 & 1 & 1 & 3 & 1 & 1 \\ 1 & 1 & 1 & 1 & 1 & 3 & 1 \\ 1 & 1 & 1 & 1 & 1 & 1 & 3 \end{pmatrix}$$

$$\begin{pmatrix} 3 & 1 & 1 & 1 & 1 & 1 & 1 \\ 1 & 3 & 1 & 1 & 1 & 1 & 1 \\ 1 & 1 & 3 & 1 & 1 & 1 & 1 \\ 1 & 1 & 1 & 3 & 1 & 1 & 1 \\ 1 & 1 & 1 & 1 & 3 & 1 & 1 \\ 1 & 1 & 1 & 1 & 1 & 3 & 1 \\ 1 & 1 & 1 & 1 & 1 & 1 & 3 \end{pmatrix}$$

□ A General Construction of a Projective Plane

There is a general construction of projective planes of order q, where q is a prime power. There are other constructions of projective planes, but they all have an order that is a prime power. It has been shown that no projective plane exists of order 6 and 10.

Let $V(3, q)$ denote a 3-dimensional vectorspace over GF (q), the finite field of q elements. Its elements are vectors $\underline{a} = (a_1, a_2, a_3)$ with a_i in GF(q). The cardinality of $V(3, q)$ is q^3. Let $\underline{0} = (0, 0, 0)$.

Each line through $\underline{0}$ can be described by a non-zero vector $\underline{a}$:

$$\{\lambda \underline{a} \mid \lambda \in \text{GF}(q)\}. \tag{13.9}$$

Of course, non-zero scalar multiples of $\underline{a}$ will give rise to the same line in $V(3, q)$. So, there are $(q^3 - 1)/(q - 1) = q^2 + q + 1$ different lines through $\underline{0}$.

Similarly, a plane through $\underline{0}$ in $V(3, q)$ can be described by a non-zero vector $\underline{u}$:

$$\{(a_1, a_2, a_3) \in V(3, q) \mid a_1 u_1 + a_2 u_2 + a_3 u_3 = 0\}. \tag{13.10}$$

Again, non-zero scalar multiples of $\underline{u}$ will give rise to the same plane in $V(3, q)$, therefore, there are $(q^3 - 1)/(q - 1) = q^2 + q + 1$ different planes through $\underline{0}$. A different way to describe a plane through $\underline{0}$ is $\{\lambda \underline{a} + \mu \underline{b} \mid \lambda \in \text{GF}(q), \mu \in \text{GF}(q)\}$.

Each non-zero point on a plane through $\underline{0}$ defines a line through $\underline{0}$. As before, non-zero scalar multiples of this point define the same line. We conclude that there are $(q^2 - 1)/(q - 1) = q + 1$ lines (through $\underline{0}$) on a plane (through $\underline{0}$).

Each line $\{\lambda \underline{a} \mid \lambda \in \text{GF}(q)\}$ can be embedded in a plane $\{\lambda \underline{a} + \mu \underline{b} \mid \lambda \in \text{GF}(q), \mu \in \text{GF}(q)\}$ by selecting any of the $q^3 - q$ points not on the line. Of course, not all these planes are different. A particular plane containing $\{\lambda \underline{a} \mid \lambda \in \text{GF}(q)\}$ can be obtained by any of the $q^2 - q$ points in the plane not on the line. It follows that each line (through $\underline{0}$) lies on exactly $(q^3 - q)/(q^2 - q) = q + 1$ planes (through $\underline{0}$).

> **Theorem 13.7**
> Let $\mathcal{P}$ be the set of lines through $\underline{0}$ in $V(3, q)$, where q is prime power, and let $\mathcal{L}$ be the set of planes through $\underline{0}$ in $V(3, q)$. Then $(\mathcal{P}, \mathcal{L})$ is a projective plane of order q.

Remark 1:

It is easy to get confused here. The projective points correspond to lines in $V(3, q)$ (through 0) and the projective lines correspond to planes in $V(3, q)$ (through 0).

Remark 2:

Note that we have already verified the properties PP-4, PP5, and PP-6 mentioned in Theorem 13.6.

Proof:

Proof of PP-1:

The four lines through $\underline{0}$ and each of the points $(1, 0, 0)$, resp. $(0, 1, 0)$, $(0, 0, 1)$, $(1, 1, 1)$ define four projective points in $\mathcal{P}$, no three of which lie on a projective line. The reason is that no three of these four points in $V(3, q)$ lie on the same plane through $\underline{0}$.

Proof of PP-2:

Let P and Q be two different projective points, and let them be defined by the lines $\{\lambda\underline{a} \mid \lambda \in \mathrm{GF}(q)\}$ and $\{\lambda\underline{b} \mid \lambda \in \mathrm{GF}(q)\}$ in $V(3, q)$. There is exactly one plane containing these two lines, namely $\{\lambda\underline{a} + \mu\underline{b} \mid \lambda \in \mathrm{GF}(q), \mu \in \mathrm{GF}(q)\}$. This plane defines the unique projective line through P and Q.

Proof of PP-3:

Let ℓ and m be two different projective lines. They correspond to two planes in $V(3, q)$ through $\underline{0}$. The line of intersection of these two planes is a line through $\underline{0}$, which defines the unique projective point on both ℓ and m.

$\square$

There are different techniques of generating a set of $q^2 + q + 1$ non-zero points in $V(3, q)$ that will give rise to different lines and planes through $\underline{0}$ in $V(3, q)$ (see (13.9) and (13.10)), i.e. to $q^2 + q + 1$ different projective points and projective planes.

A nice way, as we shall see in the following example, is to take a primitive element in $\mathrm{GF}(q^3)$, say ω, represent it as vector in $V(3, q)$, and take as points the elements $1, \omega, ..., \omega^{q^2+q}$. Indeed, let $\alpha = \omega^{(q^3-1)/(q-1)} = \omega^{q^2+q+1}$. Since ω has order $q^3 - 1$, it follows that α has order $q - 1$. It also follows that $\{0, 1, \alpha, ..., \alpha^{q-2}\} = \mathrm{GF}(q)$ (see Theorem B.29 and the Remark at the end of Subsection B.4.6). This means that for each $1 \leq j \leq q - 2$ the points ω^i and $\omega^{i+j(q^3-1)/(q-1)}$ in $V(3, q)$ give rise to the same projective point and thus we only have to consider $1, \omega, ..., \omega^{q^2+q}$.

Example 13.6

Take $q = 3$. To find a primitive polynomial of degree 3 over GF(3), we first have to load the Mathematica package Algebra`FiniteFields`. *After that we can apply the function* FieldIrreducible.

```
<< Algebra`FiniteFields`
```

```
m = 3; p = 3;
FieldIrreducible[GF[p, m], x]
```

$$1 + 2 x^2 + x^3$$

So, $GF(3^3)$ can be described by the set of ternary polynomials modulo $f(x) = x^3 + 2x^2 + 1$. Let $\omega \in GF(3^3)$ be a zero of $f(x)$. Since $f(x)$ is a primitive polynomial, it follows that ω has order 26. This can be checked with

```
f27 = GF[3, {1, 0, 2, 1}];
om = f27[{0, 1, 0}];
om^2
om^13
```

$$\{0, 0, 1\}_3$$

$$\{2, 0, 0\}_3$$

The element $\alpha = \omega^{(q^3-1)/(q-1)}$ is $\omega^{13} = 2$ in this case. Indeed, $\{0, 1, \alpha\} = GF(3)$.

So, the $3^2 + 3 + 1 = 13$ projective points can be found by computing ω^i, $0 \leq i < 13$. In this example, we take the equivalent set $1 \leq i \leq 13$ to keep the output uniform in appearance.

```
Do[Print[om^i], {i, 1, 13}]
```

$$\{0, 1, 0\}_3$$

$$\{0, 0, 1\}_3$$

$$\{2, 0, 1\}_3$$

$$\{2, 2, 1\}_3$$

$$\{2, 2, 0\}_3$$

$$\{0, 2, 2\}_3$$

$$\{1, 0, 1\}_3$$

$$\{2, 1, 1\}_3$$

$$\{2, 2, 2\}_3$$

$$\{1, 2, 1\}_3$$

$$\{2, 1, 0\}_3$$

$$\{0, 2, 1\}_3$$

$\{2, 0, 0\}_3$

To check if a projective point $\omega^i = (a_1, a_2, a_3)$ lies on the projective line defined by $\omega^j = (u_1, u_2, u_3)$ (see (13.10)), we need to check if $a_1 u_1 + a_2 u_2 + a_3 u_3 = 0$. In Mathematica this can be done as follows (the [[1]] removes the subscript in the presented output).

```
i = 5; j = 12;
a = om^i[[1]]
b = om^j[[1]]
Mod[a.b, 3] == 0
```

```
{2, 2, 0}
```

```
{0, 2, 1}
```

```
False
```

So, we are now ready to generate the projective plane of order 3. We present it by means of its incidence matrix.

```
A = Table[If[Mod[(om^i[[1]]).(om^j[[1]]), 3] == 0, 1, 0],
    {i, 1, 13}, {j, 1, 13}];
MatrixForm[
  A]
```

$$\begin{pmatrix}
0 & 1 & 1 & 0 & 0 & 0 & 1 & 0 & 0 & 0 & 0 & 0 & 1 \\
1 & 0 & 0 & 0 & 1 & 0 & 0 & 0 & 0 & 0 & 1 & 0 & 1 \\
1 & 0 & 0 & 0 & 0 & 0 & 1 & 0 & 1 & 1 & 0 & 0 & 0 \\
0 & 0 & 0 & 1 & 0 & 1 & 1 & 0 & 0 & 0 & 1 & 0 & 0 \\
0 & 1 & 0 & 0 & 0 & 0 & 0 & 1 & 0 & 1 & 1 & 0 & 0 \\
0 & 0 & 0 & 1 & 0 & 0 & 0 & 0 & 0 & 1 & 0 & 1 & 1 \\
1 & 0 & 1 & 1 & 0 & 0 & 0 & 1 & 0 & 0 & 0 & 0 & 0 \\
0 & 0 & 0 & 0 & 1 & 0 & 1 & 1 & 0 & 0 & 0 & 1 & 0 \\
0 & 0 & 1 & 0 & 0 & 0 & 0 & 0 & 1 & 0 & 1 & 1 & 0 \\
0 & 0 & 1 & 0 & 1 & 1 & 0 & 0 & 0 & 1 & 0 & 0 & 0 \\
0 & 1 & 0 & 1 & 1 & 0 & 0 & 0 & 1 & 0 & 0 & 0 & 0 \\
0 & 0 & 0 & 0 & 0 & 1 & 0 & 1 & 1 & 0 & 0 & 0 & 1 \\
1 & 1 & 0 & 0 & 0 & 1 & 0 & 0 & 0 & 0 & 0 & 1 & 0
\end{pmatrix}$$

We can check the properties PP-2, PP3 and PP4, PP5 by computing (see (13.8))

```
MatrixForm[A.Transpose[A]]
```

$$
\begin{pmatrix}
4 & 1 & 1 & 1 & 1 & 1 & 1 & 1 & 1 & 1 & 1 & 1 & 1 \\
1 & 4 & 1 & 1 & 1 & 1 & 1 & 1 & 1 & 1 & 1 & 1 & 1 \\
1 & 1 & 4 & 1 & 1 & 1 & 1 & 1 & 1 & 1 & 1 & 1 & 1 \\
1 & 1 & 1 & 4 & 1 & 1 & 1 & 1 & 1 & 1 & 1 & 1 & 1 \\
1 & 1 & 1 & 1 & 4 & 1 & 1 & 1 & 1 & 1 & 1 & 1 & 1 \\
1 & 1 & 1 & 1 & 1 & 4 & 1 & 1 & 1 & 1 & 1 & 1 & 1 \\
1 & 1 & 1 & 1 & 1 & 1 & 4 & 1 & 1 & 1 & 1 & 1 & 1 \\
1 & 1 & 1 & 1 & 1 & 1 & 1 & 4 & 1 & 1 & 1 & 1 & 1 \\
1 & 1 & 1 & 1 & 1 & 1 & 1 & 1 & 4 & 1 & 1 & 1 & 1 \\
1 & 1 & 1 & 1 & 1 & 1 & 1 & 1 & 1 & 4 & 1 & 1 & 1 \\
1 & 1 & 1 & 1 & 1 & 1 & 1 & 1 & 1 & 1 & 4 & 1 & 1 \\
1 & 1 & 1 & 1 & 1 & 1 & 1 & 1 & 1 & 1 & 1 & 4 & 1 \\
1 & 1 & 1 & 1 & 1 & 1 & 1 & 1 & 1 & 1 & 1 & 1 & 4
\end{pmatrix}
$$

□ **The Projective Plane Authentication Code**

> **Definition 13.6**
> Let $(\mathcal{P}, \mathcal{L})$ denote a projective plane. Let ℓ be one of the projective lines.
> The corresponding authentication code $(\mathcal{M}, \mathcal{K}, C)$ is defined by $\mathcal{M} = \ell$, $\mathcal{K} = \mathcal{P}\setminus\{P \mid P \text{ on } \ell\}$, $C = \mathcal{L}\setminus\{\ell\}$ and the mapping
>
> $$f_P(Q) \text{ is the unique line } c \text{ through } P \text{ and } Q, \qquad P \in \mathcal{K}, Q \in \mathcal{M}.$$

In words, the message set $\mathcal{M}$ consists of the points on ℓ, the key space $\mathcal{K}$ consists of all points not on ℓ, the code set C consists of all lines in $\mathcal{L}$, except for ℓ itself.

Finding the message back from the received codeword c is quite easy. Just intersect $c = f_P(Q)$ with ℓ. Their intersection point is the message.

That the above scheme defines an authentication code is easy to check. Its parameters are given in the following theorem.

> **Theorem 13.8**
> The A-code defined by a projective plane of order n has parameters
>
> $$|\mathcal{M}| = n+1, \ |\mathcal{K}| = n^2, \ |C| = n^2 + n.$$
>
> The probabilities of success for the impersonation and substitution attack are given by
>
> $$P_I = P_S = \frac{1}{n}.$$

The reader may want to check the above theorem on the Fano plane below. The four points not on ℓ form the key space $\mathcal{K}$, the three points on ℓ the message space $\mathcal{M}$, and the other six lines the codeword set C.

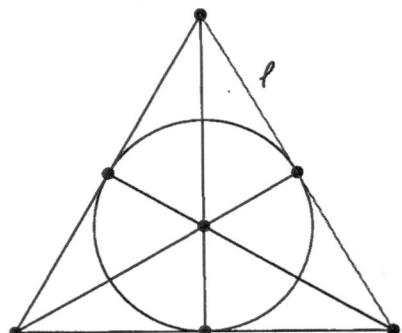

Proof of Theorem 13.8:

The parameters in this theorem follow directly from Theorem 13.6.

To compute P_I, we observe that an opponent can do no better than to select as a codeword a line c ($c \neq \ell$) that contains as many points outside ℓ (these are the possible keys) as possible. However, this number of points outside ℓ is independent of the choice of c. It is n by PP-4. So, by (13.2),

$$P_I = \frac{n}{|\mathcal{K}|} = \frac{n}{n^2} = \frac{1}{n}.$$

Similarly, if the opponent has observed codeword c (not equal to ℓ), there are still n keys (points on c but not on ℓ) possible. Let P be the intersection of c with ℓ. To replace it with another message (point Q on ℓ) the opponent can do no better than select a line d through such a point Q with as many points on c as possible. But by PP-2 this number is 1, independent of the choice of c and d, namely the unique point of intersection of c and d. So, by (13.3),

$$P_S = \frac{1}{n}.$$

$\square$

The authentication codes coming from projective planes, are perfect because P_I, P_S, and P_D are all $1/n$, which is equal to $1/\sqrt{|\mathcal{K}|}$.

Moreover, $|\mathcal{M}| = n + 1 = \sqrt{|\mathcal{K}|} + 1$, so, Theorem 13.5, tells us that the message set is of maximal size given this key set.

A construction of authentication codes by means of shift register sequences can be found in [Joha94a]. Its implementation is simpler than the projective plane construction above. For large message sets, e.g. data files, the codes discussed in Section 13.3.4 may be more practical.

13.3.3 A-Codes From Orthogonal Arrays

Definition 13.7
An *orthogonal array* $OA(n, k, \lambda)$ is a $k \times (\lambda.n^2)$ matrix of n symbols, such that in any two rows every possible pair of symbols occurs exactly λ times.
The number λ is called the *index* of the orthogonal array and k its *depth*.

Note that the above implies that each symbol occurs exactly $\lambda.n$ times in each row.

Example 13.7 (Part 1)

An example of an OA(4, 5, 1) is given by

$$U = \begin{pmatrix} 0 & 0 & 0 & 0 & 1 & 1 & 1 & 1 & 2 & 2 & 2 & 2 & 3 & 3 & 3 & 3 \\ 0 & 1 & 2 & 3 & 0 & 1 & 2 & 3 & 0 & 1 & 2 & 3 & 0 & 1 & 2 & 3 \\ 0 & 1 & 2 & 3 & 1 & 0 & 3 & 2 & 2 & 3 & 0 & 1 & 3 & 2 & 1 & 0 \\ 0 & 1 & 2 & 3 & 2 & 3 & 0 & 1 & 3 & 2 & 1 & 0 & 1 & 0 & 3 & 2 \\ 0 & 1 & 2 & 3 & 3 & 2 & 1 & 0 & 1 & 0 & 3 & 2 & 2 & 3 & 0 & 1 \end{pmatrix};$$

The following theorem shows how orthogonal arrays define A-codes in a natural way.

Theorem 13.9
Let U be an orthogonal array $OA(n, k, \lambda)$. Let the rows of U be indexed by the set $\mathcal{M}$ and the columns by the set $\mathcal{K}$. Further, put $\mathcal{T} = \{1, 2, ..., n\}$. Define the mapping $g : \mathcal{M} \times \mathcal{K} \rightarrow \mathcal{T}$ by $g_k(m) = U_{m,k}$. Then g defines an A-code with parameters: $|\mathcal{M}| = k$, $|\mathcal{K}| = \lambda.n^2$, $|\mathcal{T}| = n$.
Further
$$P_I = P_S = 1/n.$$

Proof: The parameters of this A-code follow from those of U.

The chance that an impersonation attack succeeds is $1/n$, because each symbol occurs equally often in a row of U.

The probability of a successful substitution attack is also $1/n$. The reason is that each intercepted authenticator occurs λ with each possible symbol, no matter which message was intercepted and which message one wants it to be replaced with.

□

Example 13.7 (Part 2)

For instance, in the matrix U defined above, message 4 under key 13 will be authenticated by

```
m = 4; k = 13;
U[[4, 13]]
```

```
1
```

When, message 4 is intercepted with authenticator 1, one knows that the key is among {2, 8, 11, 13}. Mathematica can find these positions with the functions _Flatten_ and _Position_.

```
l = Flatten[Position[U[[4]], 1]]
```

```
{2, 8, 11, 13}
```

Each other row has all four symbols on these four locations. This can be checked with the functions _MatrixForm_ and _Transpose_. The [[l]] below gives the restriction of the matrix to the rows indexed by the elements of the list l.

```
SubU = Transpose[U][[l]];
MatrixForm[Transpose[SubU]]
```

$$\begin{pmatrix} 0 & 1 & 2 & 3 \\ 1 & 3 & 2 & 0 \\ 1 & 2 & 0 & 3 \\ 1 & 1 & 1 & 1 \\ 1 & 0 & 3 & 2 \end{pmatrix}$$

There is a great deal of literature on orthogonal arrays. See [Hall67] or [BeJL86] for constructions, bounds and existence results. For instance, it is known that an $OA(q, q + 1, 1)$ exist for all prime powers q, because orthogonal arrays with these parameters exist if and only if projective planes of order q exist (see Theorem. 13.7 for a construction of a projective plane of order q).

Below we give a sketch of the proof of this result.

Let $(\mathcal{P}, \mathcal{L})$ be a projective plane of order q. Pick any of the lines ℓ in $\mathcal{L}$. Number the points on ℓ by $P_1, P_2, ..., P_{q+1}$ and the other points by $Q_1, Q_2, ..., Q_{q^2}$.

Let $\mathcal{L}_i$, $1 \le i \le q + 1$, be the collection of all lines through P_i except for ℓ itself. By PP-5, each $\mathcal{L}_i$ has cardinality q. Number the lines in each $\mathcal{L}_i$ from 1 to q.

Define $U_{i,j}$, $1 \le i \le q + 1$, $1 \le j \le q^2$, as k, where k, $1 \le k \le q$, is the index of the unique line in $\mathcal{L}_i$ that meets Q_j (which is the unique line in $\mathcal{L}$ through P_i and Q_j). Then U is an $OA(q, q + 1, 1)$.

Example 13.8

Consider the incidence matrix A of the projective plane of order 3 in Example 13.6.

$$A = \begin{pmatrix} 0 & 1 & 1 & 0 & 0 & 0 & 1 & 0 & 0 & 0 & 0 & 0 & 1 \\ 1 & 0 & 0 & 0 & 1 & 0 & 0 & 0 & 0 & 0 & 1 & 0 & 1 \\ 1 & 0 & 0 & 0 & 0 & 0 & 1 & 0 & 1 & 1 & 0 & 0 & 0 \\ 0 & 0 & 0 & 1 & 0 & 1 & 1 & 0 & 0 & 0 & 1 & 0 & 0 \\ 0 & 1 & 0 & 0 & 0 & 0 & 0 & 1 & 0 & 1 & 1 & 0 & 0 \\ 0 & 0 & 0 & 1 & 0 & 0 & 0 & 0 & 0 & 1 & 0 & 1 & 1 \\ 1 & 0 & 1 & 1 & 0 & 0 & 0 & 1 & 0 & 0 & 0 & 0 & 0 \\ 0 & 0 & 0 & 0 & 1 & 0 & 1 & 1 & 0 & 0 & 0 & 1 & 0 \\ 0 & 0 & 1 & 0 & 0 & 0 & 0 & 0 & 1 & 0 & 1 & 1 & 0 \\ 0 & 0 & 1 & 0 & 1 & 1 & 0 & 0 & 0 & 1 & 0 & 0 & 0 \\ 0 & 1 & 0 & 1 & 1 & 0 & 0 & 0 & 1 & 0 & 0 & 0 & 0 \\ 0 & 0 & 0 & 0 & 0 & 1 & 0 & 1 & 1 & 0 & 0 & 0 & 1 \\ 1 & 1 & 0 & 0 & 0 & 1 & 0 & 0 & 0 & 0 & 0 & 1 & 0 \end{pmatrix};$$

We define a function RowSwap to perform row exchanges in a matrix.

```
RowSwap[B_, i_, j_] :=
  Module[{U, V}, U = B; V = U[[i]]; U[[i]] = U[[j]]; U[[j]] = V; U]
```

Next we perform some column permutations on A to get a line ℓ as top row with all its points on the left. We use the <u>Transpose</u> *function.*

```
B = Transpose[A];
B = RowSwap[B, 1, 7]; B = RowSwap[B, 4, 13];
B = Transpose[B];
MatrixForm[B]
```

$$\begin{pmatrix} 1 & 1 & 1 & 1 & 0 & 0 & 0 & 0 & 0 & 0 & 0 & 0 & 0 \\ 0 & 0 & 0 & 1 & 1 & 0 & 1 & 0 & 0 & 0 & 1 & 0 & 0 \\ 1 & 0 & 0 & 0 & 0 & 0 & 1 & 0 & 1 & 1 & 0 & 0 & 0 \\ 1 & 0 & 0 & 0 & 0 & 1 & 0 & 0 & 0 & 0 & 1 & 0 & 1 \\ 0 & 1 & 0 & 0 & 0 & 0 & 0 & 1 & 0 & 1 & 1 & 0 & 0 \\ 0 & 0 & 0 & 1 & 0 & 0 & 0 & 0 & 0 & 1 & 0 & 1 & 1 \\ 0 & 0 & 1 & 0 & 0 & 0 & 1 & 1 & 0 & 0 & 0 & 0 & 1 \\ 1 & 0 & 0 & 0 & 1 & 0 & 0 & 1 & 0 & 0 & 0 & 1 & 0 \\ 0 & 0 & 1 & 0 & 0 & 0 & 0 & 0 & 1 & 0 & 1 & 1 & 0 \\ 0 & 0 & 1 & 0 & 1 & 1 & 0 & 0 & 0 & 1 & 0 & 0 & 0 \\ 0 & 1 & 0 & 0 & 1 & 0 & 0 & 0 & 1 & 0 & 0 & 0 & 1 \\ 0 & 0 & 0 & 1 & 0 & 1 & 0 & 1 & 1 & 0 & 0 & 0 & 0 \\ 0 & 1 & 0 & 0 & 0 & 1 & 1 & 0 & 0 & 0 & 0 & 1 & 0 \end{pmatrix}$$

Next we perform a number of row exchanges to get the subsets $\mathcal{L}_i$ nicely aligned ($\mathcal{L}_1$ will appear in rows 2, 3, 4, $\mathcal{L}_2$ in rows 5, 6, 7, etc).

```
BB = B;
BB = RowSwap[BB, 2, 8]; BB = RowSwap[BB, 6, 11];
BB = RowSwap[BB, 7, 13];
BB = RowSwap[BB, 8, 13];
MatrixForm[BB]
```

$$
\begin{pmatrix}
1 & 1 & 1 & 1 & 0 & 0 & 0 & 0 & 0 & 0 & 0 & 0 & 0 \\
1 & 0 & 0 & 0 & 1 & 0 & 0 & 1 & 0 & 0 & 0 & 1 & 0 \\
1 & 0 & 0 & 0 & 0 & 0 & 1 & 0 & 1 & 1 & 0 & 0 & 0 \\
1 & 0 & 0 & 0 & 0 & 1 & 0 & 0 & 0 & 0 & 1 & 0 & 1 \\
0 & 1 & 0 & 0 & 0 & 0 & 0 & 1 & 0 & 1 & 1 & 0 & 0 \\
0 & 1 & 0 & 0 & 1 & 0 & 0 & 0 & 1 & 0 & 0 & 0 & 1 \\
0 & 1 & 0 & 0 & 0 & 1 & 1 & 0 & 0 & 0 & 0 & 1 & 0 \\
0 & 0 & 1 & 0 & 0 & 0 & 1 & 1 & 0 & 0 & 0 & 0 & 1 \\
0 & 0 & 1 & 0 & 0 & 0 & 0 & 0 & 1 & 0 & 1 & 1 & 0 \\
0 & 0 & 1 & 0 & 1 & 1 & 0 & 0 & 0 & 1 & 0 & 0 & 0 \\
0 & 0 & 0 & 1 & 0 & 0 & 0 & 0 & 0 & 1 & 0 & 1 & 1 \\
0 & 0 & 0 & 1 & 0 & 1 & 0 & 1 & 1 & 0 & 0 & 0 & 0 \\
0 & 0 & 0 & 1 & 1 & 0 & 1 & 0 & 0 & 0 & 1 & 0 & 0 \\
\end{pmatrix}
$$

The last 9 columns define the orthogonal array OA(3, 4, 1). For instance, column 5 minus its first entry looks like (1, 0, 0, 0, 1, 0, 0, 0, 1, 0, 0, 1). This vector is the concatenation of four three-tuples, each containing one 1. It will be mapped to four entries in {1, 2, 3}, depending on whether the 1 is on the first coordinate, the second, or the third, therefore, column 5 will be mapped to (1, 2, 3, 3).

In this way the last 9 columns are mapped with the Mathematica functions Table, If, *and* Do *to the 4 × 9 matrix:*

```
U = Table[0, {i, 1, 4}, {j, 1, 9}];
Do[b = {BB[[2 + (i - 1) * 3, j]],
     BB[[3 + (i - 1) * 3, j]], BB[[4 + (i - 1) * 3, j]]};
            U[[i, j - 4]] = If[b == {1, 0, 0},
     1, If[b == {0, 1, 0}, 2, 3]],
            {i, 1, 4}, {j, 5, 13}];
MatrixForm[U]
```

$$
\begin{pmatrix}
1 & 3 & 2 & 1 & 2 & 2 & 3 & 1 & 3 \\
2 & 3 & 3 & 1 & 2 & 1 & 1 & 3 & 2 \\
3 & 3 & 1 & 1 & 2 & 3 & 2 & 2 & 1 \\
3 & 2 & 3 & 2 & 2 & 1 & 3 & 1 & 1 \\
\end{pmatrix}
$$

This is indeed an OA(3, 4, 1) and hence it defines an A-code with $|\mathcal{M}| = 4$, $|\mathcal{K}| = 9$, $|\mathcal{T}| = 3$ and $P_I = P_S = 1/3$.

Note that the last 9 columns in U (or A) can be further permuted to get

$$\begin{pmatrix} 1 & 1 & 1 & 2 & 2 & 2 & 3 & 3 & 3 \\ 1 & 2 & 3 & 1 & 2 & 3 & 1 & 2 & 3 \\ 1 & 3 & 2 & 3 & 2 & 1 & 2 & 1 & 3 \\ 2 & 3 & 1 & 1 & 2 & 3 & 3 & 1 & 2 \end{pmatrix} ;$$

13.3.4 A-Codes From Error-Correcting Codes

In [JohKS93] it is shown how authentication codes can be constructed from error-correcting codes (EC-codes) and vice versa. In this subsection we shall show how to convert an EC-code to an A-code. Our description is slightly different from the original one.

Let C be any $(n, |C|, d_H)$ EC-code over $GF(q)$, i.e. C is a subset of $V(n, q)$, the n-dimensional vectorspace over $GF(q)$, with minimum Hamming distance d_H. The latter means that all elements in C, which are called codewords, differ in at least d_H coordinates from each other. The dimension n of $V(n, q)$ is also called the length of C.

Let C have the additional property that

$$\underline{c} \in C \implies \underline{c} + \lambda \underline{1} \in C, \quad \text{for all } \lambda \in GF(q), \tag{13.11}$$

where $\underline{1}$ stands for the all-one vector.

For instance, any linear code containing the all-one vector satisfies (13.11). Note that (13.11) implies that q divides the cardinality of C.

The relation $\sim$ defined on C by

$$\underline{c} \sim \underline{c}' \quad \text{if and only if} \quad \underline{c} - \underline{c}' \equiv \lambda \underline{1} \quad \text{for some } \lambda \in GF(q), \tag{13.12}$$

defines an equivalence relation on C. Let M be a subcode of C, containing one representative from each equivalence class. So, M has cardinality $|C|/q$ and $C = \{\underline{m} + \lambda . \underline{1} \mid \underline{m} \in M, \lambda \in GF(q)\}$.

Let $\underline{m}_i$, $0 \le i < |C|/q$, be any enumeration of the codewords in M. As message set $\mathcal{M}$ for the authentication code that we are constructing, we take $\mathcal{M} = \{0, 1, ..., (|C|/q) - 1\}$. This means that we have a 1-1 correspondence between the subcode M and the index set $\mathcal{M}$. It is often convenient not to distinguish between these two sets. So, from now on we shall speak of message $\underline{m}_i$ instead of message i.

Example 13.9 (Part 1)

Consider the binary linear code C with generator matrix

$$G = \begin{pmatrix} 1 & 0 & 0 & 0 & 1 & 1 & 0 \\ 0 & 1 & 0 & 0 & 1 & 0 & 1 \\ 0 & 0 & 1 & 0 & 0 & 1 & 1 \\ 0 & 0 & 0 & 1 & 1 & 1 & 1 \end{pmatrix};$$

This means that C consists of the 16 vectors in the (binary) linear span of the rows. It is easy to check that different codewords in C differ in at least 3 coordinates. This makes C a (7, 16, 3) code in V(7, 2). Some readers may recognize C as a Hamming code.

That the all-one word is in C can easily be checked.

```
inf = {1, 1, 1, 1};
Mod[inf.G, 2]
```

```
{1, 1, 1, 1, 1, 1, 1}
```

It follows that C satisfies (13.11).

As subcode M of C we take all codewords in C with first coordinate equal to 0. So, M consists of the linear span of the lower three rows of G. The message set $\mathcal{M} = \{0, 1, ..., 7\}$ will be identified with M.

The key set $\mathcal{K}$ of the authentication code that we are constructing, will consist of the pairs (i, λ) with $1 \le i \le n$ and $\lambda \in GF(q)$. So, $\mathcal{K} = \{1, 2, ..., n\} \times GF(q)$ and $|\mathcal{K}| = n.q$.

The authenticator $g_k(\underline{m})$ of message $\underline{m} \in M$ under key $k = (i, \lambda)$ is simply given by

$$g_k(\underline{m}) = m_i + \lambda. \tag{13.13}$$

So, the authenticator set $\mathcal{T}$ is just $GF(q)$.

Theorem 13.10
Let C be an $(n, |C|, d_H)$ code satisfying (13.11). Let M be a subcode of C containing one element of each equivalence class under relation (13.12).
Let $\mathcal{K} = \{1, 2, ..., n\} \times GF(q)$ and $\mathcal{T} = GF(q)$. Further, $g_k(\underline{m}): M \times \mathcal{K} \rightarrow \mathcal{T}$ is defined by (13.13).
Then $(M, \mathcal{K}, \mathcal{T})$ is an A-code with parameters

$$|M| = |C|/q, \qquad |\mathcal{K}| = n.q, \qquad |\mathcal{T}| = q. \tag{13.14}$$

$$P_I = 1/q, \qquad P_S \le 1 - d_H/n. \tag{13.15}$$

Remark:

To make P_S acceptably low, one needs EC-codes with d_H close to n. For q-ary codes this is no problem, as we shall see in Example 13.10. Of course, q also needs to be large.

Proof of Theorem 13.10:

The parameters in (13.14) follow immediately from the construction.

To compute P_I, we note that an opponent who wants to impersonate the sender needs to find the right authenticator for his message $\underline{m}'$. However, for each coordinate $1 \le i \le n$ the set $\{m_i + \lambda \mid \lambda \in GF(q)\}$ is equal to $GF(q)$. In other words, each symbol occurs equally often as authenticator of $\underline{m}'$. So, the probability that the opponent will choose the correct authenticator is $1/q$, independent of the choice of the authenticator and independent of the message $\underline{m}'$ that the opponent tries to transmit. This proves that $P_I = 1/q$.

An opponent who wants to replace an authenticated message $(\underline{m}, t)$, (where $t = g_k(\underline{m})$) by another authenticated message, knows that the key in use is from a set of n possible keys (i, λ). To be more precise, for each coordinate $1 \le i \le n$ there is exactly one value of λ such that $m_i + \lambda = t$.

The optimal strategy for the opponent who wants to substitute another authenticated message for $(\underline{m}, t)$ is to find a message $\underline{m}'$, $\underline{m}' \neq \underline{m}$, such that in $g_k(\underline{m}') = t'$ for as many of those n keys as possible. This symbol t' is the authenticator for $\underline{m}'$ that will be accepted most likely.

It remains to show that t' will be accepted in at most $n - d_H$ cases, which implies that the probability of a successful substitution is at most $(n - d_H)/n = 1 - d_H/n$. This assertion follows from

$$| \{ (i, \lambda) \in \{1, 2, \ldots n\} \times GF(q) \mid (\underline{m})_i + \lambda = t \ \& \ (\underline{m}')_i + \lambda = t' \} |$$

$$= | \{ 1 \le i \le n \mid (\underline{m} - \underline{m}')_i = t - t' \} |$$

$$= n - d_H (\underline{m} - \underline{m}', (t - t')\underline{1})$$

$$\le n - d_H,$$

because $\underline{m} - \underline{m}'$ and $(t - t')\underline{1}$ are different words in the code C ($\underline{m}$ and $\underline{m}'$ are in different equivalence classes).

$\square$

Example 13.9 (Part 2)

To illustrate the second part of the proof above, we continue with the code of Example 13.9. If Alice wants to send message 7, she finds $\underline{m}$ with the Mathematica function `IntegerDigits` *from:*

```
mes = 7;
inf = IntegerDigits[mes, 2, 4]
m = Mod[inf.G, 2]
```

```
{0, 1, 1, 1}
```

```
{0, 1, 1, 1, 0, 0, 1}
```

(Remember that all messages had their first coordinate equal to 0.)

Suppose, that Alice and Bob have agreed upon key $(3, 1),$. *Then Alice will append the authenticator* $t = (\underline{m})_3 + 1 \equiv 0 \,(mod\ 2)$ *to her message, therefore, Alice will send*

```
i = 3; lam = 1;
{mes, Mod[m[[i]] + lam, 2]}
```

```
{7, 0}
```

Opponent Eve, observing this codeword, can conclude that the key is in the set $\{(i, \lambda) \mid 1 \leq i \leq 7, m_i + \lambda \equiv t \,(mod\ 2)\} = \{(1, 0), (2, 1), (3, 1), (4, 1), (5, 0), (6, 0), (7, 1)\}$. *To verify this, we use the Mathematica functions* `Table` *and* `Mod`.

```
t = 0;
T = Table[{i, Mod[t - m[[i]], 2]}, {i, 1, 7}]
```

```
{{1, 0}, {2, 1}, {3, 1}, {4, 1}, {5, 0}, {6, 0}, {7, 1}}
```

Suppose now that Alice wants to send message 5. The corresponding codeword $\underline{m}'$ *is given by*

```
mes' = 5;
inf' = IntegerDigits[mes', 2, 4];
m' = Mod[inf'.G, 2]
```

```
{0, 1, 0, 1, 0, 1, 0}
```

If Eve chooses $t' = 0$ *as authenticator she has a probability of* $4/7$ *of getting her message accepted, because exactly four of the possible keys would lead to this authenticator. With authenticator* $t' = 1$ *this probability is* $3/7$. *(We use the Mathematica functions* `Length` *and* `Intersection` *to test this.)*

```
t' = 0;
T' = Table[{i, Mod[t' - m'[[i]], 2]}, {i, 1, 7}]
Length[Intersection[T, T']]
```

```
{{1, 0}, {2, 1}, {3, 0}, {4, 1}, {5, 0}, {6, 1}, {7, 0}}
```

```
4
```

Example 13.10

The q-ary Reed-Solomon code of dimension k (see [MacW77]) has length $n = q - 1$ and minimum distance $d_H = n - k$. By multiplying each coordinate with a suitable constant, one may assume that $\underline{1} \in C$. Theorem 13.10 gives an A-code with parameters:

$$|\mathcal{M}| = q^{k-1}, \qquad |\mathcal{K}| = (q-1)q, \qquad |\mathcal{T}| = q.$$

$$P_I = 1/q, \qquad P_S \le k/(q-1).$$

The method explained in this section is certainly not the only way to make A-codes from EC-codes. It does have the property that each impersonation attack has the same probability of success (namely $1/q$).

Since every message can have each symbol in $\mathcal{T} = GF(q)$ as authenticator, it follows that the codeword set C has cardinality $|\mathcal{M}|.q$. This implies that Theorem 13.2 holds with equality.

In [JohKS93] the authors also show how to convert an A-code into an error-correcting code.

13.4 Problems

Problem 13.1
Prove that properties PP-1,PP2,PP3 in Definition 13.5 imply that a projective plane also contain four lines, no three of which go through the same point.

Problem 13.2
Prove that the Fano plane is unique (apart from a relabelling of the points and lines) .

Problem 13.3
Compare the Projective Plane Authentication Code construction (see Definition 13.6) with the authentication code with $M = \mathcal{K} = C = \mathbf{Z}_q$ defined by the one-time pad, i.e. $m \longrightarrow c$ with $c \equiv m + k \pmod{q}$. Also, answer this question when M is a random subset of $\mathbf{Z}_q$ of size $\sqrt{q}$.

Problem 13.4
Check that the rows of the incidence matrix in Example 13.6 can be permuted in such a way that the matrix becomes a circulant (each row is cyclic shift to the right of the previous row).

Problem 13.5 [M]
Use the same technique as in Example 13.6, to determine the top row of an incidence matrix of a projective plane of order 5.
Cycle this row around and check that it does define a projective plane of order 5.

Problem 13.6 [M]
Convert the orthogonal array OA(4, 5, 1) in Example 13.7 into a projective plane of order 4.

Problem 13.7
Show that condition (13.11) in Theorem 13.10 can be replaced by the requirement that C contains at least one codeword of weight n.

Problem 13.8 [M]
Repeat Example 13.9 (both parts) for the ternary $(11, 3^6, 5)$ code generated by

$$
G = \begin{pmatrix}
2 & 0 & 1 & 2 & 1 & 1 & 0 & 0 & 0 & 0 & 0 \\
0 & 2 & 0 & 1 & 2 & 1 & 1 & 0 & 0 & 0 & 0 \\
0 & 0 & 2 & 0 & 1 & 2 & 1 & 1 & 0 & 0 & 0 \\
0 & 0 & 0 & 2 & 0 & 1 & 2 & 1 & 1 & 0 & 0 \\
0 & 0 & 0 & 0 & 2 & 0 & 1 & 2 & 1 & 1 & 0 \\
0 & 0 & 0 & 0 & 0 & 2 & 0 & 1 & 2 & 1 & 1
\end{pmatrix} ;
$$

14   Zero Knowledge Protocols

Cryptographic protocols are exchanges of data between two or more parties following a precise order and format with the goal of achieving a particular security. Of course, the above definition is not very precise, but we have already seen some examples of cryptographic protocols. One is the identity verification protocol in Subsection 4.1.2, another is the Diffie-Hellman key exchange protocol in Subsection 8.1.2 and a few others are mentioned in Section 8.2.

A *zero-knowledge proof* is a technique to convince somebody else that one has certain knowledge, without having to reveal even a single bit of information (or a fraction thereof) about that knowledge. As a consequence, the verifier nor any passive eavesdropper gains any information from taking part in any number of executions of the protocol.

One may think of using a zero-knowledge protocol in the situation that one wants to use an ATM to withdraw money from a bank account. Instead of having to enter a PIN-code it should be enough to convince the teller that one knows this PIN-code. One wants to do this in such a way that no information about the PIN-code is released. In the next section, we shall give an example of how this can be done. In Section 14.2, another identity verification will be presented.

14.1   The Fiat-Shamir Protocol

As in Subsection 4.1.2, we are again in the situation that a smart card wants to convince a smart card reader that it is genuine. A trusted party that has to issue these cards selects a large composite number n, for instance n is the product of two large primes p and q, just as in the RSA system. The number n is a system parameter known to all parties.

The security of the *Fiat-Shamir protocol* [FiaS87] will be based on the assumption that taking square roots modulo a large composite number n is, in general, intractable. This is the same assumption that was made in the Rabin variant of the RSA system (Section 9.5). In Theorem 9.18, it was shown that the problem of finding a square root modulo a composite number is as hard as factoring it.

The trusted party computes an identity number ID for the smart card that should have the additional property that

$$\text{ID} \equiv s^2 \ (\text{mod } n) \tag{14.1}$$

for some integer s. The number ID may be computed from the name of the card holder and other relevant data, but a few bits should be left open for the trusted party to complete in order to make ID the square of an integer modulo n (ID has to be a quadratic residue mod p and mod q).

The trusted party computes the square root s of ID (it can do this, because it knows the factorization of n, see Subsection 9.5.3) and stores s in a segment of the memory of the smart card that is not accessible from the outside world.

One round of the Fiat-Shamir Protocol is depicted in Figure 4.1 below.

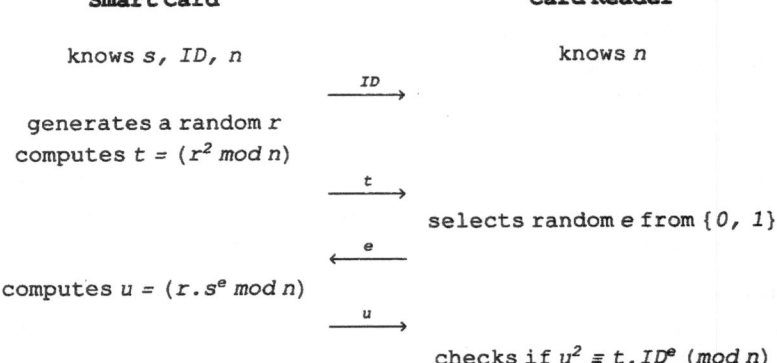

Fiat-Shamir identification protocol (one round)

Figure 14.1

The smartcard or card holder makes the identity number ID known to the card reader. To prove that the card was indeed issued by the trusted party, the card wants to convince the card reader that it knows s, the square root of ID modulo n.

To this end, the card generates a random number r, computes its square

$$t = (r^2 \bmod n) \tag{14.2}$$

and sends that to the card reader. In the jargon of this field, t is called a *witness* to the card's knowledge of r.

The card reader selects a random number e from $\{0, 1\}$ and presents that as a *challenge* to the card.

How the protocol *responds* to the challenge depends on the value of e.

If $e = 0$, the card simply sends the random number r back. The card reader then checks if its square is indeed equal to the value t that it received earlier from the card.

If $e = 1$, the card computes $u = r.s$, the product of the random number r and the secret square root s, and sends u to the card reader. The card reader checks if u^2 is indeed equal to $t \times$ ID modulo n, which should be the case, since $t \equiv r^2 \pmod{n}$ and ID $\equiv s^2 \pmod{n}$.

In Figure 14.1, these two alternatives are combined in the response $u = (r.s^e \bmod n)$. The card reader checks if

$$u^2 \equiv t.\text{ID}^e \pmod{n}. \tag{14.3}$$

It may be clear that if the card can supply r (when $e = 0$) and at the same time can supply $r.s$ (when $e = 1$), it must know the square root s of ID. It is also clear that if the smart card fails the test in (14.3), the card reader will reject the smart card.

If an unauthorized smart card knows beforehand the value of the challenge e, it can fool the card reader. This is obvious in the $e = 0$ case. In this case, the smart card takes a random r, presents $t = (r^2 \bmod n)$ as witness and later presents r itself as response. The secret square root s never played a role in these calculations.

If the illegitimate card knows that the challenge will be 1, it generates a random r, computes $t \equiv r^2 / \mathrm{ID} \pmod{n}$ and presents this value of t to the card reader. After having received the challenge $e = 1$, the smart card will present $u = r$. The card reader checks (see (14.3)) if u^2 is congruent to $t.\mathrm{ID}$ modulo n. This is obviously the case with $u = r$ and $t \equiv r^2 / \mathrm{ID} \pmod{n}$.

Note that the unauthorized card can not meet the challenge if he makes the wrong guess about the challenge. So, it will be caught with probability 1/2, if the smart card selects its challenge at random.

For this reason, smart card and card reader will run k times through the above protocol, where k is a security parameter. A smart card that does not know the value of s can guess the k random challenges with probability $(1/2)^k$, so it will be caught with probability $1 - (1/2)^k$.

The card should not use the same random number r twice, because as soon as the card reader knows both r and $r.s$ (through u), it can calculate the secret square root s.

The idea of proving certain things without revealing any information about it is counter-intuitive, but very powerful. There is a growing field of applications of zero-knowledge proofs.

Examples are electronic voting schemes that make it possible to cast votes in an anonymous way. On the other hand, the voter will be caught when attempting to vote twice. In these schemes, it can be checked that all votes have been counted in the final tally.

Another application is a payment system that allows you to withdraw money from your account in digital form and spend it anonymously. Even your own bank can no longer trace it to you. However, if you try to double spend the money, your identity can be recovered.

14.2   Schnorr's Identification Protocol

Schnorr's identity verification protocol [Schn91] is based on the difficulty of the discrete logarithm problem (Table 8.1). As in the Diffie-Hellman scheme, all participants share some parameters. First of all there is a finite field GF(q) (this could be $\mathbb{Z}_q$, if q is prime) and a prime divisor p of $q - 1$. Let ω be a primitive element of GF(q) and take $\alpha = \omega^{(q-1)/p}$. Then α is a primitive p-th root of unity. This means that $1, \alpha, ..., \alpha^{p-1}$ are all different and that $\alpha^p = 1$.

Example 14.1 (Part 1)

Let $p = 104729$ and $q = 8p + 1 = 837833$. Take $\omega = 3$ and $\alpha = \omega^{(q-1)/p} = \omega^8 = 6561$. To check that q is prime and that $\omega = 3$ is a primitive element in $\mathbb{Z}_q$ (which makes α a primitive p-th root of unity), we use the Mathematica functions Prime, PrimeQ, *and the function MultiplicativeOrder (defined in Appendix D, but standard in Mathematica 4) which computes the multiplicative order of an element..*

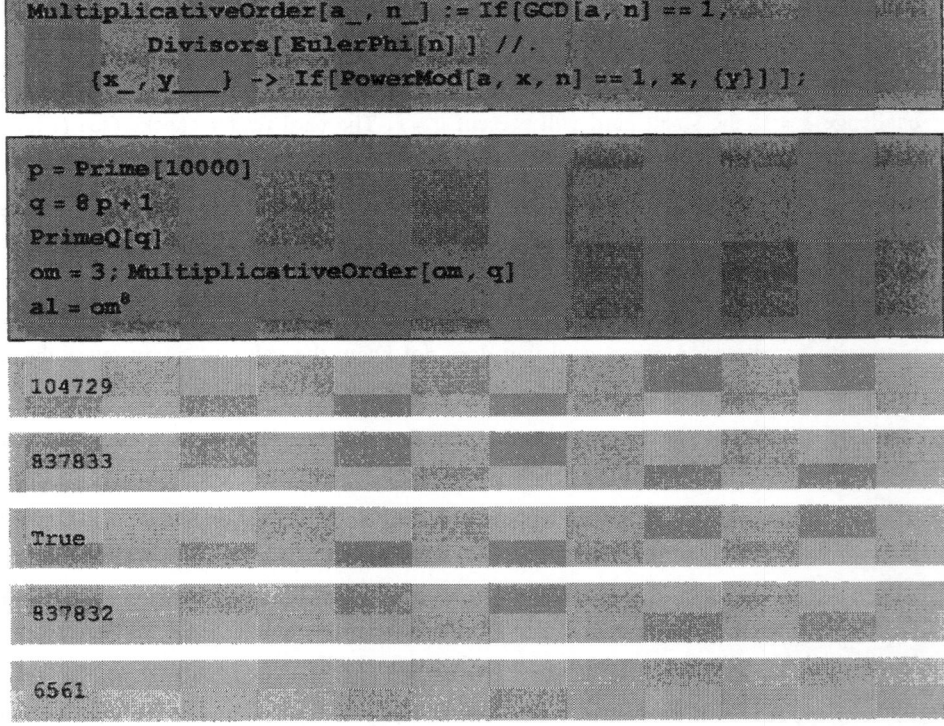

```
MultiplicativeOrder[a_, n_] := If[GCD[a, n] == 1,
      Divisors[EulerPhi[n]] //.
   {x_, y___} -> If[PowerMod[a, x, n] == 1, x, {y}] ];
```

```
p = Prime[10000]
q = 8 p + 1
PrimeQ[q]
om = 3; MultiplicativeOrder[om, q]
al = om^8
```

```
104729
```

```
837833
```

```
True
```

```
837832
```

```
6561
```

Each participant P (P for prover) selects a random secret exponent x_P, computes $y_P = \alpha^{x_P}$, and makes this value public. It is assumed that other participants are able to verify that y_P is indeed P's public parameter. This can be realized if a trusted authority signs y_P or if the public values are posted on a trusted "bulletin board". If someone else, say V for Verifier, wants to check P's identity y_P he does this by checking that P knows the corresponding x_P. Of course, P does not want release the secret value of x_P to anyone. Therefore, he uses a cryptographic protocol to convince V that he has knowledge of x_P.

Example 14.1 (Part 2)

Prover P has identity number $y_P = 693$ and secret exponent $x_P = 18126$. Indeed, $\alpha^{18126} \equiv 693 \pmod{q}$.

```
xP = 18126; yP = 693;
PowerMod[al, xP, q] == yP
```

```
True
```

Schnorr's identification protocol goes as follows. The verifier is presented with P's identity number y_P. Next, prover P generates a random exponent r, $0 \leq r < p$, computes $\varrho = \alpha^r$ and presents this value ϱ to the verifier V as a witness to his secret x_P. The verifier selects a random number s, $0 \leq s < p$, and hands this to P as challenge. Prover P responds by computing $u = r + s.x_P$ and gives this value to V. The verifier checks that $\alpha^u = \varrho.(y_P)^s$. This relation should hold, because $\alpha^u = \alpha^{r+s.x_P} = \alpha^r.(\alpha^{x_P})^s = \varrho.(y_P)^s$. This scheme is depicted in the following diagram.

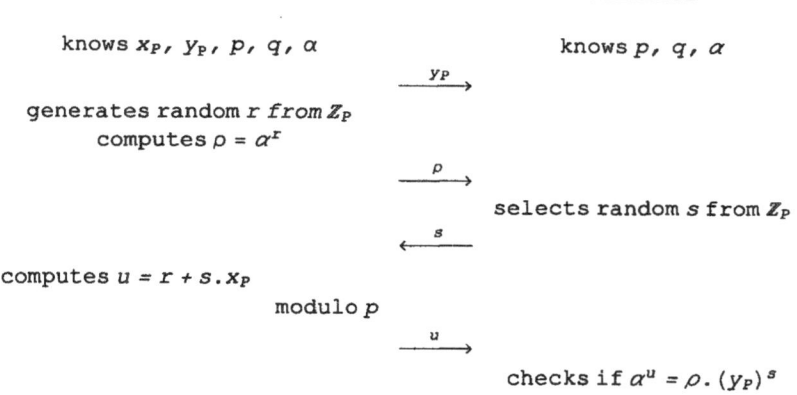

Schnorr's identification protocol

Figure 14.2

Example 14.1 (Part 3)

In the input below, the above protocol is executed. The Mathematica functions Random*,* Mod*, and* PowerMod *are used*

```
r = Random[Integer, p]; rho = PowerMod[al, r, q];
Print["witness is ", r]
s = Random[Integer, p]; Print["challenge is ", s]
u = Mod[r + s * xP, p]; Print["response is ", u]
PowerMod[al, u, q] == Mod[rho * PowerMod[yP, s, q], q]
```

```
witness is 36431
```

```
challenge is 29041

response is 65643
```

```
True
```

Of course, the prover will only be able to give the right response if he knows x_P satisfying $\alpha^{x_P} = y_P$. If he does not know x_P, he can guess the correct value of u with probability $1/p$. The value of p will be very large to make the discrete logarithm problem intractable (see Subsection 8.1.1).

Note that in the relation $u = r + s.x_P$ only the values u and s are known to V. In other words, the random value r makes sure that no information on x_P is leaked to V. This observation also shows that the prover should not use the same random number r twice. Indeed, from two relations $u_1 = r + s_1.x_P$ and $u_2 = r + s_2.x_P$ with known s_1, s_2, r_1, and r_2 the verifier can easily determine r and the secret x_P. One has $x_P = (u_1 - u_2)/(s_1 - s_2)$.

Example 14.1 (Part 4)

For the same witness, we generate a second challenge and response.

```
ss = Random[Integer, p]; Print["second challenge is ", ss]
uu = Mod[r + ss * xP, p]; Print["second response is ", uu]
PowerMod[al, u, q] == Mod[rho * PowerMod[yP, s, q], q]
```

```
second challenge is 62706

second response is 21550
```

```
True
```

To find x_P we compute $x_P = (u_1 - u_2)/(s_1 - s_2)$:

```
Mod[(u - uu) * PowerMod[s - ss, -1, p], p]
```

```
18126
```

The value 18126 is indeed the secret exponent x_P of the prover.

14.3   Problems

Problem 14.1^M
Duplicate Example 14.1 for $p = 113$. Find a suitable value for q.

15   Secret Sharing Systems

15.1   Introduction

In this chapter we shall not introduce a new cryptosystem, but we shall discuss a related topic. We start with an example from [Liu68].

> "Eleven scientists are working on a secret project. They wish to lock up the documents in a cabinet so that the cabinet can be opened if and only if six or more of the scientists are present. What is the smallest number of locks needed? What is the smallest number of keys to the locks each scientist must carry?"

Clearly, for each 5-tuple of scientists there has to be at least one lock, that can not be opened by them. Also each of the six remaining scientists has a key of that lock. More than one such lock per 5-tuple is not needed. So, $\binom{11}{5}$ locks are needed and each scientist carries $\binom{11-1}{5}$ keys. These numbers can be calculated with the *Mathematica* function `Binomial`.

```
Binomial[11, 5]
Binomial[11 - 1, 5]
```

```
462
```

```
252
```

The solution above is of course not very practical. Similarly, the described situation is not very realistic. However, there exists very real situations where one wants to share some sensitive information among a group of people, in such a way that only certain privileged coalitions are able to recover the secret information. Examples are a masterkey of a payment system or a private key that one does not want to store in a single place.

In a general setting, if P is a privileged group of people, meaning that they should be able to recover the secret data, then any other group containing P as a subgroup, should also be privileged. Also, if N is not privileged then any subset of N should not be privileged.

Definition 15.1

An *access structure* $(U, \mathcal{P}, \mathcal{N})$ consists of finite set U (of users), and two disjoint collections $\mathcal{P}$ and $\mathcal{N}$ of subsets of U ($\mathcal{P}$ for the *privileged* subsets and $\mathcal{N}$ for the *non-privileged*) with the property that

$$P \in \mathcal{P}, \ P \subset B \subset U \qquad \Longrightarrow \qquad B \in \mathcal{P}.$$
$$N \in \mathcal{N}, \ A \subset N \qquad \Longrightarrow \qquad A \in \mathcal{N}.$$

In the example above, $U = \{1, 2, ..., 11\}$, $\mathcal{P}$ consists of all subsets of U of size at least 6 and $\mathcal{N}$ of all the other subsets of U. It is a special case of what is generally called a threshold scheme.

It is often convenient to list only the set of the *minimal elements* of $\mathcal{P}$, denoted by $\mathcal{P}^-$, which can be obtained from $\mathcal{P}$ by leaving out each element of $\mathcal{P}$ that properly contains another element of $\mathcal{P}$. Similarly, one often represents $\mathcal{N}$ by the subset $\mathcal{N}^+$ consisting of its *maximal elements*.

An access structure is called *complete* or *perfect* if each subset of U is either in P or in N.

Definition 15.2

Let S be a random variable defined on a finite set S. Assume that S is uniformly distributed on S.

Let U be a collection of n participants, each having obtained a particular element S_i out of S from some trustworthy authority. Further, let $(U, \mathcal{P}, \mathcal{N})$ be an access structure. Then the collection $\{S_i\}_{i \in U}$ is called a *secret sharing scheme* for $(U, \mathcal{P}, \mathcal{N})$ if it satisfies the following two properties:

[SSS1]   each privileged group P of participants ($P \in \mathcal{P}$) can compute the secret S.

[SSS2]   each non-privileged group N of participants ($N \in \mathcal{N}$) can not compute any information on S.

The value S_i (to be called the *share* of i) should be interpreted as partial information of participant i on the secret S. In information theoretical notation (see Chapter 5), SSS1 and SSS2 can be reformulated as

[SSS1] $H(S \mid \{S_i\}_{i \in P}) = 0$ for any $P \in \mathcal{P}$.

[SSS2] $H(S \mid \{S_i\}_{i \in N}) = H(S)$ for any $N \in \mathcal{N}$.

Note that in secret sharing schemes that are not perfect, there may be coalitions C, $C \notin \mathcal{P} \bigcup \mathcal{N}$, of participants that are able to recover some information on the secret S (so, $H(S \mid \{S_i\}_{i \in C}) < H(S)$) without being privileged.

15.2   Threshold Schemes

A secret sharing scheme $\{S_i\}_{1 \le i \le n}$ is called an (n, k)-*threshold scheme*, if $\mathcal{P}$ consists of all subsets of U of cardinality $\ge k$ and $\mathcal{N}$ consists of all subsets of U of cardinality $\le k - 1$. By definition, a threshold scheme is a perfect secret sharing scheme. Properties SSS1 and SSS2 can be reformulated as

[TS1] Knowledge of k or more different S_i's makes S computable.

[TS2] Knowledge of at most $k - 1$ different S_i's leaves the secret S completely undetermined, more precisely all possible values in S are still equally likely.

Shamir describes (see [Sham79]) the following general construction of (k, n)-threshold schemes when S is a finite field GF (q), where q has to be larger than n. Here, we shall assume that q is a prime number, say $q = p$, in which case S is just $\mathbf{Z}_p$, the set of integers modulo p. The generalization to GF(q) will be immediate.

This system is based on the well known fact that a line is uniquely defined by any two points on it, that a parabola is uniquely defined by three points on it, etc. In general, a polynomial of degree $k - 1$ is uniquely determined by any k points on it.

> **Construction 15.1**
> Let the participants be labeled from 1 to n and let $S \in \mathbf{Z}_p$, $p > n$, be the secret data. Consider the polynomial
>
> $$f(x) = S + a_1 x + a_2 x^2 + \ldots + a_{k-1} x^{k-1},$$ (15.1)
>
> of degree at most $k - 1$, of which the coefficients a_j, $1 \le j \le k - 1$, are selected by some trustworthy authority in an independent, random way from $\mathbf{Z}_p$. Participant i, $1 \le i \le n$, is given as his share S_i the pair
>
> $$S_i = (i, f(i) \bmod p).$$ (15.2)

Example 15.1 (Part 1)

In order to construct a (10,4)-threshold scheme for secret $S = 17$ in $\mathbf{Z}_{19}$, we hide the secret in the polynomial $f(x)$ (note the use of the Mathematica function Mod*)*

```
Clear[f];
f[x_] := Mod[17 + 7 x + 12 x^2 + 5 x^3, 19]
```

where the coefficients of $x^j, 1 \le j \le 3$, are selected at random from $\mathbb{Z}_{19}$.

The values of the shares can be computed with the Mathematica function `Table`.

```
Table[{i, f[i]}, {i, 1, 10}]
```

```
{{1, 3}, {2, 5}, {3, 15}, {4, 6}, {5, 8},
 {6, 13}, {7, 13}, {8, 0}, {9, 4}, {10, 17}}
```

To check that the values S_i, $1 \le i \le n$, given by (15.2), form a (n, k)-threshold scheme, we have to check the two conditions TS1 and TS2.

Ad TS1:

Suppose that participants i_1, i_2, ..., i_k combine their shares $S_{i_1} = (i_1, f(i_1))$, $S_{i_2} = (i_2, f(i_2))$, ..., $S_{i_k} = (i_k, f(i_k))$. With the *LaGrange Interpolation Formula*, it is quite easy to determine $f(x)$. Indeed,

$$f(x) = \sum_{u=1}^{k} f(i_u) \prod_{l=1, l \ne u}^{k} \frac{x - i_l}{i_u - i_l}. \tag{15.3}$$

since the expression on the right hand side has degree $k - 1$, just as $f(x)$ does by (15.1), and since the right hand side takes on value $S_{i_j} = f(i_j)$ for $x = i_j$, $1 \le j \le k$, just as $f(x)$ does.

Note that by (15.1), the secret S is given by $f(0)$, therefore, in the calculation of the Lagrange Interpolation Formula, one can take $x = 0$ right from the start.

Example 15.1 (Part 2)

Suppose that participants 1, 3, 6, and 9 want to retrieve the secret S. They pool their shares $(1, 3)$, $(3, 15)$, $(6, 13)$, and $(9, 4)$.

The LaGrange Interpolation Formula can be performed with the Mathematica function `InterpolatingPolynomial`. *The function* `PolynomialMod` *is used for the reduction modulo 19.*

```
PolynomialMod[InterpolatingPolynomial[
    {{1, 3}, {3, 15}, {6, 13}, {9, 4}}, x], 19]
```

```
17 + 7 x + 12 x^2 + 5 x^3
```

The value of the secret S is the constant term in this expression. So, $S = 17$.

Ad TS2:

Suppose that shares S_{i_1}, S_{i_2}, ..., S_{i_l}, are known for some $l < k$. It follows from (15.1) and (15.3) that there are exactly q^{k-l-1} polynomials $g(x)$ satisfying $g(i_u) = S_{i_u}$, $1 \le u \le l$, and with any fixed value for $g(0)$.

Indeed, for any fixed value of $g(0)$ and any fixed group of $k - l - 1$ other participants and any given set of imaginary values of their shares, there is unique $g(x)$ meeting all requirements. This is a direct consequence of the LaGrange Interpolation Formula.

Example 15.1 (Part 3)

Suppose that participants 1, 3, and 9 attempt to retrieve secret S by pooling their shares (1, 3), (3, 15) and (9, 4).

Then the secret S can still take on any value (and each of these values is still equally likely). Indeed, adding the pair (0, S) to the above three shares leads to a unique polynomial through (0, S) and the three shares. This follows from the LaGrange Interpolation formula and can be checked as follows.

```
Clear[x]
Table[ {S, PolynomialMod[ InterpolatingPolynomial[
       {{0, S}, {1, 3}, {3, 15}, {9, 4}}, x], 19]},
                     {S, 0, 18}] // TableForm
```

0	$2 x + x^2$
1	$1 + 9 x + 5 x^2 + 7 x^3$
2	$2 + 16 x + 9 x^2 + 14 x^3$
3	$3 + 4 x + 13 x^2 + 2 x^3$
4	$4 + 11 x + 17 x^2 + 9 x^3$
5	$5 + 18 x + 2 x^2 + 16 x^3$
6	$6 + 6 x + 6 x^2 + 4 x^3$
7	$7 + 13 x + 10 x^2 + 11 x^3$
8	$8 + x + 14 x^2 + 18 x^3$
9	$9 + 8 x + 18 x^2 + 6 x^3$
10	$10 + 15 x + 3 x^2 + 13 x^3$
11	$11 + 3 x + 7 x^2 + x^3$
12	$12 + 10 x + 11 x^2 + 8 x^3$
13	$13 + 17 x + 15 x^2 + 15 x^3$
14	$14 + 5 x + 3 x^3$
15	$15 + 12 x + 4 x^2 + 10 x^3$
16	$16 + 8 x^2 + 17 x^3$
17	$17 + 7 x + 12 x^2 + 5 x^3$
18	$18 + 14 x + 16 x^2 + 12 x^3$

Remark 1:

In the generalization to arbitrary fields GF (q), the n participants are labeled by different non-zero field elements a_i, $1 \le i \le n$, and the share S_i of the i-th participant will be the pair $(a_i, f(a_i))$.

A way to realize this is to choose a primitive element (generator) $\alpha \in \text{GF}(q)$, label the participants from 1 to n and give the i-th participant as share the pair $(i, f(\alpha^i))$.

Remark 2:

The threshold scheme explained here assumes a trustworthy authority. It is also a system that can be used only once. As soon as participants have exchanged their shares to retrieve the secret, these shares are compromised. A new set of shares has to be set up for later use. In the literature one can find proposals that relax these conditions.

15.3   Threshold Schemes with Liars

In [McEl81] a variant of the construction above is proposed, that can handle the situation that some of the participants provide false information, so the share they provide does not have the correct value. Some participants may want to do this to prevent others from getting access to the secret data. It will turn out that it takes two extra shares to recover the secret for each incorrect share that is contributed. So, if $k + 2t$ participants pool their shares to recover the secret, at most t of the shares should be false.

> **Construction 15.2**
> Let S be a secret from GF(q), for some prime power q, and let $\alpha_1, \alpha_2, ..., \alpha_n$, $n \le q - 1$, be a list of n different non-zero elements in GF(q), e.g. $\alpha_i = \alpha^i$, $1 \le i \le n$, for some primitive element α in GF(q).
> Consider $f(x) = S + a_1 x + a_2 x^2 + ... + a_{k-1} x^{k-1}$,   where   the   coefficients a_j,
> $1 \le j \le k - 1$, are randomly selected from GF(q).
> The pair $(\alpha_i, f(\alpha_i))$ will be the share S_i of the i-th participant. Suppose that $k + 2t$ participants ($k + 2t \le n$) pool their shares and assume that at most t of these are incorrect. Then each of these participants can efficiently compute $f(x)$ and recover secret S.
> Moreover the incorrect shares can be identified.

Proof: The polynomial $f(x)$, used to compute the shares, is of degree $\le k - 1$ and has the additional property that at least $k + t$ of the correct shares lie on it. Could there be another polynomial, say $g(x)$, with the same properties? The answer is no. Indeed, since there are only $k + 2t$ shares, any two subsets of at least $k + t$ correct shares must have an intersection of at least k (honest) shares. These k shares lie on $f(x)$ and on $g(x)$. Since both $f(x)$ and $g(x)$ have degree at most $k - 1$, it follows that $f(x) = g(x)$.

To determine $f(x)$ the participants can try out all possible functions of degree $\le k - 1$ through k of the shares until a function passes through $\ge k + t$ of them. Of course, this is not an efficient way. For an efficient technique, the theory of error-correcting codes is needed (as in Chapter 11). The shares that are defined above in fact define codewords $(f(\alpha_1), f(\alpha_2), ..., f(\alpha_n))$ in a so-called shortened Reed-Solomon code with parameters $[n, k, n - k + 1]$.

We refer the reader, who is not familiar with this theory, to [MacWS77], Chapter 11. Both the

Berlekamp-Massey algorithm or the Euclidean algorithm give efficient ways to decode this code. In the context of our problem, where $k + 2t$ shares are known, one has to interpret the other $n - k - 2t$ shares as erasures. If the number of erasures plus twice the number of errors is less than the minimum distance of a code, one can still correct these errors and erasures. Here $(n - k - 2t) + 2.t$ is indeed less than $n - k + 1$. Efficient algorithms exists (see [Berl68], Section 10.4 and [SugK76]) to correct these errors and erasures for Reed-Solomon codes.

□

Remark 1: By taking $t = 0$ Construction 15.2 reduces to Construction 15.1.

Remark 2: If only $k + 2t - 1$ shares are available and t of them are incorrect, then $f(x)$ is not necessarily uniquely determined. For instance, it is possible that of $k + 2t - 1$ shares all of them except the first t lie on one polynomial of degree $k - 1$, while all these shares except the last t lie on another polynomial of degree $\leq k - 1$ (the intersection of the shares sets has cardinality $k - 1$).

In this case, there is however partial information on the secret.

Example 15.2

Consider $k = 3$, $t = 1$ and $p = 17$.

Of the four shares $(1, 4)$, $(2, 1)$, $(3, 5)$, $(4, 4)$, each three define a parabola, leaving the other point as incorrect value.

```
PolynomialMod[
  InterpolatingPolynomial[{{1, 4}, {2, 1}, {3, 5}}, x], 17]
PolynomialMod[InterpolatingPolynomial[
  {{1, 4}, {2, 1}, {4, 4}}, x], 17]
PolynomialMod[InterpolatingPolynomial[
  {{1, 4}, {3, 5}, {4, 4}}, x], 17]
PolynomialMod[InterpolatingPolynomial[
  {{2, 1}, {3, 5}, {4, 4}}, x], 17]
```

$14 + 12 x + 12 x^2$

$10 + x + 10 x^2$

$2 + 11 x + 8 x^2$

$12 + 8 x + 6 x^2$

Of the 17 possible secrets four are possible, all with equal probability.

15.4   Secret Sharing Schemes

Although there is a lot of literature on secret sharing schemes, there are also many central questions that still need to be answered. For this reason, we only discuss one example of a secret sharing scheme. The reader is referred to [Bric89] and [Dijk97] to find a discussion of various generalizations of the technique explained here. For a general introduction to secret sharing schemes we refer to [Stin95].

Assume that we have as access structure the set $(U, \mathcal{P}, \mathcal{N})$ with $U = \{1, 2, 3, 4\}$, $\mathcal{P}^- = \{\{1, 2\}, \{2, 3\}, \{3, 4\}\}$ and $\mathcal{N}^+ = \{\{1, 3\}, \{1, 4\}, \{2, 4\}\}$. This means that any subset of U containing both users 1 and 2, or users 2 and 3, or users 3 and 4 is a privileged set, while any other combination of users is non-privileged. Figure 15.1 depicts this situation.

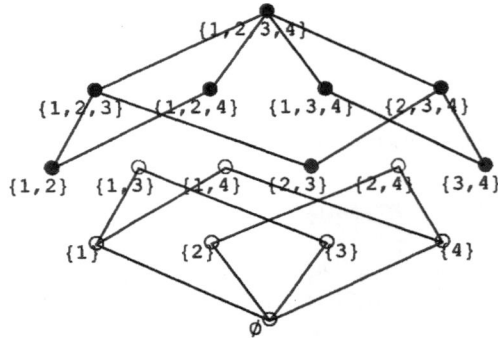

An Access Structure with Four Participants
• means privileged, o means non-privileged

Figure 15.1

The secret sharing scheme for this access structure will be set up in two steps. In the first step we want to share one bit (or byte or string) of information among the four participants.

Let s be a secret bit that we want to share among the participants of our access structure $(U, \mathcal{P}, \mathcal{N})$. The trusted authority selects two random bits a and b and gives the following shares to the participants:

participant	share
1	a
2	$s + a, b$
3	$s + b$
4	b

A Secret Sharing Scheme with One Secret Bit

Figure 15.2

The + sign stands for addition modulo 2. The reader may easily verify that this scheme meets requirements SSS1 and SSS2. For instance, participants 1 and 2 can compute s from $a + (s + a)$, where a comes from 1 and $s + a$ from 2.

Example 15.3

For instance, if the Trusted Authority wants to share secret $s = 1$ among the four participants, he may choose $a = 1$ and $b = 0$. The shares of 1, 2, 3, 4 will be 1, resp. (0,0), 1, 0.

Participants 2 and 4 can not recover s, because they only know $s + a$ and b (twice). Participants 3 and 4 can recover the secret s by adding their shares $s + b$ and $b: 1 + 0 = 1$.

We see that in the scheme of Figure 15.2 participant 2 has to store twice as many bits as is the size of the secret. This ratio can be improved by superimposing a permuted version of the scheme to itself.

Hence, now we consider a secret consisting of two bits s_1 and s_2. The trusted authority selects four random bits a, b, c, and d. He gives the following shares to the participants:

```
participant        share
     1              a, c
     2          s₁ + a, s₂ + c, b
     3          s₁ + b, s₂ + d, c
     4              b, d
```

A Secret Sharing Scheme with Two Secret Bits

Figure 15.3

In this scheme, the ratio between the size of the secret and the size of the longest share (this ratio is called *information rate*) is $2/3$. It can be shown that such a ratio is always at most 1. Secret sharing schemes that have an efficiency rate equal to 1 are called *ideal*.

There is a general matrix description of constructions of the above type. We shall explain it again for the example above.

The secret sharing system is described by the matrix G_{TA} of the trusted authority and the matrices G_i of the participants 1, 2, 3, and 4. The first two columns are labeled by the secret bits (s_1 and s_2) and the next four columns by the random variables (a, b, c, and d). Each row of G_i represents one entry of the share of participant i (expressed in terms of the secret bits and the random bits). The same holds for G_{TA}, where we view s_1, s_2 as his share.

$$GTA = \begin{pmatrix} 1 & 0 & 0 & 0 & 0 & 0 \\ 0 & 1 & 0 & 0 & 0 & 0 \end{pmatrix};$$

$$Gp1 = \begin{pmatrix} 0 & 0 & 1 & 0 & 0 & 0 \\ 0 & 0 & 0 & 0 & 1 & 0 \end{pmatrix};$$

$$Gp2 = \begin{pmatrix} 1 & 0 & 1 & 0 & 0 & 0 \\ 0 & 1 & 0 & 0 & 1 & 0 \\ 0 & 0 & 0 & 1 & 0 & 0 \end{pmatrix};$$

$$Gp3 = \begin{pmatrix} 1 & 0 & 0 & 1 & 0 & 0 \\ 0 & 1 & 0 & 0 & 0 & 1 \\ 0 & 0 & 0 & 0 & 1 & 0 \end{pmatrix};$$

$$Gp4 = \begin{pmatrix} 0 & 0 & 0 & 1 & 0 & 0 \\ 0 & 0 & 0 & 0 & 0 & 1 \end{pmatrix};$$

To see that these matrices indeed represent the secret sharing scheme we multiply them with the vector (s_1, s_2, a, b, c, d).

```
Clear[a, b, c, d, s1, s2];
vec = {s1, s2, a, b, c, d};
GTA.vec
Gp1.vec
Gp2.vec
Gp3.vec
Gp4.vec
```

```
{s1, s2}
```

```
{a, c}
```

```
{a + s1, c + s2, b}
```

```
{b + s1, d + s2, c}
```

```
{b, d}
```

We get the secret of the trusted authority and the shares of all the participants, so this is exactly the scheme that we had above.

The properties of a secret sharing scheme can now be translated as follows.

> **Theorem 15.3**
> Full rank matrices G_{TA} and G_i, $i \in U$, describe a secret sharing scheme for access structure $(U, \mathcal{P}, \mathcal{N})$ if and only if
> i) for each privileged set $A \in \mathcal{P}$ each row of G_{TA} lies in the linear span of the rows of the matrices G_i, $i \in A$,
> ii) for each non-privileged set $B \in \mathcal{N}$ no row of G_{TA} lies in the linear span of the rows of the matrices G_i, $i \in B$.

To check that the first row of G_{TA} lies in the linear span of the rows of G_1 and G_2 we use the *Mathematica* package `LinearAlgebra`MatrixManipulation`` and the functions <u>AppendColumns</u>, <u>MatrixForm</u>, <u>LinearSolve</u>, and <u>Transpose</u>.

```
<< LinearAlgebra`MatrixManipulation`;
```

```
u = GTA[[1]]
M = AppendColumns[Gp1, Gp2];
MatrixForm[M]
LinearSolve[Transpose[M], u, Modulus -> 2]
```

```
{1, 0, 0, 0, 0, 0}
```

$$\begin{pmatrix} 0 & 0 & 1 & 0 & 0 & 0 \\ 0 & 0 & 0 & 0 & 1 & 0 \\ 1 & 0 & 1 & 0 & 0 & 0 \\ 0 & 1 & 0 & 0 & 1 & 0 \\ 0 & 0 & 0 & 1 & 0 & 0 \end{pmatrix}$$

```
{1, 0, 1, 0, 0}
```

This shows that the first row of G_{TA} is the modulo-2 sum of the first row of G_1 and the first row of G_2.

Similarly, one can verify that s_2 can not be recovered by participants 1 and 3 in this way: the 2-nd row (and also the 1-st) of G_{TA} is not in the linear span of the rows of G_1 and G_3. .

```
u = GTA[[2]]
M = AppendColumns[Gp1, Gp3];
MatrixForm[M]
LinearSolve[Transpose[M], u, Modulus -> 2]
```

```
{0, 1, 0, 0, 0, 0}
```

$$\begin{pmatrix} 0 & 0 & 1 & 0 & 0 & 0 \\ 0 & 0 & 0 & 0 & 1 & 0 \\ 1 & 0 & 0 & 1 & 0 & 0 \\ 0 & 1 & 0 & 0 & 0 & 1 \\ 0 & 0 & 0 & 0 & 1 & 0 \end{pmatrix}$$

```
LinearSolve::nosol :
Linear equation encountered which has no solution.
```

```
LinearSolve[
  {{0, 0, 1, 0, 0}, {0, 0, 0, 1, 0}, {1, 0, 0, 0, 0}, {0, 0, 1, 0, 0},
   {0, 1, 0, 0, 1}, {0, 0, 0, 1, 0}}, {0, 1, 0, 0, 0, 0}, Modulus → 2]
```

We conclude this section by remarking that it is not so much a problem to make a perfect secret sharing scheme for a particular access structure, as it is to make an efficient one, i.e. with high information rate. Indeed, an inefficient secret sharing scheme for a particular access structure $(U, \mathcal{P}, \mathcal{N})$ goes as follows. Let s be the secret to be shared. For each $A \in \mathcal{P}^-$, select random bits $a_i^{(A)}$, $1 \le i \le |A|$, satisfying the binary congruence relation:

$$\sum_{i=1}^{|A|} a_i^{(A)} \equiv s \,(\mathrm{mod}\,2), \qquad A \in \mathcal{P}^-.$$

If $u \in A$, then participant u gets one of these $a_i^{(A)}$.

In the example of $U = \{1, 2, 3, 4\}$, $\mathcal{P}^- = \{\{1, 2\}, \{2, 3\}, \{3, 4\}\}$ and $\mathcal{N}^+ = \{\{1, 3\}, \{2, 4\}, \{1, 4\}\}$ we get in this way as share for secret s:

participant	share
1	$a_1^{\{1,2\}}$
2	$a_1^{\{1,2\}} + s,\ a_1^{\{2,3\}}$
3	$a_1^{\{2,3\}} + s,\ a_1^{\{3,4\}}$
4	$a_1^{\{3,4\}} + s$

A more compact way to denote this secret sharing scheme is

participant	share
1	a
2	$a + s,\ b$
3	$b + s,\ c$
4	$c + s$

This scheme has efficiency rate 1/2 and uses three random variables, as opposed to the two random variables in the scheme of Figure 15.2.

15.5   Visual Secret Sharing Schemes

In *visual secret sharing schemes* the secret to be shared consists of an image consisting of black and white (or of colored) pixels. Here we shall only discuss the black and white case, where "white" should be understood as "transparent". For instance, the number 3 can be depicted as follows.

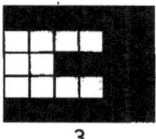

3

The shares consist of transparencies of the same shape also with black and white pixels. The idea of a visual secret sharing scheme for an access structure $(U, \mathcal{P}, \mathcal{N})$ is that privileged subsets of participants should be able to determine the secret by putting their transparencies on top of each other, while non-privileged subsets should obtain no information on the secret from their shares.

A visual secret sharing scheme can not be realized in a straightforward way. As soon as a pixel in a particular share is black, the corresponding pixel in the secret will also be black. To solve this problem, each pixel in the secret and in the shares will be subdivided in m subpixels, where m is called the *expansion factor* of the scheme. The assumption will be that two *visual threshold values* $0 \leq \alpha < \beta \leq 1$ exist such that:

- if at most $\alpha.m$ subpixels of a pixel are black, the pixel will be interpreted by the human eye as white,

- if at least $\beta.m$ subpixels of a pixel are black, the pixel will be interpreted as black.

If the number of black subpixels lies strictly between $\alpha.m$ and $\beta.m$, we assume that the human eye will not decide. The difference $\beta-\alpha$ is an indication for the level of contrast that is still present in an image if all pixels meet one of the above two requirements. There is biological evidence supporting the assumption that it is the relative difference in light intensity that is of importance to the human eye. See [VerT97] for a longer discussion.

In the context of visual secret sharing schemes, we have additional problems to face. For instance, if the shares of a non-privileged set are put on top of each other and a pixel contains more than $\alpha.m$ black subpixels, we know that the secret will be black at that place. Of course, such situations have to be avoided.

It should be clear that once we have a visual secret sharing scheme for one pixel, we can use it for the other pixels too, creating in this way a visual secret sharing scheme for the entire secret.

Here, we shall only explain a visual secret sharing scheme for a $(n, 2)$-threshold scheme. This means that any two participants should be able to recover the secret, while a single person should have no information at all about even one pixel. Before we do so, we describe the simple case where there are just two participants. We make the expansion factor $m = 2$. Let us call the following two subdivisions of a pixel L and R (for left black resp. right black):

It is clear that L and R put atop each other gives a black pixel, while both L+L and R+R are still half white and half black. Therefore, we can make a construction with threshold values $\alpha = 1/2$ and $\beta = 1$.

Construction 15.4
To share a white pixel, the trusted authority gives with equal probability either to both participants L or to both participants R.
To share a black pixel, the trusted authority gives with equal probability to one participant L and to the other R.
This gives a (2, 2)-visual threshold scheme with expansion factor $m = 2$ and threshold values $\alpha = 1/2$ and $\beta = 1$.

Below we give an example of possible shares that participants 1 and 2 have for the secret number 3 above.

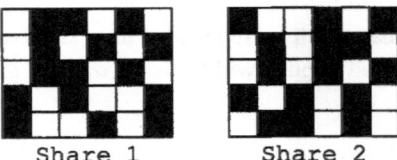

Share 1   Share 2

The reader can verify this by making transparencies of these two shares and putting them on top of each other.

There are many constructions known of (n, k)-visual threshold schemes. We shall describe a general construction for $k = 2$. Each particular implementation of the construction will lead to its own values for the expansion factor m and the threshold values α and β. It makes use of two $n \times m$ matrices, M_W and M_B, that will be used to distribute shares among the n participants for a white resp. black pixel. These matrices are further characterized by two values r and λ and have to satisfy the following properties:

VTS1: Matrix M_W consists of n identical copies of row $\overset{r}{\overline{11\ldots1}}\,\overset{m-r}{\overline{00\ldots00}}$.

VTS2: All row sums in M_B are equal to r.

VTS3: Every pair of rows in M_B has inner product λ.

The numbers m, α, β, r, and λ will be related. They can not take on any value.

Example 15.4 (Part 1)

Take n = 4 and m = 6. Let the matrices M_W and M_B be given by

$$M_W = \begin{pmatrix} 1 & 1 & 1 & 0 & 0 & 0 \\ 1 & 1 & 1 & 0 & 0 & 0 \\ 1 & 1 & 1 & 0 & 0 & 0 \\ 1 & 1 & 1 & 0 & 0 & 0 \end{pmatrix};$$

$$M_B = \begin{pmatrix} 1 & 1 & 1 & 0 & 0 & 0 \\ 1 & 0 & 0 & 1 & 1 & 0 \\ 0 & 1 & 0 & 1 & 0 & 1 \\ 0 & 0 & 1 & 0 & 1 & 1 \end{pmatrix};$$

Note that M_W and M_B satisfy properties VTS1-VTS3 for r = 3 and λ = 1.

The matrices M_W and M_B define two classes of $n \times m$ matrices:

$$\mathcal{M}_W = \{ M_W.P \mid P \text{ is a } m \times m \text{ permutation matrix} \},$$

$$\mathcal{M}_B = \{ M_B.P \mid P \text{ is a } m \times m \text{ permutation matrix} \}.$$

To distribute the shares for a particular pixel, the trusted authority takes either M_W or M_B, depending on whether the pixel is white or black, permutes the columns in a random way and gives the i-th row to participant i, $1 \le i \le n$.

Participant j makes the j-th subpixel white or black, depending on whether the j-th coordinate of his share is 0 or 1.

Example 15.4 (Part 2)

Suppose that the pixel that needs to be shared is black. The trusted authority selects a random permutation P with the Mathematica package `DiscreteMath`Permutations`` *and the function* <u>RandomPermutation</u> *as follows*

```
<<DiscreteMath`Permutations`
```

```
RP = RandomPermutation[6]
```

```
{3, 6, 4, 2, 1, 5}
```

This gives rise to the following permutation matrix (we use the functions <u>Table</u>, <u>Do</u>, *and* <u>MatrixForm</u>):

```
P = Table[0, {i, 1, 6}, {j, 1, 6}];
Do[P[[j, RP[[j]]]] = 1, {j, 1, 6}];
MatrixForm[P]
```

$$\begin{pmatrix} 0 & 0 & 1 & 0 & 0 & 0 \\ 0 & 0 & 0 & 0 & 0 & 1 \\ 0 & 0 & 0 & 1 & 0 & 0 \\ 0 & 1 & 0 & 0 & 0 & 0 \\ 1 & 0 & 0 & 0 & 0 & 0 \\ 0 & 0 & 0 & 0 & 1 & 0 \end{pmatrix}$$

Multiplying M_B on the right with P gives the matrix

```
PMB = MB.P;
MatrixForm[PMB]
```

$$\begin{pmatrix} 0 & 0 & 1 & 1 & 0 & 1 \\ 1 & 1 & 1 & 0 & 0 & 0 \\ 0 & 1 & 0 & 0 & 1 & 1 \\ 1 & 0 & 0 & 1 & 1 & 0 \end{pmatrix}$$

Putting the six subpixels in a 3×2 array in rowwise order, we get the following four shares for this black pixel:

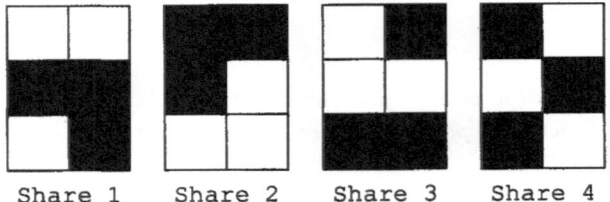

Share 1   Share 2   Share 3   Share 4

The reader can easily check that any two of these shares, when put atop of each other, will give five black subpixels and one white.

If the original pixel would have been white, we would have had

```
PMW = MW.P;
MatrixForm[PMW]
```

$$\begin{pmatrix} 0 & 0 & 1 & 1 & 0 & 1 \\ 0 & 0 & 1 & 1 & 0 & 1 \\ 0 & 0 & 1 & 1 & 0 & 1 \\ 0 & 0 & 1 & 1 & 0 & 1 \end{pmatrix}$$

This means that all four shares would have looked like

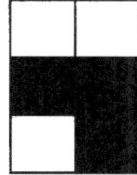

```
Each Share
```

Since each row in both M_W and M_B has the same number of ones (namely r) and since M_W and M_B are made from these by multiplying them on the right by all possible permutation matrices, it follows that each vector of length m and weight r occurs equally likely as a share for a white pixel as for a black pixel. This shows that our construction has as lower visual threshold value $\alpha = r/m$.

Because M_W is multiplied by a permutation matrix, it follows from VTS1 that when two participants have shares of a white pixel and they combine them, they do not gain anything.

On the other hand, any two rows of M_B have weight r by VTS2 and inner product λ by VTS3. This remains so if M_B is multiplied by a permutation matrix. It follows that any two shares of a black pixel have $2r - \lambda$ entries equal to one. In the example above $r = 3$ and $\lambda = 1$, giving $2r - \lambda = 5$ ones in any combination of two shares. We conclude that the construction by means of M_W and M_B has a higher visual threshold value $\beta = (2r - \lambda)/m$.

We have proved the following general construction:

> **Construction 15.5**
> Let M_B be an $n \times m$ matrix satisfying properties VTS2 and VTS3 for certain values of r and λ. Let M_W be of the form given by VTS1. Further, let M_W and M_B be the sets obtained from M_W resp. M_B by multiplying them on the right with all possible permutation matrices.
> Then a random choice of a matrix from M_W in case of a white pixel and a random choice of a matrix from M_B leads to $(n, 2)$-visual threshold scheme with expansion factor m and threshold values $\alpha = r/m$ and $\beta = (2r - \lambda)/m$.

> **Corollary 15.6**
> Take any n and let u be some value in between 2 and $n-1$. Let M_B be the matrix consisting of all columns of length n and weight u. Then M_B has $m = \binom{n}{u}$ columns.
>
> Moreover, every row of M_B has weight $r = \binom{n-1}{u-1}$ and any two rows have inner product $\lambda = \binom{n-2}{u-2}$.
>
> This defines a $(n, 2)$-visual threshold scheme with expansion factor $m = \binom{n}{u}$ and threshold values $\alpha = u/n$ and $\beta = (2n - u + 1)/n(n-1)$.

By taking $n = 4$ and $u = 2$ in the above corollary, one gets the construction of Example 15.4. Indeed, $m = \binom{n}{u} = \binom{4}{2} = 6$, $r = \binom{n-1}{u-1} = \binom{3}{1} = 3$ and $\lambda = \binom{n-2}{u-2} = \binom{2}{0} = 1$. The visual threshold values are given by $\alpha = 2/4 = 1/2$ and $\beta = 5/6$.

A disadvantage of the family of constructions described in the Corollary above, is the high expansion factor m.

A reader who is familiar with the theory of block designs and t-designs may have guessed from conditions VTS2 and VTS3 that these notions often play a role in the construction of a visual threshold scheme. We shall explain one particular construction.

Let p be any prime number. We recall from Definition A.9 that an integer u, $1 \le u < p$, is called a quadratic residue (QR) if the congruence relation $x^2 \equiv u \pmod{p}$ has a solution in $\mathbb{Z}_p$. How to determine if a number u is a quadratic residue is explained in Section A.4. With *Mathematica* one can do this with the function `JacobiSymbol`, which will output 1 if and only if u is a QR.

For instance, that $x^2 \equiv 12 \pmod{13}$ has a solution (namely ± 5) follows from

```
u = 12; m = 13; JacobiSymbol[u, m]
```

```
1
```

The Jacobi symbol is normally denoted by $\left(\frac{u}{p}\right)$ or just by $\chi(u)$, if there is no confusion about the value of p. Actually, the value of $\chi(u)$ is defined to be 0, when $u = 0$ and -1 when $1 \le u < p$ and u is not QR.

> **Corollary 15.7**
> Let p be any prime that is congruent to 3 mod 4. Define the $p \times p$ matrix M_B by
>
> $$(M_B)_{i,j} = \begin{cases} 1, & \text{if } j - i \text{ is QR,} \\ 0, & \text{otherwise.} \end{cases}$$
>
> Then every row of M_B has weight $r = (p-1)/2$ and any two rows have inner product $\lambda = (p-3)/4$.
> This defines a $(n, 2)$-visual threshold scheme with expansion factor $m = n$ and threshold values $\alpha = (p-1)/2\,p$ and $\beta = (3\,p-1)/4\,p$.

Proof:

Fixing a row index i of M_B we see that $j - i$, $0 \le j < p$, takes on all values in $\mathbb{Z}_p$. It follows from Theorem A.20 that each row in M_B has weight $(p-1)/2$.

Now consider the matrix $X = (\chi(j-i))_{0 \le i, j < p}$. Matrix M_B can be obtained from X by replacing all its -1-entries by 0. Consider two rows of X and let them be indexed by i_1 and i_2. Note that

$$\chi(i_1 - i_2) \overset{Th.A.21}{=} \chi(-1)\,\chi(i_2 - i_1) \overset{Cor.A.24}{=} -\chi(i_2 - i_1).$$

This means that the matrix X is skew-symmetric and that the i_2-th entry in row i_1 is equal to minus the i_1-th entry in row i_2. We conclude that, apart from a reordering of the coordinates, rows i_1 and i_2 will look like

$$
\begin{array}{ccccccc}
& & a & b & c & d \\ \hline
0 & +1 & +1 \ldots +1 & +1 \ldots +1 & -1 \ldots -1 & -1 \ldots -1 \\
-1 & 0 & +1 \ldots +1 & -1 \ldots -1 & +1 \ldots +1 & -1 \ldots -1
\end{array}
$$

where the two rows may have been interchanged.

The inner product of rows i_1 and i_2 in M_B is given by the value of a (since all -1's in X are replaced by 0 to get M_B). To find the values a, b, c, d we calculate first

$$\sum_{j=0}^{p-1} \chi(j - i_1)\,\chi(j - i_2) = \sum_{j=0}^{p-1} \chi(j)\,\chi\big(j - (i_2 - i_1)\big) = -1. \tag{15.4}$$

The first equality follows from the substitution $j - i_1 \to j$, the second one follows from Theorem A.22, since $i_1 \not\equiv i_2 \bmod p$.

Hence, we have the following relations:

$$
\begin{array}{ll}
2 + a + b + c + d = p, & \text{(X has } p \text{ columns)}, \\
a - b - c + d = -1, & \text{(from (15.4))}, \\
1 + a + b = (p-1)/2, & \text{(apply Thm. A.20 to the first row)}, \\
a + c = (p-1)/2, & \text{(apply Thm A.20 to the second row)}.
\end{array}
$$

These equations have a unique solution: $a = b = d = (p-3)/4$ and $c = (p+1)/4$. We conclude that the inner product of two different rows in M_B is $(p-3)/4$.

The Corollary is now a direct consequence of Construction 15.5.

□

Example 15.5

Take $p = 13$. The matrix M_B can be made with the Mathematica functions `JacobiSymbol`, `If`, *and* `Array` *as follows:*

```
p = 11;
A[i_, j_] := If[JacobiSymbol[j - i, p] == 1, 1, 0];
MB = Array[A, {p, p}];
MatrixForm[MB]
```

$$
\begin{pmatrix}
0 & 1 & 0 & 1 & 1 & 1 & 0 & 0 & 0 & 1 & 0 \\
0 & 0 & 1 & 0 & 1 & 1 & 1 & 0 & 0 & 0 & 1 \\
1 & 0 & 0 & 1 & 0 & 1 & 1 & 1 & 0 & 0 & 0 \\
0 & 1 & 0 & 0 & 1 & 0 & 1 & 1 & 1 & 0 & 0 \\
0 & 0 & 1 & 0 & 0 & 1 & 0 & 1 & 1 & 1 & 0 \\
0 & 0 & 0 & 1 & 0 & 0 & 1 & 0 & 1 & 1 & 1 \\
1 & 0 & 0 & 0 & 1 & 0 & 0 & 1 & 0 & 1 & 1 \\
1 & 1 & 0 & 0 & 0 & 1 & 0 & 0 & 1 & 0 & 1 \\
1 & 1 & 1 & 0 & 0 & 0 & 1 & 0 & 0 & 1 & 0 \\
0 & 1 & 1 & 1 & 0 & 0 & 0 & 1 & 0 & 0 & 1 \\
1 & 0 & 1 & 1 & 1 & 0 & 0 & 0 & 1 & 0 & 0
\end{pmatrix}
$$

So, we have a $(11, 2)$-visual secret sharing scheme with expansion factor $m = 11$ and threshold values $\alpha = (p - 1)/2\,p = 5/11$ and $\beta = (3\,p - 1)/4\,p = 8/11$.

15.6 Problems

Problem 15.1[M]

Set up a Shamir $(5, 3)$-threshold scheme for the secret 15 in GF(17).

Show how participants 1, 2 and 3 can recover the secret.

Show that for participants 1 and 2 together each element in GF(17) is an equally likely candidate for the secret.

Problem 15.2[M]

Consider a Shamir $(7, 4)$-threshold scheme in GF(23), where the participants 1, 3, 4, and 6 pool their shares $(1, 13)$, $(3, 19)$, $(4, 19)$, and $(6, 6)$ to retrieve the secret S. What will this secret be?

Suppose that participant 5 shows his share $(5, 3)$. Why is one of these five people lying?

Let all also participants 1 and 8 contribute there share: $(2, 4)$ and $(8, 12)$. Determine the liar and the real secret.

Problem 15.3[M]

Construct a $(7, 4)$-threshold scheme over the finite field $GF(16) = GF(2)[a]/(a^4 + a + 1)$ (see Theorem B.15).

What are the shares of the participants for secret $S = (1, 0, 1, 1)$ which stands for the field element a^{13}?

Show in detail how participants 2, 4, 5, 7 recover S.

Problem 15.4

Consider the following scheme over $\mathbb{Z}_3$:

participant	share
1	a, b, c + s_2
2	a + s_1, b, c
3	b + s_1, c - s_2, d
4	b, d + s_2

Give the matrix description of this scheme.

Prove that it is a secret sharing scheme for access structure $(U, \mathcal{P}, \mathcal{N})$ with $U = \{1, 2, 3, 4\}$, $\mathcal{P} = \{\{1, 2\}, \{2, 3\}, \{3, 1\}, \{3, 4\}\}$ and $\mathcal{N} = \{\{1, 4\}, \{2, 4\}, \{3\}\}$.

What is the information rate of this scheme? Is it perfect? Is it ideal?

Problem 15.5

Make a visualization of a set of possible shares for a black pixel in $(7, 2)$-visual threshold scheme, as constructed in Corollary 15.7.

What is the expansion factor of this scheme and what are its visual threshold values?

Appendix A   Elementary Number Theory

A.1   Introduction

Let $\mathbb{N}$ denote the set of natural numbers, $\mathbb{Z}$ the set of integers, and $\mathbb{R}$ the set of real numbers.

An integer d *divides* an integer n, if $n = k\,d$ for some $k \in \mathbb{Z}$. We shall denote this by $d \mid n$. If such an integer k does not exist, d does not divide n. This will be denoted by $d \nmid n$.

To check if the integer d divides the integer n, the *Mathematica* function `IntegerQ` can be used in the following way.

```
n = 16851; d = 123;  IntegerQ[n / d]
```

```
True
```

The *Mathematica* function `Divisor` gives a list of all divisors of a number n. For instance:

```
n = 16851; Divisors[n]
```

```
{1, 3, 41, 123, 137, 411, 5617, 16851}
```

An integer p, $p > 1$, is said to be *prime*, if 1 and p are its only positive divisors. With $p_1 = 2$, $p_2 = 3$, $p_3 = 5$, ... we introduce a natural numbering of the set of prime numbers.

Valuable *Mathematica* functions in this context are `Prime` and `PrimeQ`:

```
k = 35; Prime[k]
```

```
149
```

generating the 35-th prime number.

```
n = 1234567; PrimeQ[n]
```

```
False
```

telling if the input (here 1234567) is prime.

> **Theorem A.1**   Euclid
>
> There are infinitely many prime numbers.

Proof: Suppose the contrary. Let p_1, p_2, ..., p_k be the set of all primes. Next, we observe that the integer $(\prod_{i=1}^{k} p_i) + 1$ is not divisible by any of the primes p_1, p_2, ..., p_k. Let n be the smallest integer n that is not divisible by any of the primes p_1, p_2, ..., p_k. It can not be a prime number, because it is not in the list p_1, p_2, ..., p_k. It follows that n has a non-trivial factor d. But then this factor d is divisible by at least of the primes p_1, p_2, ..., p_k and so does n. A contradiction.

□

Between two consecutive primes there can be an arbitrary large gap of non-prime numbers. For example, the $n - 1$ elements in the sequence $n! + 2$, $n! + 3$, ..., $n! + n$ are divisible by respectively 2, 3, ..., n. Therefore none of them is prime.

> **Definition A.1**
>
> The function $\pi(n)$ counts the number of primes less than or equal to n.

In Mathematica, this function is denoted by `PrimePi[n]`.

```
n = 100; PrimePi[n]
```

```
25
```

The next theorem [see [HarW45], p.91] , which we shall not prove, tells us something about the relative frequency of the prime numbers in **N**.

> **Theorem A.2** The *Prime Number Theorem*
>
> $\lim_{n \to \infty} \frac{\pi(n)}{n/\ln n} = 1.$

```
n = 1000000; PrimePi[n] / (n / Log[n]) // N
```

```
1.08449
```

Two important definitions are those of the *greatest common divisor* and *least common multiple* of two integers.

> **Definition A.2**
>
> The *greatest common divisor* of two integers a and b, not both equal to zero, is the uniquely determined, positive integer d, satisfying

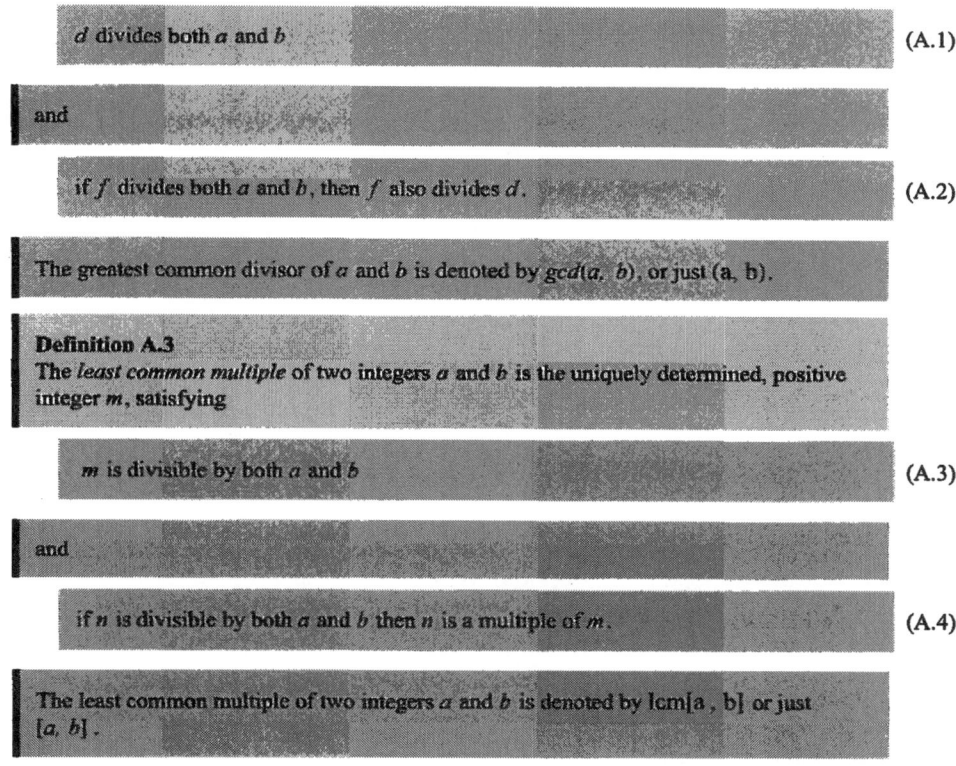

d divides both *a* and *b*   (A.1)

and

if *f* divides both *a* and *b*, then *f* also divides *d*.   (A.2)

The greatest common divisor of *a* and *b* is denoted by $\gcd(a, b)$, or just (a, b).

Definition A.3
The *least common multiple* of two integers *a* and *b* is the uniquely determined, positive
integer *m*, satisfying

m is divisible by both *a* and *b*   (A.3)

and

if *n* is divisible by both *a* and *b* then *n* is a multiple of *m*.   (A.4)

The least common multiple of two integers *a* and *b* is denoted by $\text{lcm}[a, b]$ or just
$[a, b]$.

To show the existence of gcd, we introduce the set

$$U = \{x.a + y.b \mid x \in \mathbf{Z},\ y \in \mathbf{Z},\ x.a + y.b > 0\}.$$

Let *m* denote the smallest element in *U*. We shall show that *m* satisfies (A.1) and (A.2). Clearly, if
f divides both *a* and *b* then *f* also divides *m*. So, *m* does satisfy (A.2). Now, write $a = q\,m + r$,
$0 \le r < m$ (subtract or add *m* sufficiently often from (resp. to) *a* until the remainder *r* lies in
between 0 and $m - 1$). If $r \ne 0$, then $r \in U$ (since both *a* and *m* are in *U*). This contradicts the
assumption on the minimality of *m*. So, $r = 0$, which means that *m* divides *a*. Similarly, *m* divides
b. So, *m* satisfies (A.1) too.

The uniqueness of gcd(*a*, *b*) follows from (A.1) and (A.2). Indeed, if *d* and *d'* both satisfy (A.1)
and (A.2), it follows that $d \mid d'$ and $d' \mid d$. Since both *d* and *d'* are positive, it follows that $d = d'$.

In a similar way, the existence and uniqueness of lcm[*a*, *b*] can be proved.

Alternative definitions of gcd(*a*, *b*) and lcm[*a*, *b*] are:

gcd(*a*, *b*) is the largest integer dividing both *a* and *b*
lcm[*a*, *b*] is the smallest positive integer divisible by both *a* and *b*.

The functions <u>GCD</u> and <u>LCM</u> can be evaluated by *Mathematica* as follows:

```
a = 12345; b = 67890; GCD[a, b]
```

```
15
```

```
a = 12345; b = 67890; LCM[a, b]
```

```
55873470
```

If two integers have a gcd equal to 1, we say that they are *coprime*. A consequence of the above is the following important theorem.

Theorem A.3
Let a and b be in $\mathbb{N}$. Then there exist integers u and v, such that

$$\gcd(a, b) = u.a + v.b.$$

In particular, if a and b are coprime, there exist integers u and v, such that

$$u.a + v.b = 1.$$

The following lemma seems too obvious to need a proof.

Lemma A.4
Let d divide a product ab and let the gcd of d and a be 1. Then d divides b.

Proof: Since $\gcd(d, a) = 1$, Theorem A.3 implies that $xd + ya = 1$, for some integers x and y. So, $xdb + yab = b$. Since d divides ab, it follows that d also divides $xdb + yab$ which equals b.

□

Corollary A.5
Let p be prime and let p divide $\prod_{i=1}^{k} a_i$, where a_i in $\mathbb{Z}$, $1 \le i \le k$. Then p divides at least one of the factors a_i, $1 \le i \le k$.

Proof: Use Lemma A.4 and induction on k.

□

With an induction argument the following theorem can now easily be proved.

> **Theorem A.6**   *Fundamental Theorem of Number Theory*
> Any positive integer has a unique factorization of the form
>
> $$\prod_i p_i^{e_i}, \ e_i \in \mathbb{N}.$$

Let $a = \prod_i p_i^{e_i}$, e_i in $\mathbb{N}$ and $b = \prod_i (p_i)^{f_i}$, f_i in $\mathbb{N}$. Then one easily checks that

$$\gcd(a, b) = \prod_i p_i^{\min\{e_i, f_i\}} \tag{A.5}$$

$$\operatorname{lcm}[a, b] = \prod_i p_i^{\max\{e_i, f_i\}} \tag{A.6}$$

$$\gcd(a, b) \operatorname{lcm}[a, b] = a\,b. \tag{A.7}$$

The *Mathematica* expression `FactorInteger[n]` gives the factorization of an integer n. The outcome is a list of pairs. Each pair contains a prime divisor of n and its exponent.

```
FactorInteger[123456789]
```

```
{{3, 2}, {3607, 1}, {3803, 1}}
```

```
a = 21375; b = 89775;
FactorInteger[a]
FactorInteger[b]
FactorInteger[GCD[a, b]]
FactorInteger[LCM[a, b]]
GCD[a, b] LCM[a, b] == a b
```

```
{{3, 2}, {5, 3}, {19, 1}}
```

```
{{3, 3}, {5, 2}, {7, 1}, {19, 1}}
```

```
{{3, 2}, {5, 2}, {19, 1}}
```

```
{{3, 3}, {5, 3}, {7, 1}, {19, 1}}
```

```
True
```

A.2 Euclid's Algorithm

Let a and b be two positive integers with $b \geq a$. Clearly, any divisor of a and b is a divisor of a and $b - a$ and vice versa. So, $\gcd(a, b) = \gcd(a, b - a)$. Writing $b = q.a + r$, $0 \leq r < a$, one has for the same reason that $\gcd(a, b) = \gcd(r, a)$. If $r = 0$ (and $b = q.a$), we may conclude that $\gcd(a, b) = a$, otherwise we continue in the same way with a and r. So, we write $a = q'.r + r'$, $0 \leq r' < r$, have $\gcd(a, b) = \gcd(r', r)$, etc., until one of the arguments indeed divides the other. This algorithm is an extremely fast way of computing the gcd of two integers and it is known as *Euclid's Algorithm*.

> **Algorithm A.7**   Simple Version of Euclid's Algorithm
>   **input** a, b positive integers
>   **while** $b > 0$ **do begin**
>   put r as the remainder of the a after division by b.
>   (So, write $a = q.b + r$, $0 \leq r < b$.)
>   put $a = b$
>   put $b = r$
>   **end**
>   **output** a

With the *Mathematica* functions `While`, `Floor`, and `Print`, the above algorithm runs like this

```
a = 1645; b = 861;
While[b ≠ 0, r = a - Floor[a / b] *b; {a, b} = {b, r}; Print[{a, b}]
```

{861, 784}

{784, 77}

{77, 14}

{14, 7}

{7, 0}

If one also wants to find the coefficients u and v satisfying Theorem A.3, this algorithm can be adapted as described below. Note that by leaving out the lines involving the integers u_i and v_i, this (extended) algorithm reduces to the simple version above.

```
Algorithm A.8        Extended Version of Euclid's Algorithm
     input          b ≥ a > 0
     initialize  s₀ = b; s₁ = a;
                     u₀ = 0; u₁ = 1; v₀ = 1; v₁ = 0; n = 1
     while sₙ > 0 do    begin
                     put n = n + 1;
                     write sₙ₋₂ = qₙ sₙ₋₁ + sₙ, 0 ≤ sₙ < sₙ₋₁
                     put uₙ = qₙ uₙ₋₁ + uₙ₋₂;
                     put vₙ = qₙ vₙ₋₁ + vₙ₋₂;
                     end
     put     u = (−1)ⁿ uₙ₋₁; v = (−1)ⁿ⁻¹ vₙ₋₁;
```

$$\gcd(a, b) = s_{n-1} = u.a + v.b \qquad (A.8)$$

Again Mathematica knows this extended version of Euclid's Algorithm as a standard function. It is called ExtendedGCD.

```
a = 861; b = 1645; ExtendedGCD[a, b]
```

```
{7, {107, −56}}
```

Note that in the example above one indeed has that

$$7 = \gcd(861,1645) = 107 \times 861 - 56 \times 1645$$

Proof of Algorithm A.8:

First observe that the elements s_n, $n \geq 1$, form a strictly decreasing sequence of non-negative integers. So the algorithm will terminate after at most b iterations. Later in this paragraph we shall analyze how fast Euclid's Algorithm really is.

From the recurrence relation $s_k = s_{k-2} - q_k s_{k-1}$ the algorithm it follows that

$$\gcd(a, b) = \gcd(s_0, s_1) = \gcd(s_1, s_2) = \ldots = \gcd(s_{n-1}, s_n) = \gcd(s_{n-1}, 0) = s_{n-1}.$$

This proves the first equality in (A.8). We shall now prove that for all k, $0 \leq k \leq n$,

$$(-1)^{k-1} u_k \, a + (-1)^k v_k \, b = s_k.$$

Note that substitution of $k = n - 1$ in this relation proves the second equality in (A.8).

For $k = 0$ and $k = 1$ the above relation holds by our choice of the initialization values for u_0, u_1, v_0 and v_1. We now proceed by induction. It follows from the recurrence relations in the algorithm and from the induction hypothesis, that

$$s_k = s_{k-2} - q_k s_{k-1} = \{(-1)^{k-3} u_{k-2}\, a + (-1)^{k-2} v_{k-2}\, b\} - q_k \{(-1)^{k-2} u_{k-1}\, a + (-1)^{k-1} v_{k-2}\, b\} =$$

$$= (-1)^{k-1} (u_{k-2} + q_k u_{k-1})\, a + (-1)^k (v_{k-2} + q_k v_{k-1})\, b = (-1)^{k-1} u_k\, a + (-1)^k v_k\, b.$$

$\square$

Of course there is no need to keep all the previously calculated values of s_k, u_k and v_k stored in the program. Only the last two of each together with q_k will suffice. The reason for introducing them in the algorithm was only to facilitate the readability of the proof above.

With the *Mathematica* functions While, Floor, and Print, the above algorithm runs like this:

```
b = 1645; a = 861;
n = 1;
so = b; sn = a;
uo = 0; un = 1;
vo = 1; vn = 0;
While[sn ≠ 0,
  Print[(-1)^(n-1), "x", un, "x", a, " + ", (-1)^n, "x",
    vn, "x", b, "=", sn];              q = Floor[so / sn];
  n = n + 1;              {so, sn, uo, un, vo, vn} =
  {sn, so - q * sn, un, q * un + uo, vn, q * vn + vo}]
```

```
1×1×861 + -1×0×1645=861

-1×1×861 + 1×1×1645=784

1×2×861 + -1×1×1645=77

-1×21×861 + 1×11×1645=14

1×107×861 + -1×56×1645=7
```

We would like to conclude this section by saying something about the complexity of Euclid's Algorithm. It may be clear that this algorithm is at it slowest if at each step the quotient q_k has value 1 (if possible). This is the case if $s_{k-2} = s_{k-1} + s_k$ for all $2 \le k \le n-1$ and that $s_{n-2} = 2 s_{n-1}$, $s_n = 0$. In other words, the smallest value of b (and arbitrary $0 < a < b$) such that the evaluation of $\gcd(a, b)$ takes $n-1$ steps is given by $b = F_n$ and $a = F_{n-1}$, where the $\{F_i\}_{i \ge 0}$ sequence is the famous sequence of *Fibonacci numbers* defined by $F_0 = 0$, $F_1 = 1$, $F_{i+2} = F_{i+1} + F_i$ for $i \ge 0$.

By letting *Mathematica* operate repeatedly on a list of two consecutive Fibonnacci numbers (the function Nest is used for this), one gets the following method to evaluate these numbers (in the example F_{100} and F_{101} are computed):

```
f[{u_, v_}] := {v, u + v}
n = 100; Nest[f, {0, 1}, n]
```

```
{354224848179261915075, 573147844013817084101}
```

This could also have been done directly with the function `Fibonacci`.

```
Fibonacci[100]
```

```
354224848179261915075
```

The reader may check the above analysis in the following way.

```
GCDiterations[n_Integer?Positive, m_Integer?Positive] :=
 Block[{a = n, b = m, r, t = 0},
  While[b > 0, r = Mod[a, b];
   {a, b, t} = {b, r, t + 1}];                              t]
```

```
n = 100;
GCDiterations[Fibonacci[n], Fibonacci[n - 1]]
```

```
98
```

```
Table[ GCDiterations[ Fibonacci[n], Fibonacci[n - 1] ],
 {n, 2, 100} ]
```

```
{1, 1, 2, 3, 4, 5, 6, 7, 8, 9, 10, 11, 12, 13, 14, 15, 16, 17, 18,
 19, 20, 21, 22, 23, 24, 25, 26, 27, 28, 29, 30, 31, 32, 33, 34,
 35, 36, 37, 38, 39, 40, 41, 42, 43, 44, 45, 46, 47, 48, 49, 50,
 51, 52, 53, 54, 55, 56, 57, 58, 59, 60, 61, 62, 63, 64, 65, 66,
 67, 68, 69, 70, 71, 72, 73, 74, 75, 76, 77, 78, 79, 80, 81, 82,
 83, 84, 85, 86, 87, 88, 89, 90, 91, 92, 93, 94, 95, 96, 97, 98}
```

Note that the GCDiterations algorithm above does not affect the values of a and b (contrary to our implementation of the simple version of Euclid's algorithm). It also makes use of the *Mathematica* function Mod that will be discussed in the next section.

Plugging in $F_n = c f^n$ in the defining recurrence relation of the Fibonacci numbers, so in $F_{i+2} = F_{i+1} + F_i$, leads to the quadratic equation $f^2 = f + 1$, which has as zero's: $\frac{1 \pm \sqrt{5}}{2}$. Without proof we state the following upperbound on the complexity of Euclid's Algorithm. The reader may prove it with induction on b (distinguish the cases $a \le \frac{b}{f}$ and $\frac{b}{f} < a \le b$).

Theorem A.9 Complexity of Euclid's Algorithm

Let a and b be positive integers, $b \ge a$, $b \ne 1$, and let $f = \frac{1 + \sqrt{5}}{2}$. Then the number of iterations, that Euclid's Algorithm will need to compute gcd(a, b) is at most $\log_f b$.

```
a = Fibonacci[100]; b = Fibonacci[99];
GCDiterations[a, b]
Ceiling[Log[(1 + Sqrt[5]) / 2 , b]]
```

```
98
```

```
98
```

A.3   Congruences, Fermat, Euler, Chinese Remainder Theorem

A.3.1   Congruences

Definition A.4

Two integers a and b are said to be *congruent* to each other *modulo m*, if their difference $b - a$ is divisible by m. This is denoted by
$$a \equiv b \pmod{m}.$$

The *Mathematica* function Mod[a, m] gives the unique integer r, $0 \le r < m$, such that $a \equiv r \pmod{m}$.

```
a = 12345; m = 13; Mod[a, m]
```

```
8
```

An easy test if the integers a and b are congruent of each other modulo m is given by the following example:

```
m = 13; a = 12345; b = 103579; Mod[a - b, m] == 0
```

```
True
```

Definition A.5

A set of m integers a_1, a_2, ..., a_m is called a *complete residue system* modulo m, if each integer is congruent to (exactly) one of the elements a_i, $1 \le i \le m$, modulo m.

The most commonly used complete residue systems modulo m are the sets $\{0, 1, \ldots, m-1\}$ and $\{1, 2, \ldots, m\}$. With the *Mathematica* functions `Range` and `Table` one can generate these systems.

```
m = 10;
Table[i, {i, 0, m - 1}]
Range[m]
```

```
{0, 1, 2, 3, 4, 5, 6, 7, 8, 9}
```

```
{1, 2, 3, 4, 5, 6, 7, 8, 9, 10}
```

Clearly the m integers a_i, $1 \le i \le m$, form a complete residue system modulo m if and only if for each pair $1 \le i, j \le m$ one has that

$$a_i \equiv a_j \,(\mathrm{mod}\, m) \implies i = j \tag{A.9}$$

The congruence relation $\equiv$ modulo defines an equivalence relation (see Definition B.5) on $\mathbf{Z}$. A complete residue system is just a set of representatives of the m equivalence classes.

Lemma A.10

Let $k\,a \equiv k\,b \,(\mathrm{mod}\, m)$ and $\gcd(k, m) = d$. Then

$$a \equiv b \,(\mathrm{mod}\, m/d).$$

Proof: Write $k = k'd$ and $m = m'd$ with $\gcd(k', m') = 1$. It follows from $k\,a - k\,b = x\,m$, for some $x \in \mathbf{Z}$, that $k'(a-b) = x\,m'$. Since $\gcd(m', k') = 1$, it follows from Lemma A.4 that $m' \,|\, (a-b)$, i.e. $a \equiv b \,(\mathrm{mod}\, m')$.

□

Lemma A.11

Let a_1, a_2, ..., a_m be a complete residue system modulo m and let $\gcd(k, m) = 1$. Then $k\,a_1$, $k\,a_2$, ..., $k\,a_m$ is also a complete residue system modulo m.

Proof: We use criterion (A.9). By Lemma A.10, $k\,a_i \equiv k\,a_j \,(\mathrm{mod}\, m)$ implies that $a_i \equiv a_j \,(\mathrm{mod}\, m)$. This in turn implies that $i = j$.

□

A.3.2 Euler and Fermat

Often we shall only be interested in representatives of those residue classes modulo m, whose elements have coprime with m. The number of these classes is denoted by the following function.

> **Definition A.6**
> The *Euler's Totient Function* ϕ (see Euler) is defined by
>
> $$\phi(m) = |\{ 0 \leq i < m \mid \gcd(i, m) = 1 \}|.$$
>
> In words, $\phi(m)$ is the number of integers in between 0 and $m-1$ that are coprime with m.

In *Mathematica*, this function can be evaluated with the `EulerPhi[n]` function. For instance

```
m = 15; EulerPhi[m]
```

```
8
```

corresponding to the eight elements: 1, 2, ,4, 7, 8, 11, 13, and 14. Later on in this section, we see how the function $\phi(m)$ can be efficiently computed.

> **Theorem A.12**
> For all positive integers m
> $$\sum_{d \mid m} \varphi(d) = m.$$

It is quite easy to see in an example which of the m integers in between 1 and m are contributing to which term $\phi(d)$ with $d \mid m$. When $m = 15$, we have the divisors 1, 3, 5 and 15 of m. The eight elements 1, 2, ,4, 7, 8, 11, 13, 14 all have gcd 1 with 15 (note that $\phi(15) = 8$), the four ($= \phi(5)$) elements 3, 6, 9, 12 have gcd $= 3$ with 15, the two ($= \phi(3)$) elements 5, 10 have gcd $= 5$ and the single ($= \phi(1)$) element 0 has gcd $= 15$.

Proof of Theorem A.12:

Let d divide m. By writing $r = id$ one sees immediately that the number of elements r, $0 \leq r < m$, with $\gcd(r, m) = d$ is equal to the number of integers i with $0 \leq i < \frac{m}{d}$ and $\gcd(i, \frac{m}{d}) = 1$, therefore, this number is $\phi(\frac{m}{d})$.

On the other hand, $\gcd(r, m)$ divides n for each integer r, $0 \leq r < m$. It follows that $\sum_{d \mid m} \phi(\frac{m}{d}) = m$. This statement is equivalent to what needs to be proved.

$\square$

The following non-standard *Mathematica* statement evaluates sums of function values $f[d]$ over all divisors d of a given integer m.

```mathematica
DivisorSum[f_, m_] := Plus @@ (f /@ Divisors[m])
```

One can use this function to check Theorem A.12.

```mathematica
m = 15; DivisorSum[EulerPhi, m]
```

```
15
```

Definition A.7
A set of $\phi(m)$ integers r_1, r_2, ..., $r_{\phi(m)}$ is called a *reduced residue system* modulo m if each integer j with $\gcd(j, m) = 1$, is congruent to (exactly) one of the elements r_i, $1 \le i \le \phi(m)$.

A reduced residue system can be quite easily generated by means of the following newly defined functions.

```mathematica
CoPrimeQ[n_Integer, m_Integer] := GCD[n, m] == 1
```

```mathematica
CoPrimeQ[35, 91]
CoPrimeQ[36, 91]
```

```
False
```

```
True
```

```mathematica
CoPrimes[n_Integer?Positive] :=
  Select[ Range[n], CoPrimeQ[n, #] & ]
```

```mathematica
CoPrimes[15]
```

```
{1, 2, 4, 7, 8, 11, 13, 14}
```

Analogously to Lemma A.11 one has the following lemma.

Lemma A.13
Let $r_1, r_2, \ldots, r_{\phi(m)}$ be a reduced residue system modulo m and let $\gcd(a, m) = 1$.
Then $a\,r_1, a\,r_2, \ldots, a\,r_{\phi(m)}$ is also a reduced residue system modulo m.

With the above lemma one can easily prove that the classes in a reduced residue system form a multiplicative group (see Subsection B.1.1).

Theorem A.14   (see Euler)
Let a and m be two integers that are coprime. Then

$$a^{\phi(m)} \equiv 1 \ (\mathrm{mod}\, m).$$

It is quite easy to check this theorem in concrete cases.

```
m = 12345; a = 11111; GCD[m, a]
EulerPhi[m]
Mod[a^EulerPhi[m], m]
```

```
1
```

```
6576
```

```
1
```

Exponentiations modulo some integer can be performed much faster in *Mathematica* with the PowerMod[a, b, m] function, which reduces all intermediate results in the computation of a^b modulo m.:

```
m = 123456789; a = 1111111111; GCD[m, a]
PowerMod[a, EulerPhi[m], m]
```

```
1
```

```
1
```

Proof: Let $r_1, r_2, \ldots, r_{\phi(m)}$ be a reduced residue system modulo m. By Lemma A.13

$$\prod_{i=1}^{\phi(m)} r_i \ \equiv \ \prod_{i=1}^{\phi(m)} (a\,r_i) \ \equiv \ a^{\phi(m)} \prod_{i=1}^{\phi(m)} r_i \quad (\mathrm{mod}\, m).$$

Since each factor r_i is coprime with m, one can divide both hands by $\prod_{i=1}^{\phi(m)} r_i$ by Lemma A.10. This results in $1 \equiv a^{\phi(m)}$ (mod m).

$\square$

Let p be a prime number. Since every integer i, $1 \leq i < p$, is coprime with p, it follows that $\phi(p) = p - 1$. Euler's Theorem implies the next theorem for all values of a except for a's that are a multiple of p. For these values, the statement in the next theorem is trivially satisfied.

> **Theorem A.15** Fermat's Little Theorem
> Let p be a prime number and let a be any integer. Then
> $$a^p \equiv a \pmod{p}.$$

This can easily be checked in individual cases with the *Mathematica* function `PowerMod`.

```
p = 98947; a = 12345; PrimeQ[p]
PowerMod[a, p, p] == a
```

```
True
```

```
True
```

As we have just observed, $\phi(p) = p - 1$ for prime. Because exactly one of every p consecutive integers is divisible by p, we have the following stronger result:

$$\phi(p^e) = p^e - (p^e / p) = p^{e-1}(p - 1) = p^e\left(1 - \frac{1}{p}\right). \tag{A.10}$$

> **Definition A.8**
> A function $f : \mathbb{N} \longrightarrow \mathbb{N}$ is said to be *multiplicative*, if for every pair of positive integers m and n
> $$\gcd(m, n) = 1 \implies f(m.n) = f(m) f(n).$$

> **Lemma A.16**
> Euler's Totient function $\phi(m)$ is multiplicative.

Proof: Let m and n be coprime and let $a_1, a_2, \ldots, a_{\phi(m)}$ and $b_1, b_2, \ldots, b_{\phi(n)}$ be reduced residue systems modulo m resp. n. It suffices to show that the $\phi(m) \phi(n)$ integers $n.a_i + m.b_j$, $1 \leq i \leq \phi(m)$ and $1 \leq j \leq \phi(n)$, form a reduced residue system modulo $m n$. It is quite easy to check that the integers $n.a_i + m.b_j$, $1 \leq i \leq \phi(m)$ and $1 \leq j \leq \phi(n)$, are all distinct modulo $m n$ and that they are coprime with $m n$. (Use Lemma A.15 and formula (A.9)).

It remains to verify that any integer k with $\gcd(k, m.n) = 1$, is congruent to $n.a_i + m.b_j$ modulo mn for some $1 \le i \le \phi(m)$ and $1 \le j \le \phi(n)$.

From Lemma A.13 we know that integers i and j, $1 \le i \le \phi(m)$ and $1 \le j \le \phi(n)$, exist for which

$$k \equiv n.a_i \,(\mathrm{mod}\, m) \quad \text{and} \quad k \equiv b_j \,(\mathrm{mod}\, n).$$

This implies that both m and n divide $k - n.a_i - m.b_j$. Since $\gcd(m, n) = 1$, it follows from (A.4) and (A.7), that also $m.n$ divides $k - n.a_i - m.b_j$.

$\square$

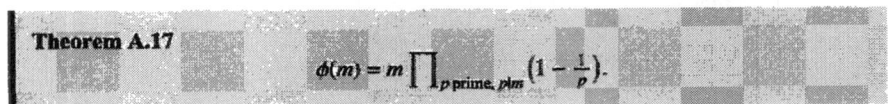

Theorem A.17
$$\phi(m) = m \prod_{p\ \mathrm{prime},\ p|m} \left(1 - \frac{1}{p}\right).$$

Proof: Combine (A.10) and Lemma A.16.

$\square$

In Section A.5 we shall see how a direct counting argument also proves Theorem A.17.

With the *Mathematica* functions `Length` and `EulerPhi` and the function CoPrimes (which makes use of CoPrimeQ) defined above one can check Theorem A.17 as follows:

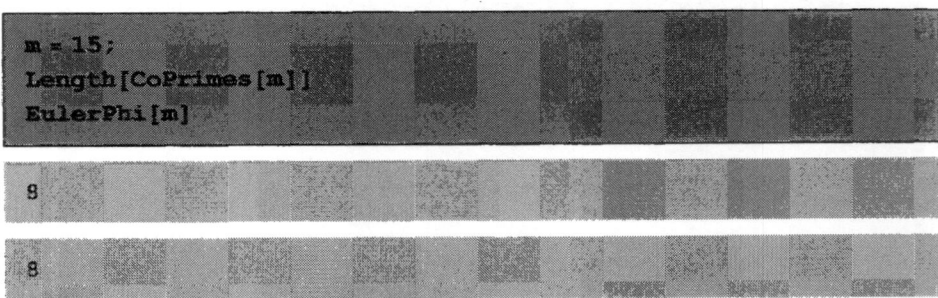

```
m = 15;
Length[CoPrimes[m]]
EulerPhi[m]
```

```
8
```

```
8
```

A.3.3 Solving Linear Congruence Relations

The simplest congruence relation, that one may have to solve, is the single, *linear congruence relation*

$$a\,x \equiv b \,(\mathrm{mod}\, m) \tag{A.11}$$

Theorem A.18
The linear congruence relation $a\,x \equiv b \,(\mathrm{mod}\, m)$ has a solution x if and only if $\gcd(a, m)$ divides b.
In this case the number of different solutions modulo m is $\gcd(a, m)$.

Proof: That $\gcd(a, m) \mid b$ is a necessary condition for (A.11) to have a solution x is trivial. We shall now prove that it is also a sufficient condition.

Let $d = \gcd(a, m)$ and write $a = a'd$, $m = m'd$ and $b = b'd$, where $\gcd(a', m') = 1$. By Lemma A.11, the congruence relation $a'x \equiv b' \pmod{m'}$ has a unique solution x' modulo m'. Clearly, a solution x of $ax \equiv b \pmod m$ satisfies $x \equiv x' \pmod{m'}$. So, each solution x modulo m can be written as $x' + im'$, $0 \le i < d$. Write $a'x' = b' + um'$, $u \in \mathbb{Z}$. Then for each $0 \le i < d$,

$$a(x' + im') = da'x' + ida'm' = db' + udm' + ia'm = b + (u + \mathrm{i}a')m.$$

Hence, the numbers $x' + im'$, $0 \le i < d$, represent all the solutions modulo m of $ax \equiv b \pmod m$.

$\square$

The solution of $ax \equiv b \pmod m$, $\gcd(a, m) = 1$, can easily be found with the extended version of Euclid's Algorithm. Indeed, from $ua + vm = 1$ (see Theorem A.3), it follows that $ua \equiv 1 \pmod m$. So, the solution x is given by $bu \pmod m$. If $\gcd(a, m) = 1$, one often writes a^{-1} for the unique element u satisfying $ua \equiv 1 \pmod m$.

Example A.1 (Method 1)

To solve $14x \equiv 26 \pmod{34}$, we note that $\gcd(14, 34) = 2$, which indeed divides 26.

We first solve $7x' \equiv 13 \pmod{17}$. With the extended version of Euclid's Algorithm we find $5 \cdot 7 + (-2)\,17 = \gcd(7, 17) = 1$. So, $7 \cdot 5 \equiv 1 \pmod{17}$ and x' can be computed from $x' \equiv 7^{-1} \cdot 13 \equiv 5 \cdot 13 \equiv 14 \pmod{17}$.

By the theorem above, $14x \equiv 26 \pmod{34}$ has the numbers 14 and 14+17=31 as solutions modulo 34.

```
ExtendedGCD[7, 17]
Mod[5 * 13, 17]

{1, {5, -2}}

14
```

Example A.2 (Method 2)

To solve $123456789\,x \equiv 135798642 \pmod{179424673}$, we first check if $\gcd(123456789, 179424673)$ divides 135798642. Next, we compute $123456789^{-1} \bmod 179424673$ and then compute $123456789^{-1} \cdot 135798642$ which gives 21562478 as solution .

Instead of using Euclid's Algorithm to compute $123456789^{-1} \bmod 179424673$, we can also use Euler's Theorem. Indeed, $a^{\phi(m)} \equiv 1 \pmod m$ implies that $a\,a^{\phi(m)-1} \equiv 1 \pmod m$ and thus that $a^{-1} \equiv a^{\phi(m)-1} \pmod m$.

```
GCD[148953050, 179424673]
PowerMod[123456789, EulerPhi[179424673] - 1, 179424673]
```

```
1
```

```
172609538
```

So, the number 172609538 is the multiplicative inverse of 123456789 modulo 179424673. The solution x of the congruence relation 123456789 x ≡ 135798642 (mod 179424673) is given by:

```
Mod[135798642 * 172609538, 179424673]
```

```
21562478
```

We can check this:

```
Mod[123456789 * 21562478, 179424673]
```

```
135798642
```

The *Mathematica* function PowerMod computes the multiplicative inverse of a number very efficiently in the following way:

```
PowerMod[123456789, -1, 179424673]
```

```
172609538
```

The *Mathematica* function `Solve` gives all the solutions of the congruence relation $a x \equiv b \pmod{m}$, if they do exist.

```
Clear[x];
Solve[{12 x == 8, Modulus == 16}, x]
```

```
{{Modulus → 16, x → 2}, {Modulus → 16, x → 6},
 {Modulus → 16, x → 10}, {Modulus → 16, x → 14}}
```

To get only the solutions, one can execute

```
x /. Solve[ {12 x == 8, Modulus == 16}, x]
```

```
{2, 6, 10, 14}
```

The reader is invited to try

```
x /. Solve[ {13 x == 1, Modulus == 16}, x]
```

```
Solve[ {12 x == 7, Modulus == 16}, x]
```

A.3.4   The Chinese Remainder Theorem

We shall now discuss the case that x has to satisfy several, linear congruence relations simultaneously, say $a_i x \equiv b_i \pmod{m_i}$ with $\gcd(a_i, m_i) \mid b_i$ for $1 \le i \le k$. Dividing the i-th relation by $d_i = \gcd(a_i, m_i)$, $1 \le i \le k$, one gets as before the congruence relation $a_i' x \equiv b_i' \pmod{m'}$, with $\gcd(a_i', m_i') = 1$. By the proof of Theorem A.18, a solution of this congruence relation is equivalent to a solution of one of the d congruence relations $a_i' x \equiv b_i' + jm_i' \pmod{m_i}$, $0 \le j < d$. In view of this, we restrict our attention to the case that $\gcd(a_i, m_i) = 1$ for all i, $1 \le i \le k$.

> **Theorem A.19**   *The Chinese Remainder Theorem*
> Let m_i, $1 \le i \le k$, be k pairwise coprime integers. Further, let a_i, $1 \le i \le k$, be integers with $\gcd(a_i, m_i) = 1$. Then the system of k simultaneous congruence relations
>
> $$a_i x \equiv b_i \pmod{m_i}, \qquad 1 \le i \le k, \qquad\qquad (A.12)$$
>
> has a <u>unique solution</u> modulo $\prod_{i=1}^{k} m_i$ for all possible k-tuples of integers $b_1, b_2, ..., b_k$.

Proof: Suppose that x' and x'' both form a solution. Then $a_i (x' - x'') \equiv 0 \pmod{m_i}$, $1 \le i \le k$. By Lemma A.4, m_i divides $x' - x''$ for all $1 \le i \le k$. It follows that $x' \equiv x'' \pmod{\prod_{i=1}^{k} m_i}$. Hence, if the k congruence relations have a simultaneous solution, it will be unique modulo $\prod_{i=1}^{k} m_i$.

On the other hand, since there are as many different values for x modulo $\prod_{i=1}^{k} m_i$ as there are possible   k-tuples of reduced right hand sides $b_1, b_2, ..., b_k$ there must be a one-to-one correspondence between them.

□

The proof above does not give an efficient algorithm to determine the solution of (A.12). We shall now explain how this can be done.

Let $1 \le i \le k$ and let u_i be the unique solution modulo $\prod_{i=1}^{k} m_i$ of

$$a_i u_i \equiv 1 \pmod{m_i}, \tag{A.13}$$

$$a_j u_i \equiv 0 \pmod{m_i}, \quad 1 \le j \le k, \quad j \ne i. \tag{A.14}$$

With Euclid's Algorithm u_i is easy to determine. Indeed from (A.14) it follows that u_i is a multiple of $m^{(i)}$ defined by $\prod_{j, \, j \ne i} m_j$, say $u_i = r \, m^{(i)}$ for some $0 \le r < m_i$. The value of r follows from (A.13). Indeed, r is the solution of $a_i \, r \, m^{(i)} \equiv 1 \pmod{m_i}$. Hence

$$u_i = \left\{ (a_i \, m^{(i)})^{-1} \pmod{m_i} \right\} m^{(i)}.$$

The numbers u_i, $1 \le i \le k$, can be stored using at most $k \log_2 m$ bits of memory space.

The solution of (A.12) is now given by

$$x = u_1 \, b_1 + u_2 \, b_2 + \ldots + u_k \, b_k.$$

Example A.3

To solve

$$3 \, x \equiv 7 \, (\mathrm{mod} \, 11) \qquad 2 \, x \equiv 9 \, (\mathrm{mod} \, 13) \qquad 12 \, x \equiv 5 \, (\mathrm{mod} \, 17)$$

we rewrite these congruences as

$$x \equiv 3^{-1} \cdot 7 \, (\mathrm{mod} \, 11) \quad x \equiv 2^{-1} \cdot 9 \, (\mathrm{mod} \, 13) \quad x \equiv 12^{-1} \cdot 5 \, (\mathrm{mod} \, 17)$$

which reduces to

$$x \equiv 4 \cdot 7 \, (\mathrm{mod} \, 11) \qquad x \equiv 7 \cdot 9 \, (\mathrm{mod} \, 13) \qquad x \equiv 10 \cdot 5 \, (\mathrm{mod} \, 17)$$

i.e.

$$x \equiv 6 \, (\mathrm{mod} \, 11) \qquad x \equiv 11 \, (\mathrm{mod} \, 13) \qquad x \equiv 16 \, (\mathrm{mod} \, 17).$$

Next we compute the solutions of

$$
\begin{array}{lll}
u_1 \equiv 1 \, (\mathrm{mod} \, 11) & u_1 \equiv 0 \, (\mathrm{mod} \, 13) & u_1 \equiv 0 \, (\mathrm{mod} \, 17) \\
u_2 \equiv 0 \, (\mathrm{mod} \, 11) & u_2 \equiv 1 \, (\mathrm{mod} \, 13) & u_2 \equiv 0 \, (\mathrm{mod} \, 17) \\
u_3 \equiv 0 \, (\mathrm{mod} \, 11) & u_3 \equiv 0 \, (\mathrm{mod} \, 13) & u_3 \equiv 1 \, (\mathrm{mod} \, 17).
\end{array}
$$

Writing $u_1 = l_1 \cdot 13 \cdot 17$, $u_2 = l_2 \cdot 11 \cdot 17$, $u_3 = l_3 \cdot 11 \cdot 13$, *we find with Theorem A.18, (or the* Solve *function) that* $l_1 \equiv 1 \, (\mathrm{mod} \, 11)$, $l_2 \equiv 8 \, (\mathrm{mod} \, 13)$, $l_3 \equiv 5 \, (\mathrm{mod} \, 17)$ *and thus that* $u_1 \equiv 221 \, (\mathrm{mod} \, 11 \cdot 13 \cdot 17)$, $u_2 \equiv 1496 \, (\mathrm{mod} \, 11 \cdot 13 \cdot 17)$, $u_3 \equiv 715 \, (\mathrm{mod} \, 11 \cdot 13 \cdot 17)$.

We conclude that $x \equiv 6 \cdot 221 + 11 \cdot 1496 + 16 \cdot 715 \equiv 50 \, (\mathrm{mod} \, 11 \cdot 13 \cdot 17)$.

To solve congruence relations $x_i \equiv b_i \pmod{m_i}$, $1 \le i \le k$, with all the m_i's mutually prime with the Chinese Remainder Theorem with *Mathematica,* we first read the package

```
NumberTheory`NumberTheoryFunctions`
```

```
<<NumberTheory`NumberTheoryFunctions`
```

Such a system can now be solved with the *Mathematica* function `ChineseRemainderTheorem` that is available in the above package. We demonstrate this by determining u_1, u_2, and u_3 in the above example.

```
ChineseRemainderTheorem[{1, 0, 0}, {11, 13, 17}]
ChineseRemainderTheorem[{0, 1, 0}, {11, 13, 17}]
ChineseRemainderTheorem[{0, 0, 1}, {11, 13, 17}]
```

```
221
```

```
1496
```

```
715
```

When considering the system of congruence relations $a_i x_i \equiv b_i \pmod{m_i}$, $1 \le i \le k$, where the m_i's are relatively prime and where $\gcd(a_i, m_i) = 1$ for $1 \le i \le k$, it is quite easy for *Mathematica* to reduce this system to the equivalent system $x_i \equiv a_i^{-1} b_i \pmod{m_i}$, $1 \le i \le k$, which can be solved with the Chinese Remainder Theorem function. We use the functions `PowerMod` and `Mod` for this reduction. They operate equally well on vectors (coordinatewise) as on numbers.

We demonstrate this with the parameters of the example above.

```
a = {3, 2, 12}; b = {7, 9, 5}; m = {11, 13, 17};
b = Mod[b * PowerMod[a, -1, m], m]
ChineseRemainderTheorem[b, m]
```

```
{6, 11, 16}
```

```
50
```

A.4   Quadratic Residues

Let p be an odd prime. The quadratic congruence relation $ax^2 + bx + c \equiv 0 \, (\text{mod } p)$, $a \not\equiv 0 \, (\text{mod } p)$, can be simplified by dividing the congruence relation by a followed by the substitution $x \rightarrow x - b/(2\,a)$. In this way, $ax^2 + bx + c \equiv 0 \, (\text{mod } p)$ reduces to a quadratic congruence relation of the type:

$$x^2 \equiv u \, (\text{mod } p) \tag{A.15}$$

> **Definition A.9**
> Let p be an odd prime and u an integer not divisible by p. Then u is called a *quadratic residue (QR)*, if (A.15) has an integer solution, and *quadratic non-residue (NQR)*, if (A.15) does not have an integer solution.

> **Definition A.10**
> Let p be an odd prime and u an integer. The *Legendre symbol* $\left(\frac{u}{p}\right)$ is defined by
> $$\left(\frac{u}{p}\right) = \begin{cases} +1 & \text{if } u \text{ is a quadratic residue mod } p, \\ -1 & \text{if } u \text{ is a quadratic nonresidue mod } p, \\ 0 & \text{if } p \text{ divides } u. \end{cases}$$
> If there is no confusion about the actual choice of the prime number p, one often writes $\chi(u)$ instead of $\left(\frac{u}{p}\right)$.

The Legendre symbol is a special case of the following function.

> **Definition A.11**
> Let $m = \prod_i (p_i)^{e_i}$ be an odd integer and let u be an integer with $\gcd(u, m) = 1$. Then the *Jacobi symbol* $\left(\frac{u}{m}\right)$ is defined by
> $$\left(\frac{u}{m}\right) = \prod_i \left(\frac{u}{p_i}\right)^{e_i}$$
> where $\left(\frac{u}{p}\right)$ denotes the Legendre symbol.

The Jacobi symbol (and a fortiori the Legendre symbol) can be evaluated with the standard *Mathematica* function `JacobiSymbol[u, m]`. So, we can check if 12 is a quadratic residue modulo 13 (indeed $5^2 \equiv 12 \, (\text{mod } 13)$) by means of the Jacobi Symbol[12, 13] which should give value 1.

```
u = 12; m = 13; JacobiSymbol[u, m]
```

```
1
```

We want to derive some properties of the Legendre symbol.

Let $a^2 \equiv u \, (\text{mod } p)$. Then, also $(p - a)^2 \equiv u \, (\text{mod } p)$. The polynomial $x^2 - u$ has at most two zeros in GF(p) (see Theorem B.15), so modulo p there can not be more than two different solutions to $x^2 \equiv u \pmod{p}$. It follows that the quadratic residues modulo p are given by the integers

$$i^2 \pmod{p}, 1 \le i \le \tfrac{p-1}{2},$$

or, alternatively, by the integers $(p-i)^2 \pmod{p}$, $1 \le i \le \tfrac{p-1}{2}$. We conclude that there are exactly $\tfrac{p-1}{2}$ QR's and $\tfrac{p-1}{2}$ NQR's. This proves the first of the following two theorems.

> **Theorem A.20**
> Let p be an odd prime. Then, exactly $\tfrac{p-1}{2}$ of the integers 0, 1, ..., $p-1$ are quadratic residue and $\tfrac{p-1}{2}$ are quadratic non-residue. In formula
>
> $$\sum_{u=0}^{p-1} \chi(u) = 0.$$

The reader can check the above theorem in concrete examples by means of the following two *Mathematica* functions.

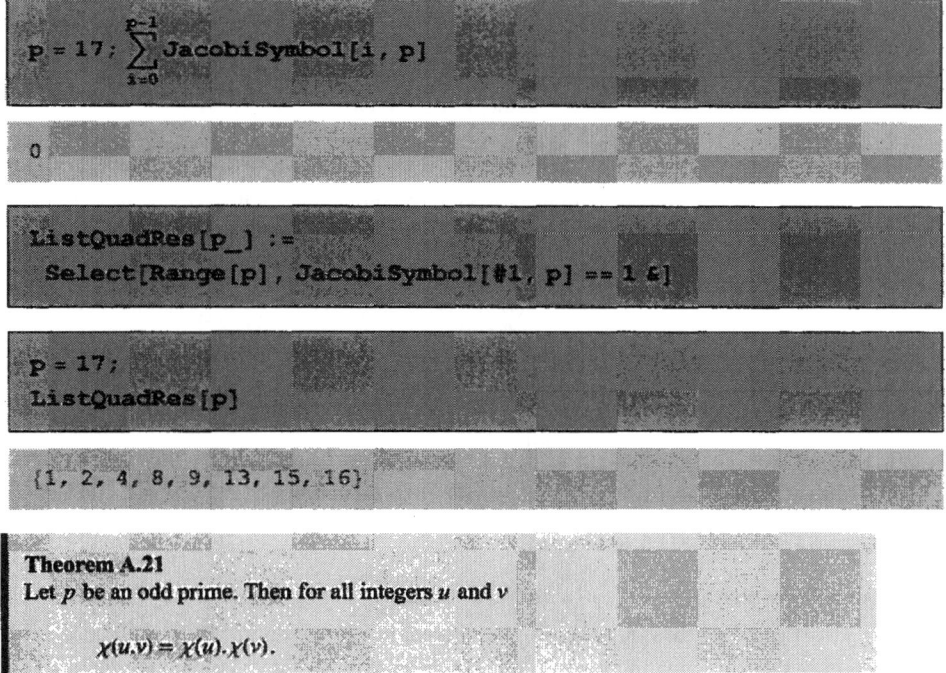

```
p = 17; ∑_{i=0}^{p-1} JacobiSymbol[i, p]
```

```
0
```

```
ListQuadRes[p_] :=
  Select[Range[p], JacobiSymbol[#1, p] == 1 &]
```

```
p = 17;
ListQuadRes[p]
```

```
{1, 2, 4, 8, 9, 13, 15, 16}
```

> **Theorem A.21**
> Let p be an odd prime. Then for all integers u and v
>
> $$\chi(u.v) = \chi(u).\chi(v).$$

Proof: This theorem will be a trivial consequence of Theorem A.23 later on. We shall present here a more elementary proof.

If p divides u or v the assertion is trivial, because both hands are equal to zero. The proof in case that p does not divide u or v is split up in three cases.

Case 1: u and v are both QR.

Then $u \equiv a^2 \pmod p$ and $v \equiv b^2 \pmod p$, for some integers a and b. It follows that $u.v \equiv (a.b)^2 \pmod p$. So $u.v$ is QR.

Case 2: Exactly one of u and v is QR, say u is QR and v is NQR.

Suppose that also $u.v$ is QR. Then there exist integers a and b such that $u \equiv a^2 \pmod p$ and $u.v \equiv b^2 \pmod p$. Since $a \not\equiv 0 \pmod p$, it follows that $v \equiv (b/a)^2 \pmod p$. A contradiction!

Case 3: Both u and v are NQR.

From Lemma A.11 we know that $i \cdot u$, $i = 1, 2, \ldots, p-1$, runs through all non-zero elements modulo p. For the $\frac{p-1}{2}$ values of i for which i is QR, we have by Case 2 that $i.u$ is NQR. So, for the $\frac{p-1}{2}$ values of i for which i is NQR, it follows that $i.u$ is QR. So $u.v$ is QR.

$\square$

Although the next theorem will never be used in this textbook, we do mention it, because it is often needed in related areas in Discrete Mathematics.

> **Theorem A.22**
> Let p be an odd prime. Then, for every integer v
> $$\sum_{u=0}^{p-1} \chi(u).\chi(u+v) = \begin{cases} p-1, & \text{if } p \text{ divides } v, \\ -1, & \text{otherwise.} \end{cases}$$

Proof: If p divides v, the statement is trivial. When p does not divide v, one has by Theorem A.21 and Theorem A.20 that

$$\sum_{u=0}^{p-1} \chi(u) \chi(u+v) = \sum_{u=1}^{p-1} \chi(u) \chi(u+v) = \sum_{u=1}^{p-1} \chi(u) \chi(u) \chi(1+v/u) =$$
$$\sum_{u=1}^{p-1} \chi(1+v/u) = \sum_{w \neq 1} \chi(w) = -1 + \sum_{w=0}^{p-1} \chi(w) = -1$$

$\square$

Let u be QR, say $u \equiv a^2 \pmod p$. By Fermat's Theorem $u^{\frac{p-1}{2}} \equiv a^{p-1} \equiv 1 \pmod p$. So, the $\frac{p-1}{2}$ QR's are zero of the polynomial $x^{\frac{p-1}{2}} - 1$ over GF(p). Since a polynomial of degree $\frac{p-1}{2}$ over GF(p) has at most $\frac{p-1}{2}$ different zeros in GF(p) (see Theorem B.15), one has in GF(p):

$$x^{(p-1)/2} - 1 = \prod_{u \text{ is QR}} (x - u). \tag{A.16}$$

It also follows that $u^{\frac{p-1}{2}} \neq 1$, if u is NQR. Since $\left(u^{\frac{p-1}{2}}\right)^2 \equiv 1 \pmod p$ by Fermat's Theorem and since $y^2 \equiv 1 \pmod p$ has only 1 and -1 as roots, it follows that $u^{\frac{p-1}{2}} \equiv -1 \pmod p$, if u is NQR. This proves the following theorem for all u coprime with p. For $p \mid u$ the theorem is trivially true.

> **Theorem A.23**
> Let p be an odd prime. Then for all integers u,
> $$\left(\frac{u}{p}\right) \equiv u^{(p-1)/2} \pmod p.$$

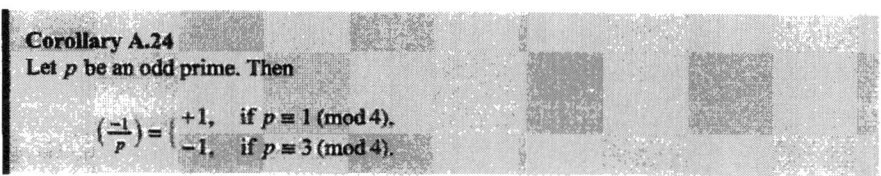

Corollary A.24
Let p be an odd prime. Then
$$\left(\frac{-1}{p}\right) = \begin{cases} +1, & \text{if } p \equiv 1 \ (\mathrm{mod}\ 4), \\ -1, & \text{if } p \equiv 3 \ (\mathrm{mod}\ 4). \end{cases}$$

Proof: $(-1)^{\frac{p-1}{2}} = 1$ if and only if $p \equiv 1 \ (\mathrm{mod}\ 4)$.

□

Another value of the Legendre symbol that we shall need later on is $\left(\frac{2}{p}\right)$.

Theorem A.25
Let p be an odd prime. Then
$$\left(\frac{2}{p}\right) = \begin{cases} +1, & \text{if } p \equiv \pm 1 \ (\mathrm{mod}\ 8), \\ -1, & \text{if } p \equiv \pm 3 \ (\mathrm{mod}\ 8). \end{cases}$$

Proof:

$$2^{\frac{p-1}{2}} \prod_{k=1}^{\frac{p-1}{2}} k \equiv \prod_{k=1}^{\frac{p-1}{2}} (2k) \equiv \left(\prod_{k=1}^{\lfloor \frac{p-1}{4}\rfloor}(2k)\right)\cdot\left(\prod_{k=1+\lfloor\frac{p-1}{4}\rfloor}^{\frac{p-1}{2}}(2k)\right) \equiv$$

$$(-1)^{\frac{p-1}{2}-\lfloor\frac{p-1}{4}\rfloor}\cdot\left(\prod_{k=1}^{\lfloor\frac{p-1}{4}\rfloor}(2k)\right)\left(\prod_{k=1+\lfloor\frac{p-1}{4}\rfloor}^{\frac{p-1}{2}}(p-2k)\right) \equiv (-1)^{\frac{p-1}{2}-\lfloor\frac{p-1}{4}\rfloor}\cdot\left(\prod_{k=1}^{\frac{p-1}{2}} k\right)(\mathrm{mod}\ p).$$

Dividing both hands in the above relation by $\prod_{k=1}^{\frac{p-1}{2}} k$ yields

$$2^{\frac{p-1}{2}} \equiv (-1)^{\frac{p-1}{2}-\lfloor\frac{p-1}{4}\rfloor}(\mathrm{mod}\ p).$$

The assertion now follows from Theorem A.23.

□

We recall the definition of the Jacobi symbol in terms of the Legendre symbol

$$\left(\frac{u}{m}\right) = \prod_i \left(\frac{u}{p_i}\right)^{e_i}, \text{ where } m = \prod_i p_i^{e_i}. \tag{A.17}$$

Theorem A.26
Let m and n be odd integers. Then the following relations hold for the Jacobi symbol

i) $\left(\frac{u}{m}\right) = \left(\frac{u-m}{m}\right)$,

ii) $\left(\frac{uv}{m}\right) = \left(\frac{u}{m}\right)\left(\frac{v}{m}\right)$,

iii) $\left(\frac{u}{mn}\right) = \left(\frac{u}{m}\right)\left(\frac{u}{n}\right)$,

iv) $\left(\frac{-1}{m}\right) = 1$ if and only if $m \equiv 1 \pmod 4$,

v) $\left(\frac{2}{m}\right) = 1$ if and only if $m \equiv \pm 1 \pmod 8$.

Proof: The first two relations hold for the Legendre symbol and, by (A.17), also for the Jacobi symbol. The third relation is a direct consequence of (A.17).

To see that the fourth relation is a direct consequence of (A.17) and Corollary A.24, it suffices to observe that a product of an odd number of integers, each congruent to 3 modulo 4, is also congruent to 3 modulo 4, while for an even number the product will be 1 modulo 4. The proof of the last relation goes analogously (now use Theorem A.25).

□

One more relation is needed to be able to compute $\left(\frac{u}{m}\right)$ fast. We shall not give its proof, because the theory goes beyond the scope of this book. The interested reader is referred to Theorem 99 in [HarW45] or Theorem 7.2.1 in [Shap83].

Theorem A.27   (Quadratic Reciprocity Law by Gauss)
Let m and n be odd coprime integers. Then
$$\left(\frac{m}{n}\right)\left(\frac{n}{m}\right) = (-1)^{\frac{(m-1)(n-1)}{4}}$$

With the relations in Theorem A.25, Theorem A.26, and Theorem A.27 one can evaluate the Jacobi symbol very quickly.

Example A.4

$$\left(\frac{12703}{16361}\right) \overset{A.27}{=} \left(\frac{16361}{12703}\right) \overset{A.26\,i)}{=} \left(\frac{3658}{12703}\right) \overset{A.26\,ii)}{=} \left(\frac{2}{12703}\right)\cdot\left(\frac{1829}{12703}\right) \overset{A.26\,v)\ \&\,A.27}{=}$$

$$= \left(\frac{12703}{1829}\right) \overset{A.26\,i)}{=} \left(\frac{1729}{1829}\right) \overset{A.27}{=} \left(\frac{1829}{1729}\right) \overset{A.26\,i)}{=} \left(\frac{100}{1729}\right) \overset{A.26\,ii)}{=} \left(\frac{2}{1729}\right)^2 \cdot \left(\frac{25}{1729}\right)$$

$$= \left(\frac{25}{1729}\right) \overset{A.27}{=} \left(\frac{1729}{25}\right) \overset{A.26\,i)}{=} \left(\frac{4}{25}\right) \overset{A.26\,ii)}{=} \left(\frac{2}{25}\right)^2 = 1.$$

It should be easy for the reader to verify that the above method has roughly the same complexity as Euclid's Algorithm.

Of course we could have evaluated $\left(\frac{12703}{16361}\right)$ directly with *Mathematica*, as we have seen before.

```
JacobiSymbol[12703, 16361]
```

```
1
```

A.5   Continued Fractions

Quite often one wants to approximate a real number by means of a rational number. For instance, many people use 22/7 as an approximation of π. A better approximation of π is already given by 333/106 and again better is 355/113. One has to increase the denominator to 33102 to get the next improvement.

```
N[Pi - 22/7 ]

N[Pi - 333/106 ]

N[Pi - 355/113 ]

N[Pi - 103993/33102 ]
```

```
-0.00126449
```

```
0.0000832196
```

```
-2.66764 × 10^-7
```

```
5.77891 × 10^-10
```

It is the theory of continued fractions that explains how to get such good approximations.

Definition A.12

A *finite continued fraction* is an expression of the form

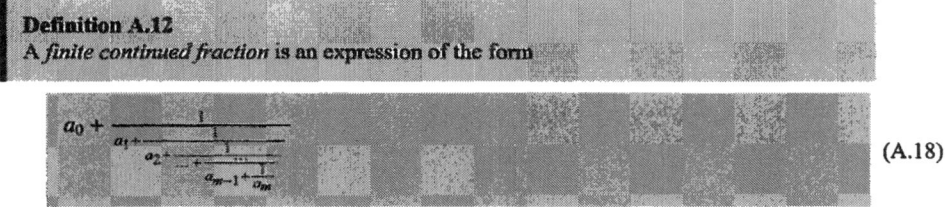

$$a_0 + \cfrac{1}{a_1 + \cfrac{1}{a_2 + \cfrac{1}{\cdots + \cfrac{1}{a_{n-1} + \cfrac{1}{a_n}}}}}$$

(A.18)

where $a_0 \in \mathbb{Z}$ and $a_i \in \mathbb{N}, 1 \leq i \leq m$.
It will often be denoted by the sequence $[a_0, a_1, \ldots, a_m]$.

If $m \to \infty$ in (A.18), we speak of an *infinite* continued fraction. It has the form

$$a_0 + \cfrac{1}{a_1 + \cfrac{1}{a_2 + \cfrac{1}{a_3 + \cfrac{1}{\cdots}}}}$$

and will be shortened to $[a_0, a_1, a_2, \ldots]$.

Clearly, each finite continued fraction represents a rational number. One can find it by simplifying the continued fraction step by step, starting with $a_{m-1} + \frac{1}{a_m} = \frac{a_{m-1} a_m + a_m}{a_m}$, $\frac{1}{a_{m-1} + \frac{1}{a_m}} = \frac{a_m}{a_{m-1} a_m + a_m}$, etc.

In *Mathematica* this can be achieved with the function `Normal`.

$$\texttt{Normal}\left[3 + \cfrac{1}{7 + \cfrac{1}{15 + \cfrac{1}{1 + \frac{1}{292}}}}\right]$$

$$\frac{103993}{33102}$$

We shall now show that the opposite is also true: each rational number has a finite continued fraction.

Lemma A.28
Each rational number has a finite continued fraction.

Proof: Let a/b, $b > 0$, represent a rational number. We apply the simple version of Euclid's Algorithm (Alg. A.7) to the pair (a, b), so we put $s_0 = a$, $s_1 = b$, and compute recursively $s_i = q_i s_{i+1} + s_{i+2}$, with $0 \leq s_{i+2} < s_{i+1}$, until $s_{m+2} = 0$ (and thus $s_m = q_m s_{m+1}$) for some integer m. Then

$$\frac{a}{b} = \frac{s_0}{s_1} = \frac{q_0 s_1 + s_2}{s_1} = q_0 + \frac{1}{s_1/s_2} = q_0 + \cfrac{1}{\frac{q_1 s_2 + s_3}{s_2}} = q_0 + \cfrac{1}{q_1 + \frac{1}{s_2/s_3}} = \ldots$$

$$\ldots = q_0 + \cfrac{1}{q_1 + \cfrac{\cdots}{\cdots + \cfrac{1}{q_{m-1} + \frac{1}{s_m/s_{m+1}}}}} = q_0 + \cfrac{1}{q_1 + \cfrac{\cdots}{\cdots + \cfrac{1}{q_{m-1} + \frac{1}{q_m}}}}.$$

We conclude that a/b has $[q_0, q_1, \ldots, q_m]$ as continued fraction.

$\square$

It is important to observe that the representation of a rational number as a finite simple continued fraction, where all the q_i's ($i \geq 1$) are positive, is not completely unique. Although the manner in which the q_i's are calculated with the simple version of Euclid's Algorithm (see proof above) gives a unique value of the q_i's, it is clear that in the last step we have $q_m \geq 2$, since $s_{m+1} < s_m$.

As the last term in the expansion is a positive integer, and not equal to one, we can therefore rewrite the last term as follows:

$$\frac{1}{q_m} = \frac{1}{(q_m - 1) + \frac{1}{1}} .$$

This shows that $[q_0, q_1, ..., q_m]$ has the same value as $[q_0, q_1 ..., q_m - 1, 1]$.

The last term in a continued fraction can be chosen in such a way as to make the number of terms in the expansion either even or odd, if that would be convenient.

Formula (A.18) suggests the following way of computing a continued fraction of a number α.

> **Algorithm A.29**
> The continued fraction of a number α can be computed by
>
>   **initialize** $\alpha_0 = \alpha$
>   **compute** recursively $a_i = \lfloor \alpha_i \rfloor$ and
> $\alpha_{i+1} = 1/(\alpha_i - a_i)$.   for $i \geq 0$,
>   **output** $[a_0, a_1, a_2, ...]$.

Example A.5

Consider $\alpha = 11/9$. Then we get

```
Clear[a];
alpha = 11 / 9; α[0] = alpha;
a[0] = ⌊α[0]⌋
α[1] = 1 / (α[0] - a[0]);
```

```
1
```

To get the next term, we compute

```
a[1] = ⌊α[1]⌋
α[2] = 1 / (α[1] - a[1]);
```

```
4
```

We continue with

```
a[2] = ⌊α[2]⌋
α[3] = 1 / (α[2] - a[2]);
```

```
2
```

Power::infy : Infinite expression $\frac{1}{0}$ encountered.

We conclude that $\alpha_2 = a_2$ and thus that the continued fraction is given by [1, 4, 2]. We can check this quite easily:

$$\text{Normal}\left[1 + \frac{1}{4 + \frac{1}{2}}\right]$$

$$\frac{11}{9}$$

To let *Mathematica* compute the continued fraction of a number, first the package `NumberTheory`ContinuedFractions`` has to be loaded.

```
<<NumberTheory`ContinuedFractions`
```

To find the continued fraction of a rational number, one can use the function `ContinuedFraction`.

```
ContinuedFraction[135/159]
```

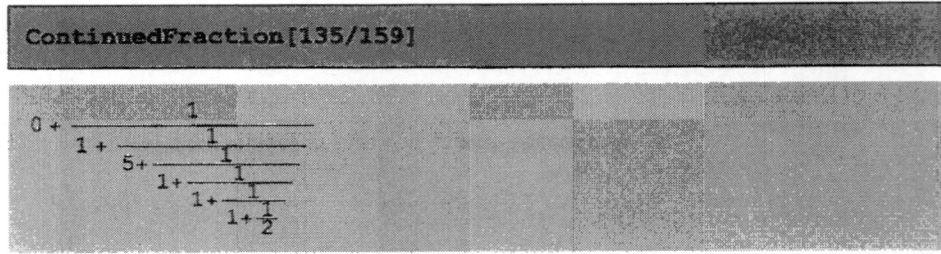

If α is not rational, one has to include the number of terms that one wants to see.

```
ContinuedFraction[Pi, 11]
```

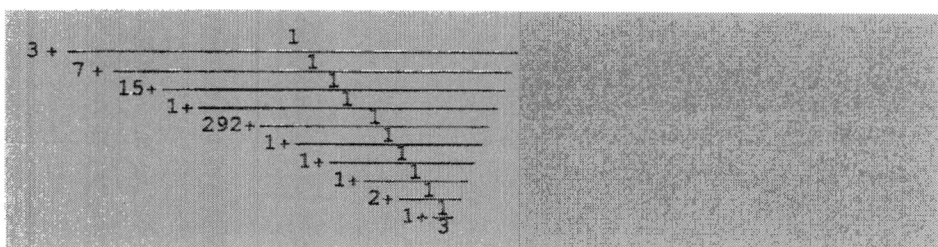

To express such a continued fraction as a regular fraction, one can use the *Mathematica* function `Normal` again.

```
Normal[ContinuedFraction[Pi, 11]]
```

$$\frac{4272943}{1360120}$$

If a continued fraction is given in the form $[a_0, a_1, \ldots, a_m]$, one gets the regular continued fraction by means of the function `ContinuedFractionForm`. The reader should know that in *Mathematica* the numbering of the indices starts with 1, 2, etc.

```
AA={3,7,15,1,292};
ContinuedFractionForm[AA]
```

$$3 + \cfrac{1}{7 + \cfrac{1}{15 + \cfrac{1}{1 + \cfrac{1}{292}}}}$$

To obtain the continued fraction of a number α in the form $[a_0, a_1, \ldots, a_m]$, one can just appends [[1]] to the function ContinuedFraction[α, n].

```
ContinuedFraction[Pi, 11][[1]]
```

```
{3, 7, 15, 1, 292, 1, 1, 1, 2, 1}
```

Definition A.13
The k-th *convergent* C_k of a continued fraction $[a_0, a_1, \ldots, a_m]$, $0 \le k \le m$, is defined by $[a_0, a_1, \ldots, a_k]$.

These convergents can be quite easily evaluated with the functions `Table`, `Normal`, `Take`, `ContinuedFractionForm`, and `Length`.

```
AA={3,7,15,1,292};
Table[Normal[ContinuedFractionForm[Take[AA,i]]],{i,1,Length
[AA]}]
```

$$\left\{3, \frac{22}{7}, \frac{333}{106}, \frac{355}{113}, \frac{103993}{33102}\right\}$$

Each convergent, being a rational number, can be written as p_k/q_k. The values of p_k and q_k can be found with the *Mathematica* functions Numerator and Denominator.

```
C5=Normal[ContinuedFraction[Pi,5]]
p5=Numerator[C5]
q5=Denominator[C5]
```

$$\frac{103993}{33102}$$

103993

33102

The next theorem gives a nice relation between a continued fraction and its convergents. To be able to shorten the proof, we shall relax our usual restriction of the integrality of the a_i's.

Theorem A.30

Let $\{a_i\}_{i \geq 0}$ be a finite or infinite sequence of reals, all positive with the possible exception of a_0.

Let $C_k = p_k/q_k$ be defined by $[a_0, a_1, \ldots, a_k]$ as in (A.18). Then, the numbers p_k and q_k satisfy the recurrence relation

$$p_0 = a_0, \qquad p_1 = a_0 a_1 + 1,$$
$$q_0 = 1, \qquad q_1 = a_1,$$
$$p_k = a_k p_{k-1} + p_{k-2}, \quad k \geq 2,$$
$$q_k = a_k q_{k-1} + q_{k-2}, \qquad k \geq 2.$$

Proof: The proof is by induction on k.

For $k = 0$, we have $\frac{p_0}{q_0} = C_0 = a_0 = \frac{a_0}{1}$, so indeed $p_0 = a_0$ and $q_0 = 1$.

For $k = 1$, we have $\frac{p_1}{q_1} = C_1 = [a_0, a_1] = a_0 + \frac{1}{a_1} = \frac{a_0 a_1 + 1}{a_1}$, so indeed $p_1 = a_0 a_1 + 1$ and $q_1 = a_1$.

Assume that the theorem has been proved up to a certain value of k. So,

$$C_k = [a_0, a_1, \ldots, a_k] = \frac{p_k}{q_k} = \frac{a_k p_{k-1} + p_{k-2}}{a_k q_{k-1} + q_{k-2}}.$$

Now substitute $a_k \rightarrow a_k + 1/a_{k+1}$ above. Then

$$C_{k+1} \stackrel{\text{def.}}{=} [a_0, a_1, \ldots, a_k, a_{k+1}] \stackrel{\text{Def.} A.12}{=} \left[a_0, a_1, \ldots, a_k + \frac{1}{a_{k+1}}\right]$$

$$\stackrel{\text{induct.}}{=} \frac{\left(a_k + \frac{1}{a_{k+1}}\right) p_{k-1} + p_{k-2}}{\left(a_k + \frac{1}{a_{k+1}}\right) q_{k-1} + q_{k-2}} = \frac{a_{k+1}(a_k \, p_{k-1} + p_{k-2}) + p_{k-1}}{a_{k+1}(a_k \, q_{k-1} + q_{k-2}) + q_{k-1}}$$

$$\stackrel{\text{rec.rel.}}{=} \frac{a_{k+1} \, p_k + p_{k-1}}{a_{k+1} \, q_k + q_{k-1}} \stackrel{\text{rec.rel.}}{=} \frac{p_{k+1}}{q_{k+1}}.$$

$\square$

A small result, that we need later, is the inequality

$$q_k \geq F_k, \tag{A.19}$$

where F_k is the k-th Fibonnaci number, defined by $F_0 = 0$, $F_1 = 1$, and the recurrence relation $F_k = F_{k-1} + F_{k-2}$, $k \geq 2$. The inequality $q_k \geq F_k$ follows with an easy induction argument from $q_0 > 0$, $q_1 \geq 1$, and the recurrence relation $q_k = a_k \, q_{k-1} + q_{k-2}$ in which $a_k \geq 1$ (use $q_k \geq q_{k-1} + q_{k-2}$).

> **Lemma A.31**
> Let $C_k = p_k / q_k$ be the k-th convergent of a continued fraction. Then
> $$p_k \, q_{k-1} - p_{k-1} \, q_k = (-1)^{k-1}$$

Proof: The proof is again by induction on k. For $k = 1$ we have by Theorem A.30 that $p_1 \, q_0 - p_0 \, q_1 = (a_0 \, a_1 + 1) \times 1 - a_0 \times a_1 = 1$.

To prove the step from k to $k+1$ we use the recurrence relation in Theorem A.30:

$$p_{k+1} \, q_k - p_k \, q_{k+1} \stackrel{\text{Thm.} A.30}{=} (a_{k+1} \, p_k + p_{k-1}) q_k - p_k(a_{k+1} \, q_k + q_{k-1}) =$$

$$p_{k-1} \, q_k - p_k \, q_{k-1} \stackrel{\text{ind.}}{=} (-1)(-1)^{k-1} = (-1)^k.$$

$\square$

> **Corollary A.32**
> Let $C_k = p_k / q_k$ be the k-th convergent of a continued fraction. Then
> $$\gcd(p_k, q_k) = 1.$$

Proof: This is an immediate consequence of $p_{k-1} \, q_k - p_k \, q_{k-1} = (-1)^{k-1}$. Indeed, each number dividing p_k and q_k must also divide -1.

$\square$

Theorem A.33

Let $C_k = p_k / q_k$ be the k-th convergent of a finite or infinite continued fraction $[a_0, a_1, \ldots]$. Then

$$C_k - C_{k-1} = \frac{(-1)^{k-1}}{q_{k-1} q_k}, \qquad k \geq 1, \tag{A.20}$$

$$C_k - C_{k-2} = \frac{a_k(-1)^k}{q_{k-2} q_k}, \qquad k \geq 2. \tag{A.21}$$

$$C_0 < C_2 < C_4 < \ldots\ldots < C_5 < C_3 < C_1. \tag{A.22}$$

For an infinite continued fraction, the strictly increasing bounded sequence of the even convergents has the same limit as the strictly decreasing bounded sequence of the odd convergents.

Proof: By Lemma A.31 and Theorem A.30

$$C_k - C_{k-1} = \frac{p_k}{q_k} - \frac{p_{k-1}}{q_{k-1}} = \frac{p_k q_{k-1} - p_{k-1} q_k}{q_{k-1} q_k} = \frac{(-1)^{k-1}}{q_k q_{k-1}}$$

$$C_k - C_{k-2} = \frac{p_k}{q_k} - \frac{p_{k-2}}{q_{k-2}} = \frac{p_k q_{k-2} - p_{k-2} q_k}{q_{k-2} q_k}$$

$$= \frac{(a_k p_{k-1} + p_{k-2}) q_{k-2} - p_{k-2}(a_k q_{k-1} + q_{k-2})}{q_{k-2} q_k}$$

$$= \frac{a_k p_{k-1} q_{k-2} - a_k p_{k-2} q_{k-1}}{q_{k-2} q_k} = \frac{a_k(-1)^k}{q_{k-2} q_k}.$$

This proves (A.20) and (A.21). That the even convergents form a strictly increasing sequence follows from (A.21), which implies that $C_{2k} - C_{2k-2} > 0$ (the a_i's are positive). For the same reason, the odd convergents are strictly decreasing.

To show that each even convergent, say C_{2i}, is less that any odd convergent, say C_{2j+1}, we first observe that $C_{2k+1} - C_{2k} > 0$ by (A.20). We combine this with the above to get

$$C_{2i} < C_{2i+2j} < C_{2i+2j+1} < C_{2j+1}.$$

Finally, by (A.19) and (A.20), for $k \geq 2$

$$(|C)_k - C_{k-1}| = \frac{1}{q_{k-1} q_k} \leq \frac{1}{F_{k-1} F_k} \leq \frac{1}{(k-1)^2}$$

thus, the difference between two consecutive convergents tends to zero as k tends to infinity. This shows that the limit of the even convergents must be the same as the limit of the odd convergents.

$\square$

Example A.6

Below we have listed the first 10 convergents of π in their natural ordering.

```
<<NumberTheory`ContinuedFractions`

Do[Print[k - 1, "   ",
  N[Normal[ContinuedFraction[Pi, k]], 16]], {k, 1, 9, 2}]
Print[π, "   ", N[Pi, 16]]
Do[Print[k - 1, "   ",
  N[Normal[ContinuedFraction[Pi, k]], 16]], {k, 10, 2, -2}]
```

```
0   3.

2   3.141509433962264

4   3.141592653011902

6   3.141592653467437

8   3.141592653581078

π   3.141592653589793

9   3.141592653591404

7   3.141592653618936

5   3.141592653921421

3   3.141592920353983

1   3.142857142857143
```

The next two theorems will be stated without their proofs. These can be found in any introduction to continued fractions, e.g. [Rose84], but the arguments are too technical for our purposes.

Theorem A.34
Let $C_k = p_k/q_k$ be the k-th convergent of a finite or infinite continued fraction $\alpha = [a_0, a_1, \ldots]$ and suppose that $|\alpha - r/s| < |\alpha - p_k/q_k|$.
Then $s > q_k$.

For instance, since $\frac{355}{113}$ is a convergent of π, we now know that only rationals with a denominator greater than 113 may lie closer to π than $\frac{355}{113}$ does.

Theorem A.35
Let $\alpha \in \mathbb{R}$ and let r/s (with $\gcd(r, s) = 1$) be a rational such that $|\alpha - r/s| < 1/2s^2$.
Then r/s is a convergent of the continued fraction expansion of α.

This theorem says that a rational number r/s that lies at distance at most $1/2s^2$ from a number α will appear as convergent in the continued fraction of that number.

A.6   Möbius Inversion Formula, the Principle of Inclusion and Exclusion

A.6.1   Möbius Inversion Formula

Often in Discrete Mathematics a function f is defined in terms of another function, say g. The question is, how g can be expressed in terms of f. With the theory of partially ordered sets and the (generalized) Möbius Inversion Formula one can frequently solve this problem (see Chapter IV in [Aign79]). In this section we shall discuss two important special cases.They both follow from the theory, mentioned above, but it turns out that they can also be proved directly.

Often we shall need an explicit factorization of an integer n. We no longer want the strict ordering of the prime numbers given by $p_1 = 2$, $p_2 = 3$, etc.. However, different subscripts will still denote different prime numbers.

> **Definition A.14**
> Let $n = \prod_{i=1}^{k} (p_i)^{e_i}$, $e_i > 0$, $1 \le i \le k$, where the p_i's are different primes. Then the *Möbius function* $\mu(n)$ (Möbius) is defined by
> $$\mu(n) = \begin{cases} 1 & \text{if } n = 1, \\ 0 & \text{if } e_i \ge 2 \text{ for some } 1 \le i \le k, \\ (-1)^k & \text{if all } e_i \text{ are equal to } 1. \end{cases}$$

In other words, $\mu(n)$ is the multiplicative function satisfying $\mu(1) = 1$, $\mu(p) = -1$, and $\mu(p^i) = 0$, $i \ge 2$, for any prime p. Mathematica has the standard function $\underline{\texttt{MoebiusMu}[n]}$ to evaluate $\mu(n)$.

```
n = 30; MoebiusMu[n]
```

```
-1
```

The Möbius function is defined in this peculiar way to have the following property.

> **Lemma A.36**
> Let n be a positive integer. Then
> $$\sum_{d|n} \mu(d) = \begin{cases} 1 & \text{if } n = 1, \\ 0 & \text{if } n > 1. \end{cases}$$

Proof: For $n = 1$ the assertion is trivial. For $n > 1$ we write as above $n = \prod_{i=1}^{k} p_i^{e_i}$, $e_i > 0$, $1 \le i \le k$. Then $k > 0$ and thus

$$\sum_{d|n} \mu(d) = \sum_{d | p_1^{e_1} p_2^{e_2} \dots p_k^{e_k}} \mu(d) = \sum_{d | p_1 p_2 \dots p_k} \mu(d) =$$

$$= 1 + \sum_{l=1}^{k} \sum_{1 \le i_1 < i_2 < \dots < i_l \le k} \mu(p_{i_1} p_{i_2} \cdots p_{i_l})$$

$$= \sum_{l=1}^{k} \binom{k}{l} (-1)^l = (1 - (1))^k = 0.$$

$\square$

The reader may want to check the above lemma by means of:

```
DivisorSum[f_, m_] := Plus @@ (f /@ Divisors[m])
```

```
m = 100; DivisorSum[MoebiusMu, m]
```

```
0
```

Lemma A.37
Let m and n be two positive integers such that m divides n. Then

$$\sum_{d, m|d|n} \mu(n/d) = \begin{cases} 1 & \text{if} \quad m = n, \\ 0 & \text{if otherwise.} \end{cases}$$

Proof: Let $n = n'm$. For each d with $m|d|n$, we write $d = d'm$. Then $\sum_{d, m|d|n} \mu(n/d) = \sum_{d'|n'} \mu(n'/d')$, which by Lemma A.36 is 1 for $n' = 1$, (i.e. $m = n$), and is 0 for $n' > 1$.

$\square$

Theorem A.38   Möbius Inversion Formula
Let f be a function defined on $\mathbb{N}$ and let the function g on $\mathbb{N}$ be defined by

$$g(n) = \sum_{d|n} f(d), \qquad n \in \mathbb{N},$$

Then, for all $n \in \mathbb{N}$

$$f(n) = \sum_{d|n} \mu(d) g(n/d) = \sum_{d|n} \mu(n/d) g(d).$$

Proof: By the definition of $g(n)$ and Lemma A.37

$$\sum_{d|n} \mu(n/d) g(d) = \sum_{d|n} \mu(n/d) \sum_{e|d} f(e) = \sum_{e|n} f(e) \sum_{d, e|d|n} \mu(n/d) = f(n).$$

$\square$

> **Corollary A.39**   Multiplicative Möbius Inversion Formula
> Let F be a function defined on $\mathbb{N}$ and let the function G on $\mathbb{N}$ be defined by
>
> $$g(n) = \prod_{d|n} f(d), \qquad n \in \mathbb{N},$$
>
> Then for all n in $\mathbb{N}$
>
> $$f(n) = \prod_{d|n} \mu(d) \, g(n/d) = \prod_{d|n} \mu(n/d) \, g(d).$$

Proof: Substitute $g(n) = \log(G(n))$ and $f(n) = \log(F(n))$ in the Möbius Inversion Formula. ∎

Example A.7

From Theorem A.12 we know that Euler's Totient Function satisfies

$$\sum_{d|n} \phi(d) = n.$$

It follows from the Möbius Inversion Formula (Thm. A.38) that for $n = \prod_{i=1}^{k} (p_i)^{e_i}$, $e_i > 0$, $1 \le i \le k$,

$$\phi(n) \;=\; \sum_{d|n} \mu(d) \frac{n}{d} \;=$$

$$= \frac{n}{1} - \sum_{1 \le i \le k} \frac{n}{p_i} + \sum_{1 \le i < j \le k} \frac{n}{p_i p_j} - \dots + (-1)^k \frac{n}{p_1 p_2 \cdots p_k} \;=$$

$$= n \left(1 - \frac{1}{p_1}\right)\left(1 - \frac{1}{p_2}\right) \cdots \left(1 - \frac{1}{p_k}\right).$$

This proves Theorem A.17 in a different way.

Theorem B.17 in Section B.3 will show a nice application of the Multiplicative Möbius Inversion Formula.

A.6.2   The Principle of Inclusion and Exclusion

We shall conclude this section with another useful principle. To develop some intuition, consider the integers in between 0 and $p.q - 1$, where p and q are different primes. We want to evaluate $\phi(p.q)$ directly, i.e. we want to count the number of integers i, $0 \le i < p.q$, that are coprime with $p.q$. Of course, this number is pq minus the number of integers i, $0 \le i < p.q$, that have a nontrivial factor in common with $p.q$, i.e. that are divisible by p or q. There are q multiples of p in the range 0, 1, ..., $p.q - 1$ and similarly p multiples of q. However, one of the multiples of p is also a multiple of q, namely 0 itself. We conclude that

$$\phi(p.q) = p.q - p - q + 1 = (p-1)(q-1) = p.q\left(1 - \frac{1}{p}\right)\left(1 - \frac{1}{q}\right),$$

as it should be according to Theorem A.17.

Theorem A.40 The Principle of Inclusion and Exclusion
Let S be a finite set with N elements. Suppose that the elements in S can satisfy certain properties $P(i)$, $1 \le i \le k$.
Let $N(i_1, i_2, \ldots, i_s)$ be the number of elements in S that satisfy properties $P(i_1), P(i_2), \ldots, P(i_s)$, where $1 \le i_1 < i_2 < \cdots < i_s \le k$, $1 \le s \le k$, (and possibly also some of the other properties).
Let $N(\emptyset)$ denote the number of elements in S that satisfy none of the properties $P(i)$, $1 \le i \le k$.
Then

$$N(\emptyset) = N - \sum_{1 \le i \le k} N(i) + \sum_{1 \le i < j \le k} N(i, j) - \ldots + (-1)^k N(1, 2, \ldots, k).$$

Proof: An element s in S that satisfies exactly r of the k properties is counted

$$1 - \binom{r}{1} + \binom{r}{2} - \ldots + (-1)^r \binom{r}{r} = (1-1)^r = \begin{cases} 1 & \text{if } r = 0, \\ 0 & \text{if } r \ne 0. \end{cases}$$

times in the right hand side, just as in the left hand side.

$\square$

We leave it as an exercise to the reader to prove Theorem A.17 directly from the definition of the Euler Totient Function and the above principle (Hint: Let p_i, $1 \le i \le k$, denote the prime numbers that divide n, take $S = \{0, 1, \ldots, n-1\}$, and say that element $s \in S$ has property $P(i)$, $1 \le i \le k$, if s is divisible by p_i.)

A.7   Problems

Problem A.1^M

Let $\prod_{i=1}^{k} p_i^{a_i}$ be the prime factorization of an integer n. How many different divisors does n have? For $n = 1000$, check your answer with the *Mathematica* function `DivisorSigma[k, n]` which computes $\sum_{d|n} d^k$ (use $k = 0$).

Problem A.2^M

Compute u and v such that $\gcd(455, 559) = 455\,u + 599\,v$.

Problem A.3

Prove that $\gcd(a^m - 1, a^n - 1) = a^{\gcd(m,n)} - 1$ for every positive integer a. (Hint: reduce the pair $\{m, n\}$, $m \geq n$, to $\{m - n, n\}$ and then follow the simple version of Euclid's Algorithm).

Problem A.4^M

a) Check that 563 is a prime number.
b) Use Euclid's algorithm to compute 11^{-1} (mod 563).
c) Solve $11\,x \equiv 85$ (mod 563).

Problem A.5

Find the solutions of $33\,x \equiv 255$ (mod 1689). Note that $1689 = 3 \times 563$ and use the results of Problem A.4.

Problem A.6

a) Determine $\phi(100)$. Check the result with the `EulerPhi` function.
b) Compute the two least significant digits of 2004^{2004} without using the computer.

Problem A.7^M

Solve the system of congruence relations (hint: use Theorem A.19):
$$3\,x \equiv 2 \ (\mathrm{mod}\ 11), \qquad 7\,x \equiv 9 \ (\mathrm{mod}\ 13), \qquad 4\,x \equiv 14 \ (\mathrm{mod}\ 15).$$

Problem A.8^M

Determine the Jacobi Symbol (7531, 3465).

Problem A.9

Use the Chinese Remainder Theorem to solve $x^2 \equiv 56$ (mod 143). (Hint: first reduce it to several systems of linear congruence relations).
How many different solutions are there modulo 143?

Problem A.10

Determine the first five terms of the continued fraction of f, the largest zero of $f^2 = f + 1$. Determine also the first five convergents.
What do you conjecture about the other terms in the continued fraction of f? Prove this conjecture (hint: use Algorithm A.29 and the definition of f).

Problem A.11

Prove Theorem A.17 with the Principle of Inclusion and Exclusion (Thm. A.40) and the definition of the Euler function $\varphi(n)$.

Appendix B   Finite Fields

Introductory Remarks

Most readers will be familiar with the algebraic structure of the sets of rational, real, and complex numbers. These sets have all the properties with respect to addition and multiplication that one may want them to have. They are called *fields*.

In discrete mathematics, in particular in the context of cryptology and coding theory, fields of finite cardinality play a crucial role. In this chapter, an introduction will be given to the theory of finite fields.

The outline of this is as follows:

In Section B.1, we recapitulate the basic definitions and properties of abstract algebra and of linear algebra. In particular, we shall show that the set of integers modulo a prime number from a finite field. In Section B.2, a general construction of finite fields will be given. In Section B.3 a formula is derived for the number of irreducible polynomials over a given finite field. This shows that finite fields exist whenever the size is a power of a prime. An analysis of the structure of finite fields will be given in Section B.4. In particular, it will be shown that a finite field of size q exists if and only if q is a prime power. Moreover, such a field is unique, its additive group has the structure of a vector space and its multiplicative group has a cyclic structure.

B.1   Algebra

Although we assume that the reader is already familiar with all notions discussed in this and the next subsection. we offer this summary as a service to the reader.

B.1.1   Abstract Algebra

□ **Set operations**

Let S be a nonempty set. An operation $*$ defined on S is a mapping from $S \times S$ into S. The image of the pair (s, t) under $*$ is denoted by $s * t$. Examples of operations are the addition $+$ in R and the multiplication $\times$ in C. The operation $*$ is called commutative if for all s and t in S:

S.1 $s * t = t * s$ for all s and t in S.

An element e in S that satisfies

S.2 $s * e = e * s$ for all s in S.

will be called a unit-element of $(S, *)$.

If $(S, *)$ has a unit-element, it will be unique. Indeed, suppose that e and e' both satisfy S.1. Then, by using S.2 twice one gets

$$e = e * e' = e'.$$

Example B.1

Take S as the set of integers $\mathbb{Z}$ and $+$ (i.e. addition) as operation. This operation is commutative and $(\mathbb{Z}, +)$ has 0 as unit-element.

Example B.2

Let S be the set of 2×2 real matrices with matrix multiplication as operation. This operation is not commutative, e.g.

$$\begin{pmatrix} 1 & 1 \\ 0 & 1 \end{pmatrix} \cdot \begin{pmatrix} 0 & 1 \\ 1 & 1 \end{pmatrix} == \begin{pmatrix} 0 & 1 \\ 1 & 1 \end{pmatrix} \cdot \begin{pmatrix} 1 & 1 \\ 0 & 1 \end{pmatrix}$$

```
False
```

On the other hand, this set S does have a unit-element, namely $\begin{pmatrix} 1 & 0 \\ 0 & 1 \end{pmatrix}$. Compute for instance:

$$\texttt{MatrixForm}\left[\begin{pmatrix} a & b \\ c & d \end{pmatrix} \cdot \begin{pmatrix} 1 & 0 \\ 0 & 1 \end{pmatrix}\right]$$

$$\begin{pmatrix} a & b \\ c & d \end{pmatrix}$$

□ **Group**

Definition B.1
Let G be a non-empty set and $*$ an operation defined on G. Then, the pair $(G, *)$ is called a group, if
G1: $(g * h) * k = g * (h * k)$ for all g, h, $k \in G$ (associativity),
G2: G contains a unit element, say e,
G3:   for each g in G an element h in G exists such that $g * h = h * g = e$.
         This element is called the inverse of g and often denoted by g^{-1}.

Property G1 tells us that there is no need to write brackets in strings like $g * h * k$. The element h in Property G3 is unique. Indeed, if h and h' both satisfy G3, then $h = h * e = h * (g * h') = (h * g) * h' = e * h' = h'$. In the same way one can show that for each $a, b \in G$ the equations

$$a.x = b \qquad \text{and} \qquad x.a = b$$

have a unique solution in G, namely

$$x = a^{-1} b, \qquad \text{resp.} \qquad x = b.a^{-1}.$$

The reader easily checks that $(\mathbb{Z}, +)$ in Example B.1 shows a commutative group. Other well-known examples of commutative groups are: $(\mathbb{Q}, +)$, $(\mathbb{Q} \setminus \{0\}, \cdot)$, and $(\mathbb{R}, +)$.

Example B.2 does not yield a group because not all matrices have an inverse (e.g. the all-zero matrix).

Let $(G, *)$ be a group and H a subset of G with the property that $(H, *)$ is also a group, then H will be called a subgroup of G. It can be shown (see Problem B.3) that H is a subgroup of G if and only if

$$h_1 h_2^{-1} \in H, \qquad \text{for every } h_1, h_2 \in H.$$

Let $m \in \mathbb{Z} / \{0\}$ and define $m\mathbb{Z} = \{m k \mid k \in \mathbb{Z}\}$. Then $(m\mathbb{Z}, +)$ is a commutative subgroup of $(\mathbb{Z}, +)$, as one can easily check.

Example B.3

Let $m \in \mathbb{Z} / \{0\}$ and define $\mathbb{Z}_m^*$ as the reduced residue system

$$\mathbb{Z}_m^* = \{0 \le i < m \mid gcd(i, m) = 1\}.$$

The cardinality of set $\mathbb{Z}_m^*$ is $\varphi(m)$ by Definition A.6.

It follows from Lemma A.13 that the product of two elements in $\mathbb{Z}_m^*$ can again be represented by an element in $\mathbb{Z}_m^*$. Clearly, 1 is en element of $\mathbb{Z}_m^*$ which is the unit element under this multiplication. That each element in $\mathbb{Z}_m^*$ has a multiplicative inverse follows from Theorem A.18 (note that with a $\in \mathbb{Z}_m^*$ one has that gcd(a, m) = 1 and thus the equivalence relation $a x \equiv 1 \pmod{m}$ has a unique solution).

We conclude that the multiplicative group $(\mathbb{Z}_m^*, \times)$ is a commutative group of cardinality $\varphi(m)$.

Commutative groups are also called Abelian groups. Quite often, Abelian groups are represented in an additive way: the operation is denoted by a plus sign and the unit-element is called the zero element (denoted with a zero). An abelian group in this notation is called an additive group.

The most commonly used additive group in this introduction will be $(\mathbb{Z}_m, 0)$, but in Chapter 10, we shall see another example (see Theorem 10.2).

We shall now consider the more interesting situation that two operations are defined on a set. The first will be denoted by $g + h$, the second by $g \cdot h$.

□ **Ring**

> **Definition B.2**
> The triple $(R, +, \cdot)$ is called a ring, if
> R1: $(R, +)$ is a commutative group.
>   Its unit-element will be denoted by 0.
> R2:   The operation $\cdot$ is associative.
> R3:   Distributivity holds, i.e. for all $r, s, t \in R$
> $r \cdot (s + t) = r \cdot s + r \cdot t$ and $(r + s) \cdot t = r \cdot t + s \cdot t$.

From now on we shall often simply write $g\,h$ instead of $g \cdot h$. The (additive) inverse of an element g in the group $(R, +)$ will simply be denoted by $-g$, just as we write $2\,g$ for $g + g$, and $3\,g$ for $g + g + g$, etc. Note that 0 really behaves like a zero-element, because for every $r \in R$ one has that $0\,r = (r - r)\,r = r^2 - r^2 = 0$ and similarly that $r\,0 = 0$.

Suppose that the operation $\cdot$ is commutative on $R \backslash \{0\}$. Then the ring $(R, +, \cdot)$ is called commutative. Examples of commutative rings are $(\mathbb{R}, +, \cdot)$, $(\mathbb{Q}, +, \cdot)$, $(\mathbb{Z}, +, \cdot)$, but also $(m\,\mathbb{Z}, +, \cdot)$, when $m \neq 0$.

Let $(R, +, \cdot)$ be a ring and S a subset of R with the property that $(S, +, \cdot)$ is itself a ring, then S will be called a subring of R. Note that $(6\,\mathbb{Z}, +, \cdot)$ is a subring of $(2\,\mathbb{Z}, +, \cdot)$, which in turn is a subring of $(\mathbb{Z}, +, \cdot)$.

□ **Ideal**

> **Definition B.3**
> A subring $(S, +, \cdot)$ of a ring $(R, +, \cdot)$ is called an ideal if
> I: for all $r \in R$ and $s \in S$ $[r\,s \in S$ and $s\,r \in S]$.

Let $m \in \mathbb{Z} \backslash \{0\}$. It is easy to check that any integer multiple of an m-tuple, is also an m-tuple. It follows that $(m\,\mathbb{Z}, +, \cdot)$ is an ideal in $(\mathbb{Z}, +, \cdot)$.

Now suppose that $(R, \cdot)$ has a unit-element, say e, then some elements in R may have an inverse in R i.e. an element b such that $a\,b = b\,a = e$. This inverse, which is again unique, is called the multiplicative inverse of a and will be denoted by a^{-1}. Clearly, the element 0 will not have a multiplicative inverse. Indeed, suppose that $r\,0 = e$ for some $r \in R$. Then for each $a \in R$ one has that $a = a\,e = a(r\,0) = (a\,r)\,0 = 0$, i.e. $R = 0$.

It follows from the above that $(R, \cdot)$, when $R \neq \{0\}$, can not be a group. However, $(R \backslash \{0\}, \cdot)$ may very well have the structure of a group.

□ **Field**

> **Definition B.4**
> A triple $(F, +, \cdot)$ is called a field, if
> F1: $(F, +)$ is a commutative group. Its unit-element is denoted by 0.
> F2: $(F, \cdot)$ is a group. The multiplicative unit-element is denoted by e.
> F3: Distributivity holds.

Unlike some rings, a field can not have so-called zero-divisors, i.e. elements f and g, both unequal to 0, whose product $f\,g$ is equal to 0. Indeed, suppose that $f\,g = 0$ and $f \neq 0$. Then, $g = e\,g = (f^{-1}\,f)\,g = f^{-1}(f\,g) = f^{-1}\,0 = 0$, so every element in F is zero.

If a subring $(K, +, \cdot)$ of a field $(F, +, \cdot)$ has the structure of a field, we shall call it a subfield of $(F, +, \cdot)$.

Examples of fields are the rationals $(\mathbb{Q}, +, \cdot)$, the reals $(\mathbb{R}, +, \cdot)$, and the complex numbers $(\mathbb{C}, +, \cdot)$, each one being a subfield of the next one.

We speak of a finite group $(G, *)$, ring $(R, +, \cdot)$, or field $(F, +, \cdot)$ of order n, if G, resp. R, and F are finite sets of cardinality n. For finite fields it is customary to denote the cardinality by q.

In this chapter, we shall study the structure of finite fields. It will turn out that finite fields of order q only exist when q is a prime power. Moreover, these finite fields are essentially unique for a fixed prime power q. This justifies the widely accepted notation $\mathbb{F}_q$ or $GF(q)$ (where GF stands for Galois Field after the Frenchman Galois) for a finite field of order q. Examples of finite fields will follow in Section B.2.

Analogously to commutative rings, we define a commutative field $(F, +, \cdot)$ to be a field, for which $(F \setminus \{0\}, \cdot)$ is commutative. The following theorem will not be proved, but is very important [Cohn77, p. 196].

> **Theorem B.1**   Wedderburn
> Every finite field is commutative.

□ **Equivalence Relations**

> **Definition B.5**
> Let U be a set. Corresponding to any subset P of $U \times U$, one can define a relation $\sim$ on U by
> $$\text{for all } u, v \in U \ [u \sim v \Longleftrightarrow (u, v) \in P].$$
> An equivalence relation is a relation with the additional properties:
> E1: for all $u \in U$ $[u \sim u]$ (reflexivity).
> E2: for all $u, v \in U$ $[u \sim v \Longrightarrow v \sim u]$ (symmetry).
> E3: for all $u, v, w \in U$ $[(u \sim v \land v \sim w) \Longrightarrow u \sim w]$ (transitivity).

Let U be the set of straight lines in the (Euclidean) plane. Then "being parallel or equal" defines an equivalence relation.

In Section A.3 we have seen another example. There $U = \mathbb{Z}$ and for a fixed m, $m \neq 0$, the relation $\equiv$ was defined by $a \equiv b \,(\mathrm{mod}\, m)$ if and only if m divides $a - b$.

Let $\sim$ be an equivalence relation defined on a set U. A non-empty subset W of U is called an equivalence class, if

E1) $\forall_{v,w \in W} \, [\, v \sim w \,]$,

E2) $\forall_{w \in W} \, \forall_{u \in U \backslash W} \, [\, \neg \,(u \sim w) \,]$.

It follows from the properties above, that an equivalence class consists of all elements in U, that are in relation $\sim$ with a fixed element in U. Clearly, the various equivalence classes of U form a partition of U. The equivalence class containing a particular element w, will be denoted by $< w >$.

Let $(R, +, \cdot)$ be a commutative ring with (multiplicative) unit-element e and let $(S, +, \cdot)$ be an ideal in $(R, +, \cdot)$. We define a relation $\equiv$ on R by

$$a \equiv b \,(\mathrm{mod}\, S) \iff (a - b \in S) \qquad\qquad (B.1)$$

The reader can easily verify that (B.1) defines an equivalence relation. Let R/S (read: R modulo S) denote the set of equivalence classes. On R/S we define two operations by:

$$< a > \; + \; < b > \; := \; < a + b >, \quad a, b \in R,$$

$$< a > \; \cdot \; < b > \; := \; < ab >, \quad a, b \in R.$$

It is easy to verify that these definitions are independent of the particular choice of the elements a and b in the equivalence class $< a >$ and $< b >$. We leave it as an exercise to the reader to prove the following theorem.

Theorem B.2
Let $(R, +, \cdot)$ be a commutative ring and let $(S, +, \cdot)$ be an ideal in $(R, +, \cdot)$. With the above definitions $(R/S, +, \cdot)$ is a commutative ring with unit-element.

The ring $(R/S, +, \cdot)$ is called a residue class ring of R modulo S. In the next section we will see applications of Theorem B.2.

□ **Cyclic Groups**

Before we conclude this section, there is one more topic that needs to be discussed. Let $(G, \cdot)$ be a finite group and let a be an element in $G\backslash\{e\}$. Let a^2, a^3, ..., denote $a\,a$, $a\,a\,a$, etc. Consider the sequence of elements e, a, a^2, ..., in G. Since G is finite, there exists a unique integer n such that the elements e, a, a^2, ..., a^{n-1} are all different, while $a^n = a^j$ for some j, $0 \le j < n$. It follows that $a^{n+1} = a^{j+1}$, etc.. We shall now show that $j = 0$, i.e. that $a^n = e$. Suppose that $j > 0$. Then it would follow from $a^n = a^j$ that $a^{n-1} = a^{j-1}$. However, this contradicts our definition of n. We conclude that the n elements a^i, $0 \le i < n$, are all distinct and that $a^n = e$.

It is now clear that the elements e, a, a^2, ..., a^{n-1} form a subgroup H in G. Such a (sub)group H is called a cyclic subgroup of order n. We say that the element a generates H and that a has (multiplicative) order n.

Since all elements in a cyclic group are a power of the same element, it follows that a cyclic group is commutative.

> **Lemma B.3**
> Let $(G, \cdot)$ be a group and a an element in G of order n. Then, for all $m \in \mathbb{Z}$
> $$a^m = e \quad \Longleftrightarrow \quad n \mid m.$$

Proof:

Write $m = q\,n + r$, $0 \le r < n$. Then, $a^m = e$, iff $a^r = e$, i.e. iff $r = 0$, i.e. iff $n \mid m$.

□

It follows that an element a in G has order d if and only if $a^d = e$ and $a^{d/p} \ne e$ for every prime divisor p of d.

To find the multiplicative order of an integer a in $\mathbb{Z}_m^*$ (so $\gcd(a, m) = 1$), it follows from Euler's Theorem (Thm. A.14) and Lemma B.3 that one only has to check the divisors of $\varphi(m)$. The following module does this in an efficient way. It makes use of the *Mathematica* functions GCD, Divisors, EulerPhi, and PowerMod. In *Mathematica* 4, MultiplicativeOrder is a standard function.

```
MultiplicativeOrder[a_, m_] :=
    If[GCD[a, m] == 1,          Divisors[EulerPhi[m]] //.
       {x_, y___} -> If[PowerMod[a, x, m] == 1, x, {y}] ];
```

```
a = 2; m = 123456789;
n = MultiplicativeOrder[a, m]
```

Lemma B.4
Let $(G, \cdot)$ be a group and a an element in G of order n. For $k > 0$, element a^k has order

$$\frac{n}{\gcd(k, n)} \cdot$$

Proof:

Let m be the order of a. Since $k / \gcd(k, n)$ is an integer, it follows that

$$(a^k)^{n/\gcd(k, n)} = (a^n)^{k/\gcd(k, n)} = e^{k/\gcd(k, n)} = e.$$

From Lemma B.3, we conclude that m divides $n / \gcd(k, n)$. To prove the converse, we observe that $(a^k)^m = e$. Lemma B.3 implies that n divides $k\, m$. Hence, $n / \gcd(k, n)$ divides m.

$\square$

Continuing with the same parameters as above, we have for instance:

```
k = 3;
MultiplicativeOrder[a^k, m]
n / GCD[k, n]
```

```
2285002
```

```
2285002
```

Analogous to (B.1), one can define for every subgroup $(H, \cdot)$ of a finite group $(G, \cdot)$ an equivalence relation $\sim$ by

$$a \sim b \text{ iff } ab^{-1} \in H.$$

The equivalence classes are of the form

$$\{ h a \mid h \in H \}$$

as one can easily check. They all have the same cardinality as H. It follows that the number of equivalence classes is $\frac{|G|}{|H|}$. As a consequence $|H|$ divides $|G|$. This proves the following theorem.

Theorem B.5
Let $(G, \cdot)$ be a finite group of order n. Then every subgroup $(H, \cdot)$ of $(G, \cdot)$ has an order dividing n. Also every element a, $a \neq e$, in G has an order dividing n.

B.1.2   Linear Algebra

□ **Vector Spaces and Subspaces**

Let $\mathbb{F}$ denote an arbitrary field.

> **Definition B.6**
> A vector space over $\mathbb{F}$ is a set V of objects which can be added and multiplied by elements of $\mathbb{F}$ such that the result is again in V. Besides, the following properties must be satisfied:
> 1. $(u + v) + w = u + (v + w)$ for all $u, v, w \in V$,
> 2. there is a zero-element in V, i.e. an element o such that $v + o = o + v = v$ for all $v \in V$,
> 3. for every $v \in V$ there is an element $-v$ in V such that $v + (-v) = (-v) + v = o$,
> 4. $u + v = v + u$ for all $u, v \in V$,
> 5. $\alpha(u + v) = \alpha u + \alpha v$ for all $u, v \in V$ and $\alpha \in \mathbb{F}$,
> 6. $(\alpha + \beta) v = \alpha v + \beta v$ for all $\alpha, \beta \in \mathbb{F}$ and $v \in V$,
> 7. $(\alpha \beta) v = \alpha(\beta v)$ for all $\alpha, \beta \in F$ and $v \in V$,
> 8. $1 \cdot v = v$ for all $v \in V$, where 1 denotes the unit-element of the field $\mathbb{F}$.

It is customary to call the elements of a vector space vectors although they need not be vectors in the heuristic sense.

Examples of vector spaces over $\mathbb{F}$ are:

i) $\mathbb{F}^n$, the set of n-tuples over $\mathbb{F}$

ii) $\{f(x) \in \mathbb{F}[x] \mid \deg(f(x)) < n\}$, the set of polynomials over $\mathbb{F}$ of degree less than n.

Often, it is clear from the context over which field a vector space is defined. In that case, the field will no longer be mentioned.

> **Definition B.7**
> A subset W of a given vector space V is called a linear subspace of V if W itself is a vector space with the operations already defined in V.

In order to determine whether a given subset of a vector space is a subspace, it is not necessary to check all eight vector space properties. For instance property 1 holds for all $u, v, w \in W$ because it is satisfied a fortiori by all elements in V. We have

> **Theorem B.6**
> A subset W of a vector space V is a linear subspace of V if and only if
> (i) $o \in W$,
> (ii) $u + v \in W$ for all $u, v \in W$,
> (iii) $\alpha u \in W$ for all $u \in W$ and $\alpha \in \mathbb{F}$.

Every vector space V has two so-called trivial subspaces: $\{o\}$ and V.

Let V be a vector space and let $v_1, v_2, \dots, v_n$ be elements of V. An expression of the type

$$\alpha_1 v_1 + \alpha_2 v_2 + \dots + \alpha_n v_n \quad \text{with } \alpha_i \in \mathbb{F}$$

is called a linear combination of $v_1, v_2, \dots, v_n$.

The set of all linear combinations of $v_1, v_2, \dots, v_n$ is a subspace of V, which is called the subspace spanned by $v_1, v_2, \dots, v_n$, and will be denoted by $< v_1, v_2, \dots, v_n >$.

□ Linear Independence, Basis and Dimension

Probably the most important concept when dealing with vector spaces is the concept of linear (in)dependency.

> **Definition B.8**
> A set of vectors $v_1, v_2, \dots, v_n$ in a vector space V is linearly independent if the equation $\alpha_1 v_1 + \alpha_2 v_2 + \dots + \alpha_n v_n = o$ has only the trivial solution $\alpha_1 = 0, \alpha_2 = 0, \dots, \alpha_n = 0$. If the set of vectors is not linearly independent it is linearly dependent.

Suppose that the set of vectors $v_1, v_2, \dots, v_n$ is linearly dependent. Then, there is a linear combination $\alpha_1 v_1 + \dots + \alpha_n v_n = o$ where at least one $\alpha_i \neq 0$. This enables us to write

$v_i = \alpha_i^{-1} (\alpha_1 v_1 + \dots + \alpha_{i-1} v_{i-1} + \alpha_{i+1} v_{i+1} + \dots + \alpha_n v_n)$. Thus, we get a different description of linear dependency.

> **Theorem B.7**
> A set of vectors $v_1, v_2, \dots, v_n$ in a vector space V is linearly dependent if and only if at least one of these vectors can be expressed as a linear combination of the other vectors.

This implies in particular that any set of vectors that includes the zero-vector o is linearly dependent.

> **Theorem B.8**
> Suppose that the vectors $v_1, v_2, \dots, v_n$ are linearly independent. If we replace one of these vectors by the sum of this vector and a linear combination of the other vectors, the resulting set of vectors is again linearly independent.

Now let W be a subspace of a vector space V, and let $\{w_1, w_2, \dots, w_n\} \subset W$.

> **Definition B.9**
> The set $\{w_1, w_2, \dots, w_n\}$ is a basis for W if
> (i)   this set of vectors is linearly independent,
> (ii) $< w_1, \dots, w_n > = W$, i.e. any $w \in W$ is a linear combination of $w_1, w_2, \dots, w_n$.

In particular, if $W = V$ we have a basis for the vector space V itself.

For instance, if $V = \mathbb{F}^n$ the following set of vectors is a basis for V:

$$e_1 = (1, 0, \ldots, 0), \quad e_2 = (0, 1, 0, \ldots, 0), \ldots, \quad e_n = (0, \ldots, 0, 1).$$

This basis is usually called the standard basis.

In the definition we considered only a finite basis. Not every vector space is spanned by a finite number of vectors. Take for example $\mathbb{F} = \mathbb{R}$, and V is the vector space of all real-valued functions on $\mathbb{R}$.

It can be proved that in every vector space a basis exists. Here we will be concerned only with vector spaces which are spanned by a finite number of vectors. The following theorem is very important.

> **Theorem B.9**
> Suppose one basis of a subspace W of a vector space V has n vectors, and another basis has m vectors. Then $n = m$.

A basis for a vector space is not uniquely determined; however, in the case of a finite basis the number of vectors in a basis is uniquely determined.

> **Definition B.10**
> If a vector space has a basis with n vectors we call n the dimension of this vector space. The dimension of the zero vector space $\{o\}$ is defined to be 0.

□ **Inner Product, Orthogonality**

Let V be a vector space over the field $\mathbb{F}$.

> **Definition B.11**
> An inner product on V is a bilinear map $V \times V \to \mathbb{F}$. It is denoted by (u, v), where u and v are vectors in V.

Bilinear means that the following properties should hold for all $u, v, w \in V$ and $\alpha \in \mathbb{F}$.

$$(u+v, w) = (u, w) + (v, w) \text{ and } (u, v+w) = (u, v) + (u, w)$$

$$(\alpha u, v) = \alpha(u, v) = (u, \alpha v)$$

This is a very general definition of an inner product. If in particular $\mathbb{F} = \mathbb{R}$ or $\mathbb{F} = \mathbb{C}$ usually additional properties are required. For instance, in real vector spaces one wants (u, u) to be positive definite, i.e. $(u, u) > 0$ for all vectors $u \neq o$. In this case, the length or norm of u is defined by $\sqrt{(u, u)}$ and often denoted by $\| u \|$.

If $V = \mathbb{F}^n$ then the standard inner product is defined by

$$(u, v) = u_1 v_1 + u_2 v_2 + \ldots + u_n v_n. \tag{B.2}$$

> **Definition B.12**
> (i) Two vectors u and v in V are called orthogonal if $(u,v) = 0$.
> (ii) Two subspaces U and W of V are called orthogonal if $(u,w) = 0$ for all $u \in U$ and $w \in W$.

If the field $\mathbb{F}$ is finite then there may exist nonzero vectors u such that $(u,u) = 0$. For instance, in the vector space $\mathbb{F}^n$, where $\mathbb{F} = \{0, 1\}$, with standard inner product, any vector u with an even number of nonzero coordinates is orthogonal to itself.

Let U be a subspace of V. In many applications it is useful to consider the set of all vectors orthogonal to U.

> **Definition B.13**
> The orthogonal complement of a subspace U of V, denoted by $U^{\perp}$, is the set of all vectors which are orthogonal to all vectors of U.

In formula:

$$U^{\perp} = \{v \in U \mid (u, v) = 0 \text{ for all } u \in U\}.$$

The following properties hold for subspaces U and W of a finite dimensional vector space V.

> **Theorem B.10**
> i)   The orthogonal complement of a subspace is a subspace itself, i.e. $(U^{\perp})^{\perp} = U$
> ii) $\dim(U^{\perp}) = \dim(V) - \dim(U)$.
> iii) If $U \subset W$, then $W^{\perp} \subset U^{\perp}$
> iv) $(U \cap V)^{\perp} = U^{\perp} + V^{\perp}$.

In the case where $V = \mathbb{F}^n$, with standard inner product, we have a simple representation of $U^{\perp}$. Let $\{u_1, u_2, ..., u_m\}$ be a basis for U, and let A be the $m \times n$-matrix with rows $u_1, ..., u_m$. Then we have:

$$v \in U^{\perp} \iff Av^T = o^T,$$

where the superscript T denotes the transpose of a vector, i.e. the column vector with the same coordinates as v has.

> **Definition B.14**
> A basis $\{v_1, v_2, ..., v_m\}$ of a vector space V is called self-orthogonal if all the inner products (v_i, v_j), $i \neq j$, are zero.
> It is called self-orthonormal, if in addition $(\| v \|_i \| = 0$ for $1 \leq i \leq m$.

B.2   Constructions

The set of integers modulo m, $m \in \mathbb{N} \setminus \{0\}$, that was introduced in Section A.3, can also be described as the residue class ring $(\mathbb{Z}/m\mathbb{Z}, +, \cdot)$ (see Theorem B.2), since $(m\mathbb{Z}, +, \cdot)$ is an ideal in the commutative ring $(\mathbb{Z}, +, \cdot)$. This residue class ring is commutative and has $<1>$ as multiplicative unit-element. The ring $(\mathbb{Z}/m\mathbb{Z}, +, \cdot)$ is often denoted by $(\mathbb{Z}_m, +, \cdot)$.

> **Theorem B.11**
> Let m be a positive integer. The ring $(\mathbb{Z}_m, +, \cdot)$ is a finite field with m elements if and only if m is prime.

Proof:

$\Rightarrow$ Suppose that m is composite, say $m = ab$, $a > 1$, and $b > 1$. Then $<0> = <ab> = <a><b>$, while $<a> \neq <0>$ and $<b> \neq <0>$. So the ring $(\mathbb{Z}_m, +, \cdot)$ has zero-divisors and thus it can not be a field.

$\Leftarrow$ Now suppose that m is prime (See also the Example B.3). We have to prove that for every equivalence class $<a>$, $<a> \neq <0>$, there exists an equivalence class $<b>$, such that $<a><b> = <1>$. For this it is sufficient to show that for any a with $m \nmid a$, there exists an element b, such that $ab \equiv 1 \pmod{m}$. This however follows from Lemma A.13 or Theorem A.18.

$\square$

For convenience, one often leaves out the brackets around the representatives of equivalence classes, therefore with a one really means $<a>$.

Later we shall see that for p prime, $(\mathbb{Z}_p, +, \cdot)$ is essentially the only finite field with p elements. We shall denote it by $(\mathbb{F}_p, +, \cdot)$. In information and communication theory one often works with $\mathbb{F}_2$, which just consists of the elements 0 and 1.

We are now going to construct finite fields $\mathbb{F}_q$ for $q = p^m$, p prime.

Let $(F, +, \cdot)$ be a commutative field (not necessarily finite) and let $F[x]$ be the set of polynomials over F, i.e. the set of expressions

$$f(x) = f_0 + f_1 x + f_2 x^2 + \ldots + f_n x^n.$$

where $f_i \in F$, $0 \leq 0 \leq n$, and $n \in \mathbb{N}$. The largest value of i for which $f_i \neq 0$ is called the degree of $f(x)$.

Addition and multiplication of polynomials is defined in the natural way.

$$\sum_i f_i x^i + \sum_i g_i x^i = \sum_i (f_i + g_i) x^i. \tag{B.3}$$

$$\left(\sum_i f_i x^i\right)\left(\sum_j g_j x^j\right) = \sum_k \left(\sum_{i+j=k} f_i g_j\right) x^k. \tag{B.4}$$

Example B.4

Let $F = F_2$ and consider $f(x) = 1 + x^2 + x^3$ and $g(x) = 1 + x + x^3$. Then $f(x) + g(x) = x + x^2$ and $f(x) g(x) = 1 + x + x^2 + x^3 + x^4 + x^5 + x^6$.

In *Mathematica* we can perform these calculations the function `PolynomialMod` as follows

```
p = 2; f = 1 + x^2 + x^3; g = 1 + x + x^3; PolynomialMod[f + g, p]
PolynomialMod[f * g, p]
```

```
x + x^2
```

```
1 + x + x^2 + x^3 + x^4 + x^5 + x^6
```

It is now straightforward to verify the next theorem.

Theorem B.12
Let $(F, +, \cdot)$ be a commutative field. Then $(F[x], +, \cdot)$ is a commutative ring with unit-element.

Analogously to the concepts defined in Appendix A for the set of integers, one can define the following notions in $(F[x], +, \cdot)$: divisibility, reducibility (if a polynomial can be written as the product of two polynomials of lower degree), irreducibility (which is the analog of primality), gcd, lcm, the unique factorization theorem (the analog of the fundamental theorem in number theory), Euclid's Algorithm, congruence relations, etc. We leave the details to the reader.

The following *Mathematica* functions can be helpful here: `PolynomialMod` (which also reduces one polynomial modulo another), `Factor`, `PolynomialGCD`, `PolynomialLCM`. Their usage is demonstrated in the following examples:

```
p = 2; f = 1 + x + x^2 + x^7; g = 1 + x + x^3;
PolynomialMod[f, g, Modulus -> 2]
```

```
x + x^2
```

```
Factor[x^11 - 1, Modulus -> 3]
```

```
(2 + x) (2 + 2 x + x^2 + 2 x^3 + x^5) (2 + x^2 + 2 x^3 + x^4 + x^5)
```

```
PolynomialGCD[1 + x^3, 1 + x^2, Modulus -> 2]
```

```
1 + x
```

```
PolynomialLCM[1 + x^3, 1 + x^2, Modulus -> 2]
```

```
(1 + x^2) (1 + x + x^2)
```

With the package `Algebra`PolynomialExtendedGCD`` one can use the *Mathematica* function PolynomialExtendedGCD:

```
<< Algebra`PolynomialExtendedGCD`
```

```
PolynomialExtendedGCD[1 + x^3, 1 + x^2, Modulus -> 2]
```

```
{1 + x, {1, Mod[x, 2]}}
```

One particular consequence of Theorem B.12 is stated in the following theorem and its corollary.

Theorem B.13
Let $a(x)$ and $b(x)$ be two polynomials in $F[x]$. Then there exists polynomials $u(x)$ and $v(x)$ in $F(x)$ such that
$$u(x)\,a(x) + v(x)\,b(x) = \gcd(a(x),\,b(x)).$$

Corollary B.14
Let $a(x)$ and $f(x)$ be two polynomials in $F[x]$, such that $\gcd(a(x),\,f(x)) = 1$. Then, the congruence relation
$$a(x)\,u(x) \equiv 1 \pmod{f(x)}$$
has a unique solution modulo $f(x)$.

The solution of the above congruence relation can again be found with PolynomialExtendedGCD. Indeed, from

```
PolynomialExtendedGCD[1 + x^2, 1 + x + x^4, Modulus -> 2]
```

```
{1, {1 + x + x^3, x}}
```

we can conclude that the congruence relation $(1 + x^2)\,u(x) \equiv 1 \pmod{1 + x + x^4}$ has the solution $1 + x + x^3$, as one can easily check with:

```
PolynomialMod[(1 + x^2) (1 + x + x^3), 1 + x + x^4, Modulus -> 2]
```

```
1
```

Another important property of $F[x]$ is given in the following theorem.

> **Theorem B.15**
> Any polynomial of degree n, $n > 0$, in $F[x]$ has at most n zeros in F.

Proof: For $n = 1$ the statement is trivial. We proceed by induction on n.

Let $u \in F$ be a zero of a polynomial $f(x)$ of degree n over F (if no such u exists, there is nothing to prove). Write $f(x) = (x - u) q(x) + r(x)$, degree$(r(x)) <$ degree$(x - u) = 1$. It follows that $r(x)$ is a constant, say r. Substitution of $x = u$ in the relation above shows that $t = 0$. We conclude that $f(x) = (x - u) q(x)$.

Now $q(x)$ has degree $n - 1$, thus, by the induction hypothesis, $q(x)$ has at most $n - 1$ zeros in F. Since a field can not have zero-divisors, we know that each zero of $f(x)$ is either a divisor of $x - u$ or a zero of $q(x)$. It follows that $f(x)$ has at most n zeros in F.

□

Let $s(x)$ be a non-zero polynomial in $F[x]$. It is easy to check that the set

$$\{ a(x) s(x) \mid a(x) \in F \}.$$

forms an ideal in the ring $(F[x], +, \cdot)$. We denote this ideal by $(s(x))$ and say that $s(x)$ generates the ideal $(s(x))$.

Conversely, let $(S, +, \cdot)$ be any ideal in $(F[x], +, \cdot)$, with $S \neq F[x]$. Further, let $s(x)$ be a polynomial of lowest degree in S. Take any other polynomial $f(x)$ in S and write $f(x) = q(x) s(x) + r(x)$, degree$(r(x)) <$ degree$(s(x))$. With properties I and R1, we then have that also $r(x)$ is also an element of S. From our assumption on $s(x)$ we conclude that $r(x) = 0$ and thus that $s(x)$ divides $f(x)$.

It follows from the above discussion that any ideal in the ring $(F[x], +, \cdot)$ is generated by a single element! A ring with this property is called a principal ideal ring.

From now on we shall restrict ourselves to finite fields. Up to now we have only seen examples of finite fields $\mathbb{F}_p$, with p prime.

Let $f(x) \in \mathbb{F}_p[x]$ of degree n. We shall say that f is a p-ary polynomial. Let $f(x)$ be the ideal generated by $f(x)$. From Theorem B.2 we know that $(\mathbb{F}_p[x] / (f(x)), +, \cdot)$ is a commutative ring with unit-element $< 1 >$. It contains p^n elements, represented by the p-ary polynomials of degree $< n$.

Theorem B.16
Let $(\mathbb{F}_p, +, \cdot)$ be a finite field with p elements. Let $f(x)$ be a polynomial of degree n over $\mathbb{F}_p$. Then, the commutative ring $(\mathbb{F}_p[x]/(f(x)), +, \cdot)$ is a finite field with p^n elements if and only if $f(x)$ is irreducible in $\mathbb{F}_p[x]$.

Proof: (Compare with Theorem B.11 and its proof.)

$\implies$ Suppose that $f(x) = a(x)\, b(x)$, with degree($a(x)$) > 0 and degree($b(x)$) > 0. Then $<a(x)> <b(x)> = <a(x)\, b(x)> = <f(x)> = <0>$, while $<a(x)> \neq <0>$ and $<b(x)> \neq <0>$. So, $(\mathbb{F}_p[x]/(f(x)), +, \cdot)$ is a ring with zero-divisors. Hence it can not be a field.

$\impliedby$ On the other hand, if $f(x)$ is irreducible, any non-zero polynomial $a(x)$ of degree $< n$ will have a multiplicative inverse $u(x)$ modulo $f(x)$ by Corollary B.14. For this $u(x)$ one has $<a(x)> <u(x)> = <1>$. It follows that $(\mathbb{F}_p[x]/(f(x)), +, \cdot)$ is a field. We know already that it contains p^n elements.

$\square$

Example B.5

Let $q = 2$. The field $\mathbb{F}_2$ consists of the two elements 0 and 1. Let $f(x) = 1 + x + x^3$. Then

$(\mathbb{F}_2[x]/(1 + x + x^3), +, \cdot)$ is a finite field with $2^3 = 8$ elements. These eight elements can be represented by the eight binary polynomials of degree < 3. Addition and multiplication have to be performed modulo $1 + x + x^3$. For instance

$$(1 + x + x^2)x^2 \equiv x^2 + x^3 + x^4 \equiv (x+1)(1 + x + x^3) + 1 \equiv 1 \pmod{1 + x + x^3}.$$

Thus, x^2 is the multiplicative inverse of $1 + x + x^2$ in the field $(\mathbb{F}_2[x]/(1 + x + x^3), +, \cdot)$.

In *Mathematica* one can find an irreducible polynomial over $\mathbb{F}_p$, p prime, with the function `IrreduciblePolynomial` for which the package `Algebra`FiniteFields`` needs to be loaded first.

```
<< Algebra`FiniteFields`
```

```
p = 3; deg = 11; IrreduciblePolynomial[x, p, deg]
```

```
1 + x^5 + 2 x^10 + x^11
```

In *Mathematica* the field defined by the p-ary polynomial $f(x)$ of degree can be described by GF[p, \{f_0, f_1, ..., f_m\}]. Addition, subtraction, multiplication, and division can be performed as follows:

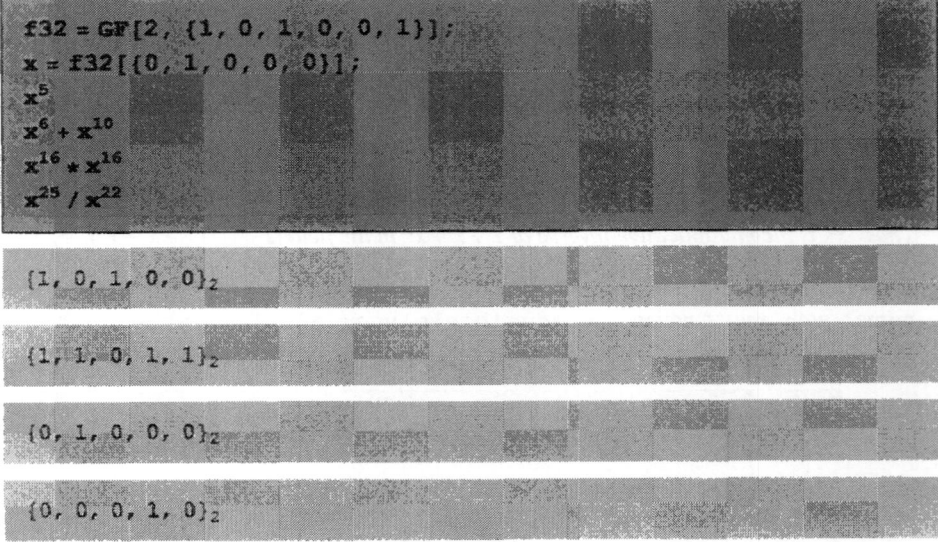

```
f32 = GF[2, {1, 0, 1, 0, 0, 1}];
f32[{1, 0, 1, 0, 0}] + f32[{0, 1, 1, 0, 1}]
f32[{1, 0, 1, 0, 0}] - f32[{0, 1, 1, 0, 1}]
f32[{1, 0, 1, 0, 0}] * f32[{0, 1, 1, 0, 1}]
f32[{1, 0, 1, 0, 0}] / f32[{0, 1, 1, 0, 1}]
```

$\{1, 1, 0, 0, 1\}_2$

$\{1, 1, 0, 0, 1\}_2$

$\{0, 0, 1, 0, 0\}_2$

$\{1, 0, 1, 1, 0\}_2$

or as follows:

```
f32 = GF[2, {1, 0, 1, 0, 0, 1}];
x = f32[{0, 1, 0, 0, 0}];
x^5
x^6 + x^10
x^16 * x^16
x^25 / x^22
```

$\{1, 0, 1, 0, 0\}_2$

$\{1, 1, 0, 1, 1\}_2$

$\{0, 1, 0, 0, 0\}_2$

$\{0, 0, 0, 1, 0\}_2$

Two questions that arise naturally at this moment are:

1) Does an irreducible, p-ary polynomial $f(x)$ of degree n exist for every prime number p and every integer n? If so, then we have proved the existence of finite fields $\mathbb{F}_q$ for all prime powers q

2) Do other finite fields exist?

The first question gets an affirmative answer in the next section. The second question gets a negative answer in Section B.4.

B.3   The Number of Irreducible Polynomials over GF(q)

In this section we want to count the number of irreducible polynomials over a finite field $\mathbb{F}_q$. Clearly, if $f(x)$ is irreducible, then so is $\alpha\,f(x)$, for $\alpha \in \mathbb{F}_q \setminus \{0\}$. Also the ideals $(f(x))$ and $(\alpha\,f(x))$ are the same, when $\alpha \in \mathbb{F}_q \setminus \{0\}$, therefore, we shall only count so-called monic polynomials of degree n, i.e. polynomials, whose leading coefficient (the coefficient of x^n) is equal to 1.

> **Definition B.15**
>
> $I_q(n) = \#\, q$-ary, irreducible, monic polynomials of degree n.
>
> $I(n) = I_2(n) = \#$ binary, irreducible polynomials of degree n.

To develop some intuition for our counting problem, we start with a brute force attack for the special case that $q = 2$. We shall try therefore to determine $I(n)$.

There are only two binary polynomials of degree 1, namely

$$x \text{ and } x + 1.$$

By definition, both are irreducible. Thus, $I(1) = 2$.

By taking all possible products of x and $x + 1$, one finds three reducible polynomials of degree 2:

$$x \cdot x = x^2, \qquad x \cdot (x + 1) = x^2 + x, \quad \text{and} \quad (x + 1)^2 = x^2 + x.$$

Since there are $2^2 = 4$ binary polynomials of degree 2, it follows that there exists only one irreducible

polynomial of degree 2, namely

$$x^2 + x + 1.$$

So, $I(2) = 1$.

Each 3-rd degree, reducible, binary polynomial can be written as a product of the lower degree irreducible polynomials x, $x + 1$ and $x^2 + x + 1$. In this way, one gets $x^i(x + 1)^{3-i}$, $0 \le i \le 3$, $(x^2 + x + 1)x$, and $(x^2 + x + 1)(x + 1)$. Since there are $2^3 = 8$ binary polynomials of degree 3, we conclude that there are

$8 - 4 - 2 = 2$ irreducible, binary polynomials of degree 3. So, $I(3) = 2$.

The two binary, irreducible polynomials of degree 3 are:

$$x^3 + x + 1 \qquad \text{and} \qquad x^3 + x^2 + 1.$$

At this moment it is important to note that for the counting arguments above, we do not have to know the actual form of the lower degree, irreducible polynomials. We only have to know how many there are of a

certain degree.

Indeed, to find $I(4)$ we can count the number of reducible, 4-th degree polynomials as follows:

			number
- product of four 1 - st degree polynomials			5
- product of one 2 - nd degree polynomial and	1×3	=	3
two 1 - st degree polynomials			
- product of two 2 - nd degree polynomials			1
- product of one 3 - rd degree polynomial and	2×2	=	4
one 1 - st degree polynomial			
		total =	13

It follows that there are $2^4 - 13 = 3$ irreducible, binary polynomials of degree 4. So, $I(4) = 3$.

With some additional work one can find these three irreducible, 4-th degree polynomials:

$$x^4 + x + 1, \quad x^4 + x^3 + 1, \quad \text{and} \quad x^4 + x^3 + x^2 + x + 1.$$

Continuing in this way one finds with the necessary perseverance and precision that $I(5) = 6$ and $I(6) = 9$, etc.

The above method does not lead to a proof that $I(n) > 0$ for all $n \in \mathbb{N}$, let alone to an approximation of the actual value of $I(n)$.

We start all over again.

Let $p_i(x)$, $i = 1, 2, \ldots$, be an enumeration of all q-ary, irreducible, monic polynomials, such that the degrees form a non-decreasing sequence. So, the first $I_q(1)$ polynomials have degree 1, the next $I_q(2)$ polynomials have degree 2, etc..

Any q-ary, monic polynomial $f(x)$ has a unique factorization of the form

$$\prod_{i=1}^{\infty} (p_i(x))^{e_i}, \ e_i \in \mathbb{N}, \ i \geq 1.$$

where only finitely many e_i's are unequal to zero. It follows that $f(x)$ can uniquely be represented by the sequence $(e_1, e_2, \ldots)$. Let a_i be the degree of $p_i(x)$ and let n be the degree of $f(x)$. Then

$$e_1 a_1 + e_2 a_2 + \ldots = n.$$

So, the polynomial $f(x)$ is in a unique correspondence with the term

$$(z^{a_1})^{e_1} (z^{a_2})^{e_2} \ldots$$

in the expression

$$(1 + z^{a_1} + z^{2 a_1} + \ldots)(1 + z^{a_2} + z^{2 a_2} + \ldots) \ldots$$

i.e. in

$$\prod_{i=1}^{\infty} (1 - z^{a_i})^{-1}.$$

```
Factor[1 + x + x² + x³ + x⁶, Modulus -> 2]
```

```
1 + x + x² + x³ + x⁴
```

B.4   The Structure of Finite Fields

B.4.1   The Cyclic Structure of a Finite Field

It follows from Theorem B.11, Theorem B.16 and Theorem B.18, that finite fields $(\mathbb{F}_q, +, \cdot)$ exist for all prime powers q. If q is a prime number $\mathbb{F}_q$ can be represented by the integers modulo p. If q is a power of a prime, say $q = p^m$, $\mathbb{F}_q$ can be represented by p-ary polynomials modulo an irreducible polynomial of degree m. We state the above as a theorem.

> **Theorem B.20**
> Let p be a prime and $q = p^m$, $m \geq 1$. Then a finite field of order q exists.

Later in this section we shall see that every finite field can be described by the construction of Theorem B.16. But first we shall prove an extremely nice property of finite fields, namely that their multiplicative group is cyclic! By Theorem B.5, we know that every non-zero element in $\mathbb{F}_q$ has a multiplicative order dividing $q - 1$.

> **Definition B.16**
> An element ω in a finite field of order q is called an n-th root of unity if $\omega^n = e$.
> An element ω is called a primitive n-th root of unity if it has order n.
> If ω is a primitive $(q - 1)$-st root of unity, then ω is called a primitive element or generator of $\mathbb{F}_q$.

> **Theorem B.21**
> Let $(\mathbb{F}_q, +, \cdot)$ be a finite field and let d be an integer dividing $q - 1$. Then $\mathbb{F}_q$ contains exactly $\phi(d)$ elements of order d.
> In particular, $(\mathbb{F}_q \setminus \{0\}, \cdot)$ is a cyclic group of order $q - 1$, which contains $\phi(q - 1)$ primitive elements.

Proof: By Theorem B.5, every non-zero element in $\mathbb{F}_q$ has a multiplicative order d, which divides $q - 1$. On the other hand, suppose that $\mathbb{F}_q$ contains an element of order d, $d \mid (q - 1)$, say ω. Then all d distinct powers of ω are a zero of $x^d - e$. It follows from Theorem B.15 that every d-th root of unity in $\mathbb{F}_q$ is a power of ω. It follows from Lemma B.4 that under the assumption that $\mathbb{F}_q$ contains an element of order d, $\mathbb{F}_q$ will contain exactly $\phi(d)$ elements of order d, namely ω^i, with $GCD[i, d] = 1$.

Let $a(d)$ be the number of elements of order d in $\mathbb{F}_q$. Then the above implies that

    i) $a(d) = 0$ or $a(d) = \phi(d)$

and also that

    ii) $\sum_{d|(q-1)} a(d) = q - 1$.

On the other hand, Theorem A.12 states that $\sum_{d|(q-1)} \phi(d) = q - 1$. So, we conclude that $a(d) = \phi(d)$ for all $d \mid (q - 1)$.

In particular, $a(q - 1) = \phi(q - 1)$ which means that $\mathbb{F}_q$ contains $\phi(q - 1)$ primitive elements and that $\mathbb{F}_q \setminus \{0\}$ is a cyclic group.

                                                                       □

To check if a particular element ω in $GF(q)$ has order d, $d \mid (q-1)$, it suffices to check that $\omega^d = 1$ and that $\omega^{d/p} \neq 1$ for every prime divisor of d. See also the discussion below Lemma B.3.

To find a primitive element in $\mathbb{Z}_p$, p prime, the *Mathematica* function `PowerList` can be used. It finds a primitive element in $\mathbb{Z}_p$ and generates all its powers (starting with the 0-th). The second element in this list is the primitive element itself. First, the package `Algebra`FiniteFields`` needs to be loaded.

```
<< Algebra`FiniteFields`
```

```
p = 17; PrimeQ[p]
PowerList[GF[p, 1]][[2]]
```

```
True
```

```
{3}
```

Problems B.6 and B.10 indicate an efficient way (due to Gauss) to find a primitive element in a finite field.

Corollary B.22
Every element ω in $\mathbb{F}_q$ satisfies

$$\omega^{q^n} = \omega, \qquad n \geq 1.$$

Proof: For $\omega = 0$ the statement is trivially true. By Theorem B.5 or Theorem B.21, any ω, $\omega \neq 0$, has an order dividing $q - 1$. So, it satisfies $\omega^{q-1} = e$ and thus also $\omega^q = \omega$. Since $\omega^{q^n} = (\omega^q)^{q^{n-1}}$, the proof now follows with an easy induction argument.

□

> **Corollary B.23**
> Let $\mathbb{F}_q$ be a finite field. Then
> $$x^q - x = \prod_{\omega \in \mathbb{F}_q} (x - \omega).$$

Proof: Every element ω in $\mathbb{F}_q$ is a zero of $x^q - x$ by Corollary B.22, therefore, the right hand side above divides the left hand side. Equality now follows because the expressions on both sides are monic and of the same degree.

□

Corollary B.23 will be used later as a tool to check if a certain element in fields containing $\mathbb{F}_q$ is actually in $\mathbb{F}_q$ itself.

Example B.6

Consider the finite field $(\mathbb{F}_2[x] / (f(x)), +, \cdot)$ with $f(x) = x^4 + x^3 + x^2 + x + 1$. It contains $2^4 = 16$ elements, which can be represented by binary polynomials of degree <4. The element x, representing the class $< x >$, is not a primitive element, since $x^5 \equiv (x + 1) f(x) + 1 \equiv 1 \pmod{f(x)}$. So x has order 5 instead of 15. With Mathematica this can be checked as follows:

```
f = 1 + x + x^2 + x^3 + x^4;
PolynomialMod[x^2, f, Modulus -> 2]
PolynomialMod[x^3, f, Modulus -> 2]
PolynomialMod[x^4, f, Modulus -> 2]
PolynomialMod[x^5, f, Modulus -> 2]
```

x^2

x^3

$1 + x + x^2 + x^3$

1

The element $x + 1$ is primitive element (its order is 15), as one can see in Table B.1. It is also easy to verify. Indeed, $x + 1$ has an order dividing 15. So, one only has to check that $(x + 1)$ raised to the power 3 or 5 does not reduce to 1 modulo $f(x)$.

```
f := 1 + x + x^2 + x^3 + x^4;
PolynomialMod[(x + 1)^3, f, Modulus -> 2]
PolynomialMod[(x + 1)^5, f, Modulus -> 2]
PolynomialMod[(x + 1)^15, f, Modulus -> 2]
```

$1 + x + x^2 + x^3$

$1 + x^2 + x^3$

1

Multiplication is easy to perform with Table B.1. For instance

$$(1 + x + x^2 + x^3)(x + x^3) \equiv (x + 1)^3 (x + 1)^{14} \equiv$$
$$(x + 1)^{17} \equiv (x + 1)^2 \equiv x^2 + 1 \ (mod \ f(x)).$$

The element $x + 1$ *is a zero of the irreducible polynomial* $y^4 + y^3 + 1$ *since*

$$(x + 1)^4 + (x + 1)^3 + 1 \equiv 0 \ (mod \ f(x)).$$

```
f := 1 + x + x^2 + x^3 + x^4;
PolynomialMod[(x + 1)^4 + (x + 1)^3 + 1, f, Modulus -> 2]
```

0

Therefore, in $(\mathbb{F}_2[x]/(g(x), +, \cdot)$ *with* $g(x) = x^4 + x^3 + 1$, *the element* x *is a primitive element. See Table B.2.*

	1	x	x^2	x^3
0	0	0	0	0
$(1 + x)^0$	1	0	0	0
$(1 + x)^1$	1	1	0	0
$(1 + x)^2$	1	0	1	0
$(1 + x)^3$	1	1	1	1
$(1 + x)^4$	0	1	1	1
$(1 + x)^5$	1	0	1	1
$(1 + x)^6$	0	0	0	1
$(1 + x)^7$	1	1	1	0
$(1 + x)^8$	1	0	0	1
$(1 + x)^9$	0	0	1	0
$(1 + x)^{10}$	0	0	1	1

$(1+x)^{11}$	1	1	0	1
$(1+x)^{12}$	0	1	0	0
$(1+x)^{13}$	0	1	1	0
$(1+x)^{14}$	0	1	0	1

Table B.1 $(\mathbb{F}_2[x]/(1+x+x^2+x^3+x^4),\ +,\ \cdot\ )$ with primitive element $1+x$.

	1	x	x^2	x^3
0	0	0	0	0
1	1	0	0	0
x	0	1	0	0
x^2	0	0	1	0
x^3	0	0	0	1
x^4	1	0	0	1
x^5	1	1	0	1
x^6	1	1	1	1
x^7	1	1	1	0
x^8	0	1	1	1
x^9	1	0	1	0
x^{10}	0	1	0	1
x^{11}	1	0	1	1
x^{12}	1	1	0	0
x^{13}	0	1	1	0
x^{14}	0	0	1	1

Table *B*.2 $(\mathbb{F}_2[x]/(1+x^3+x^4),\ +,\ \cdot)$ with primitive element x

B.4.2 The Cardinality of a Finite Field

Consider the elements e, $2\,e$, $3\,e$, etc. in $\mathbb{F}_q$. Since $\mathbb{F}_q$ is finite, not all these elements can be different. Also, if $i\,e = j\,e$, with $i < j$, also $(j-i)\,e = 0$. These observations justify the following definition.

Definition B.17
The characteristic of a finite field $\mathbb{F}_q$ with unit-element e, is the smallest positive integer c such that $c\,e = 0$.

Theorem B.24
The characteristic of a finite field $\mathbb{F}_q$ is a prime.

Proof: Suppose that the characteristic c can be written as $c'c''$, where $c' > 1$ and $c'' > 1$. Then $0 = c e = (c' e)(c'' e)$, while $c' \neq 0$ and $c'' e \neq 0$. So, $c' e$ and $c'' e$ are zero-divisors. This contradicts the assumption that F_q is a field.

□

> **Definition B.18**
> Two finite fields $(F_q, +, \times)$ and $(F_{q'}, \oplus, \otimes)$ are said to be isomorphic, if there exists a one-to-one mapping ψ from F_q onto $F_{q'}$ (so $q = q'$), such that for all ω_1 and ω_2 in F_q:
> i) $\psi(\omega_1 + \omega_2) = \psi(\omega_1) \oplus \psi(\omega_2)$,
> ii) $\psi(\omega_1 \times \omega_2) = \psi(\omega_1) \otimes \psi(\omega_2)$.

In words, two fields are isomorphic if after renaming the elements in them they behave exactly the same with respect to the operations addition and multiplication.

> **Lemma B.25**
> Let $(F_q, +, \cdot)$ be a finite field with characteristic p. Then $(F_q, +, \cdot)$ contains a subfield which is isomorphic to $(\mathbb{Z}_p, +,)$, i.e. to the integers modulo p.

Proof: The subset $\{i e \mid i = 0, 1, ..., p-1\}$ forms a subfield of $(F_q, +, \cdot)$ which is isomorphic to $(\mathbb{Z}_q, +, \cdot)$ under the isomorphism $\psi(i e) = i$, $0 \leq i < p$.

□

In view of the lemma above, we can and shall from now on identify the subfield in $(F_q, +, \cdot)$ of order p with the field $(\mathbb{Z}_p, +, \cdot)$. The subfield F_p is often called the ground field of F_q. Conversely, the field F_q is called an extension field of F_p.

> **Theorem B.26**
> Let F_q be a finite field of characteristic p. Then F_q can be viewed as a vectorspace over F_p, and $q = p^m$ for some integer m, $m \geq 1$.

Proof: Let $u_1, u_2, ..., u_m$ be a basis of F_q over F_p, i.e. every element ω in F_q can be written as

$$\omega = a_1 u_1 + a_2 u_2 + ... + a_m u_m,$$

where $a_i \in F_p$, $1 \leq i \leq m$, and there is no dependency of the field elements u_i over F_p. It follows that this representation is unique and thus $q = (\mid F)_q \mid = p^m$.

□

At this moment we know that finite fields F_q can only exist for prime powers q. Theorem $B.20$ states that F_q indeed does exist for prime powers q. That all finite fields with the same value of q are isomorphic to each other will be proved later.

B.4.3 Some Calculus Rules over Finite Fields; Conjugates

> **Theorem B.27**
> Let ω be an element in a finite field $\mathbb{F}_q$ of characteristic p. Then in $\mathbb{F}_q[x]$
> $$(x - \omega)^p = x^p - \omega^p.$$

Proof: Let $0 < i < p$. Then $\gcd(p, i!) = 1$, so

$$\binom{p}{i} \equiv \frac{p(p-1)\dots(p-i+1)}{i!} \equiv 0 \pmod{p}$$

and with the binomial theorem, we have that

$$(x - \omega)^p = x^p + (-\omega)^p = x^p - \omega^p.$$

where the last equality is obvious for odd p, while for $p = 2$ this equality follows from $+1 = -1$.

$\square$

To demonstrate this we use again the *Mathematica* function `PolynomialMod`.

```
Clear[a, x];
p = 2; m = 3;
PolynomialMod[ (x - a)^p^m, p]
```

```
a^8 + x^8
```

> **Corollary B.28**
> Let a_i, $1 \le i \le k$, be elements in a finite field $\mathbb{F}_q$ of characteristic p. Then for every n
> $$\left(\sum_{i=1}^{k} a_i\right)^{p^n} = \sum_{i=1}^{k} a_i^{p^n}.$$

```
a =.; b =.; c =.
p = 3; m = 3; PolynomialMod[ (a + b + c)^p^m, p]
```

```
a^27 + b^27 + c^27
```

Proof: Use an induction argument on k and on n. Start with $(a_1 + a_2)^p = a_1^p + a_2^p$.

$\square$

The following theorem often gives a powerful criterion to determine, whether an element in a field $\mathbb{F}_q$ of characteristic p, actually lies in the ground field $\mathbb{F}_p$.

> **Theorem B.29**
> Let $\mathbb{F}_q$ be a finite field of characteristic p. So, $q = p^m$, $m > 0$, and $\mathbb{F}_q$ contains $\mathbb{F}_p$ as a subfield. Let ω be an element in $\mathbb{F}_q$. Then
> $$\omega \in \mathbb{F}_p \iff \omega^p = \omega.$$

Proof: The p elements in the subfield $\mathbb{F}_p$ satisfy $x^p = x$ by Corollary B.23. On the other hand, the polynomial $x^p - x$ has at most p zeros in $\mathbb{F}_q$ by Theorem B.15.

□

Let ω be an element in $\mathbb{F}_q$, a field of characteristic p, but ω not in $\mathbb{F}_p$. Then $\omega^p \neq \omega$ by the previous theorem. Still there is relation between ω^p and ω.

> **Theorem B.30**
> Let ω be an element in a finite field $\mathbb{F}_q$ of characteristic p. Let $f(x)$ be a polynomial over $\mathbb{F}_p$, such that $f(\omega) = 0$. Then for all $n \in \mathbb{N}$
> $$f(\omega^{p^n}) = 0.$$

Proof: Write $f(x) = \sum_{i=0}^m f_i x^i$, Since $f_i \in \mathbb{F}_p$, $o \leq i \leq m$, one has by Corollary B.22 and Theorem B.29 that

$$0 = (f(\omega))^{p^n} = (\sum_{i=0}^m f_i \omega^i)^{p^n} = \sum_{i=0}^m (f_i \omega^i)^{p^n} =$$

$$= \sum_{i=0}^m f_i^{p^n} \omega^{i p^n} = \sum_{i=0}^m f_i (\omega^{p^n})^i = f(\omega^{p^n}).$$

□

In $\mathbb{R}$ and $\mathbb{C}$ a similar thing happens. If $f(x)$ is a polynomial over the reals and $f(\omega) = 0$, $\omega \in \mathbb{C}$, then also $f(\bar{\omega}) = 0$, where $\bar{\omega}$ is the complex conjugate of ω.

The following theorem states that the number of different elements ω^{p^i}, $i = 0, 1, \ldots$, only depends on p and the (multiplicative) order of ω.

> **Theorem B.31**
> Let ω be an element of order n in a finite field of characteristic p. Let m be the multiplicative order of p modulo n, i.e. $p^m \equiv 1 \pmod{n}$, with $m > 0$. Then, the m elements
> $$\omega, \omega^p, \omega^{p^2}, \ldots, \omega^{p^{m-1}}$$
> are all different and $\omega^{p^m} = \omega$.
> The m elements ω^{p^i}, $0 \leq i \leq m - 1$, are called the conjugates of ω.

Proof: By Lemma B.3 (twice), one has that $\omega^{p^i} = \omega^{p^j}$ if and only if $p^i \equiv p^j \pmod{n}$, and thus if and only if $p^{i-j} \equiv 1 \pmod{n}$, i.e. if and only if $i \equiv j \pmod{m}$.

$\square$

Example B.7

Consider $(\mathbb{F}_q[x]/(f(x)), +, \cdot)$ with $f(x) = x^4 + x^3 + x^2 + x + 1$ (see Example B.6). The field element x has order 5. The multiplicative order of 2 modulo 5 is 4. So, x, x^2, x^{2^2}, and x^{2^3} are all different, while $x^{2^4} = x$. Indeed, $x^4 \equiv x^3 + x^2 + x + 1 \pmod{f(x)}$, $x^8 \equiv x^3 \pmod{f(x)}$, while $x^{16} \equiv x \pmod{f(x)}$, as can be checked with the Mathematica functions `Table` and `PolynomialMod`:

```
p = 2; m = 4; f = 1 + x + x^2 + x^3 + x^4;
Table[ PolynomialMod[x^2^i, f, Modulus -> p], {i, 0, m} ] //
  TableForm
```

```
x
x^2
1 + x + x^2 + x^3
x^3
x
```

B.4.4 Minimal Polynomials, Primitive Polynomials

Theorem B.32
Let $\mathbb{F}_q$ be a finite field of characteristic p. Take $n \mid (q - 1)$ and let ω be an element of order n in $\mathbb{F}_q$. Further, let m be the multiplicative order of p modulo n.
Then the polynomial

$$m(x) = \prod_{i=0}^{m-1} \left(x - \omega^{p^i} \right) \tag{B.7}$$

has its coefficients in $\mathbb{F}_p$ and it is irreducible over $\mathbb{F}_p$. It is called the minimal polynomial of ω over $\mathbb{F}_p$

Proof: Clearly, $m(x)$ is a polynomial over $\mathbb{F}_q$. Write $m(x) = \sum_{i=0}^{m} m_i x^i$. We have to show that the coefficients m_i are in the ground field $\mathbb{F}_p$. To this end we shall use the powerful criterion of Theorem B.29.

It follows from Theorem B.27 and Corollary B.22 (with $n = 1$) that

$$(m(x))^p = \prod_{i=0}^{m-1} \left(x - \omega^{p^i}\right)^p = \prod_{i=0}^{m-1} \left(x^p - \omega^{p^{i+1}}\right) =$$

$$= \prod_{i=1}^{m} \left(x^p - \omega^{p^i}\right) = \prod_{i=0}^{m-1} \left(x^p - \omega^{p^i}\right) = m(x^p).$$

Hence

$$\sum_{i=0}^{m} m_i x^{p\,i} = m(x^p) = (m(x))^p = \left(\sum_{i=0}^{m} m_i x^i\right)^p = \sum_{i=0}^{m} m_i^p x^{p\,i}.$$

Comparing the coefficients of $x^{p\,i}$ on both hands yields $m_i = m_i^p$. It follows from **Theorem B.29** that $m_i \in \mathbb{F}_p$, $0 \le i \le m$. So, $m(x)$ is a polynomial in $\mathbb{F}_p[x]$.

From **Theorem B.30** and **Theorem B.31** it follows that no polynomial in $\mathbb{F}_p[x]$ of degree less than m can have ω as a zero. So, $m(x)$ is irreducible over $\mathbb{F}_p$.

$\square$

> **Corollary B.33**
> Let ω be an element of order n in a finite field of characteristic p. Let $m(x)$ be defined as in **Theorem B.32** and let $f(x)$ be any p-ary polynomial that has ω as zero.
> Then $f(x)$ is divisible by $m(x)$.

Proof: Combine Theorem B.30, Theorem B.31, and Theorem B.32.

$\square$

So, $m(x)$, as defined in Theorem B.32, is the monic polynomial of lowest degree over $\mathbb{F}_p$, having ω as a zero. That is the reason why $m(x)$ is called the minimal polynomial of ω over p. It has ω and all the conjugates of ω as zeros. The degree of the minimal polynomial $m(x)$ of an element ω is often simply called the degree of ω over $\mathbb{F}_p$.

If $m(x)$ is the minimal polynomial of a primitive element, then $m(x)$ is called a primitive polynomial. *Mathematica* finds a primitive polynomial of degree m over $\mathbb{F}_p$ in the variable z by means of the `FieldIrreducible` function.

```
<< Algebra`FiniteFields`
```

```
m = 6; p = 2;
FieldIrreducible[GF[p, m], z]
```

```
1 + z^5 + z^6
```

Let $f(x)$ be a primitive polynomial over $\mathbb{F}_p$ of degree m. A table (like Table B.2) in which each non-zero element in the finite field $(\mathbb{F}_p[x]/(f(x), +, \cdot)$ is represented as a polynomial in x of degree $< m$ and as a power of x is called a log table of that field. These tables are very practical to have when extensive calculations need to be done in the field.

These logarithm tables can be made quite easily by *Mathematica*. Depending on whether one wants *Mathematica* to select a suitable primitive polynomial or enter one's own, one can type :

```
p = 2;
TableForm[PowerList[GF[p, 4]]]
```

```
1    0    0    0
0    1    0    0
0    0    1    0
0    0    0    1
1    0    0    1
1    1    0    1
1    1    1    1
1    1    1    0
0    1    1    1
1    0    1    0
0    1    0    1
1    0    1    1
1    1    0    0
0    1    1    0
0    0    1    1
```

or

```
p = 2;
TableForm[PowerList[GF[p, {1, 1, 0, 0, 1}]]]
```

```
1    0    0    0
0    1    0    0
0    0    1    0
0    0    0    1
1    1    0    0
0    1    1    0
0    0    1    1
1    1    0    1
1    0    1    0
0    1    0    1
1    1    1    0
0    1    1    1
1    1    1    1
1    0    1    1
1    0    0    1
```

To determine x^i in a field GF[p, m] or, conversely, to find i such that x^i is equal to a particular element in GF[p, m], one can use the *Mathematica* functions FieldExp[GF[p, m], i], resp. FieldInd[GF[p, m] [{list}]] (essential for this calculation is the assignment True to PowerListQ).

```
PowerListQ[GF[2, {1, 1, 0, 0, 1}]] = True;
f16 = GF[2, {1, 1, 0, 0, 1}];
FieldExp[f16, 5]
FieldInd[f16[{0, 1, 1, 0}]]
```

```
{0, 1, 1, 0}₂
```

```
5
```

There are several ways to find the minimal polynomial of a field element. We shall demonstrate two methods.

Method 1:

Let α be a zero of the binary primitive polynomial $x^5 + x^2 + 1$. So, α has order 31 and the conjugates of α^3 are α^6, α^{12}, α^{24}, and α^{17}. Then the minimal polynomial of α^3 can be found by:

```
f := 1 + a² + a⁵;
PolynomialMod[
  (x - a³) (x - a⁶) (x - a¹²) (x - a²⁴) (x - a¹⁷), f, Modulus -> 2]
```

```
1 + x² + x³ + x⁴ + x⁵
```

Method 2:

Let α be a zero of the binary primitive polynomial $x^5 + x^2 + 1$. To find the minimal polynomial of $\beta = \alpha^3$, we first compute $1, \beta, \beta^2, \beta^3, \beta^4$, and β^5, using $\alpha^5 + \alpha^2 + 1 = 0$.

```
f := 1 + a² + a⁵; b = a³;
u0 = PolynomialMod[1, f, Modulus -> 2]
u1 = PolynomialMod[b, f, Modulus -> 2]
u2 = PolynomialMod[b², f, Modulus -> 2]
u3 = PolynomialMod[b³, f, Modulus -> 2]
u4 = PolynomialMod[b⁴, f, Modulus -> 2]
u5 = PolynomialMod[b⁵, f, Modulus -> 2]
```

```
1
```

```
a³
```

$a + a^3$

$a + a^3 + a^4$

$a + a^2 + a^3$

$1 + a + a^2 + a^3 + a^4$

We use the *Mathematica* function <u>CoefficientList</u> to convert the coefficients into vectors. Note that we use the <u>Join</u> function to pad the output with zeros to make all vectors of length 5.

```
M = {Join[CoefficientList[u0, a], {0, 0, 0, 0}],
    Join[CoefficientList[u1, a], {0}],
    Join[CoefficientList[u2, a], {0}],
    CoefficientList[u3, a], Join[CoefficientList[u4, a], {0}],
    CoefficientList[u5, a]};
MatrixForm[
    M]
```

$$\begin{pmatrix} 1 & 0 & 0 & 0 & 0 \\ 0 & 0 & 0 & 1 & 0 \\ 0 & 1 & 0 & 1 & 0 \\ 0 & 1 & 0 & 1 & 1 \\ 0 & 1 & 1 & 1 & 0 \\ 1 & 1 & 1 & 1 & 1 \end{pmatrix}$$

We need to find a linear dependency between 1, β, β^2, β^3, β^4, and β^5, say $\sum_{i=0}^{5} g_i \beta^i = 0$ with $g_i \in GF(2)$. To this end we use the *Mathematica* functions <u>NullSpace</u> and <u>Transpose</u>. This leads to the minimal polynomial $g(x)$ of β.

```
NullSpace[Transpose[M], Modulus -> 2]
```

```
{{1, 0, 1, 1, 1, 1}}
```

We conclude that β has minimal polynomial $1 + x^2 + x^3 + x^4 + x^5$.

B.4.5 Further Properties

Let $m(x)$ be the minimal polynomial of an element ω of degree m. It follows from Corollary B.33 that the p^m expressions $\sum_{i=0}^{m-1} f_i \omega^i$, $f_i \in \mathbb{F}_p$, $0 \le i \le m$, take on p^m different values. For these expressions addition and multiplication can be performed just as in (B.3) and (B.4), where the relation $m(\omega) = 0$ has to be used to reduce the degree of the outcome to a value less than m. It is quite easy to check that one obtains a field, that is isomorphic to $(\mathbb{F}_q[x] \ / \ (m(x)), \ +, \ \cdot)$.

If $m(x)$ is primitive, one has that the elements $1, x, \ldots, x^{(p^m-2)}$ are all different modulo $m(x)$, just as the elements $1, \omega, \ldots, \omega^{(p^m-2)}$ are all different. See for instance, Example B.6, where the primitive element $\omega = 1 + x$ has minimal polynomial $m(y) = 1 + y^3 + y^4$. Table B.2 shows the field $(\mathbb{F}_q[x] \ / \ (m(x)), \ +, \ )$.

> **Lemma B.34**
> Let $m(x)$ be an irreducible polynomial of degree m over a field with p elements and let n be a multiple of m.
> Then $m(x)$ divides $x^{p^n} - x$.

Proof: Consider the residue class ring $(\mathbb{F}_p[x]/(m(x)), +, \cdot)$. This ring is a field with $q = p^m$ elements by Theorem B.16. The field element $<x>$ is a zero of $m(x)$, since $m(<x>) = <m(x)> = <0>$. It follows from Corollary B.22 ($n = 1$) that $<x>$ is a zero of $x^{p^n} - x$, $n \ge 1$. By Corollary B.33 we conclude that $m(x)$ divides $x^{p^n} - x$.

$\square$

Also the converse of Lemma B.34 is true.

> **Theorem B.35**
> The polynomial $x^{p^n} - x$ is the product of all irreducible, monic, p-ary polynomials of a degree dividing n.

Proof: Let $m \mid n$. There are $I_p(m)$ irreducible polynomials of degree m over $\mathbb{F}_p$, all of which divide $x^{p^n} - x$ by Lemma B.34. The sum of their degrees is $m I_p(m)$. Since $\sum_{m|n} m I_p(m) = p^n = \text{degree}(x^{p^n} - x)$ by (B.5), it follows that the irreducible, monic, p-ary polynomials of degree m, $m \mid n$, form the complete factorization of $x^{p^n} - x$.

$\square$

Example B.8

$p = 2, n = 4$,

$I_2(1) = 2$, $I_2(2) = 1$, $I_2(4) = 3$ (see Section B.3).

$x^{16} - x = x(x + 1)(x^2 + x + 1)(x^4 + x^3 + x^2 + x + 1)(x^4 + x^3 + 1)(x^4 + x + 1)$

```
p = 2; m = 4;
Factor[x^p^m - x, Modulus -> p]
```

$$x (1 + x) (1 + x + x^2) (1 + x + x^4) (1 + x^3 + x^4) (1 + x + x^2 + x^3 + x^4)$$

Corollary B.36
Let $f(x)$ be an irreducible polynomial in $\mathbb{F}_p[x]$ of degree m. Let $m \mid n$. Then, a finite field with p^n elements contains m roots of $f(x)$.

Proof: By Theorem B.35, $f(x)$ divides $x^q - x$, $q = p^n$. On the other hand, $x^q - x = \prod_{\omega \in \mathbb{F}_q} (x - \omega)$ by Corollary B.23.

□

Theorem B.37
Let p be a prime and $m \in \mathbb{N}$. Then, the finite field $\mathbb{F}_{p^m}$ is unique, up to isomorphism.

Proof: Write $q = p^m$ and let $\mathbb{F}_q$ be any finite field of order q. Let $f(x)$ be any irreducible, p-ary polynomial of degree m. We shall show that $\mathbb{F}_q$ is isomorphic to $\mathbb{F}_p[x]/(f(x))$. By Corollary B.36, $\mathbb{F}_q$ contains m zeros of $f(x)$. Let ω be one of these m zeros. Since $f(x)$ is irreducible in $\mathbb{F}_p[x]$, there is no lower degree polynomial over $\mathbb{F}_p$ with ω as zero. This implies that the m elements $1, \omega, \ldots, \omega^{m-1}$ are independent over $\mathbb{F}_p$, thus, any element in $\mathbb{F}_q$ can be written as $\sum_{i=0}^{m-1} f_i \omega^i$, $f_i \in \mathbb{F}_p, 0 \leq i \leq m-1$.

The isomorphism between $\mathbb{F}_q$ and $\mathbb{F}_p[x]/(f(x))$ is now obvious.

□

Corollary B.38
$\mathbb{F}_{p^m}$ is (isomorphic to) a subfield of $\mathbb{F}_{p^n}$ if and only if m divides n.

Proof: The following assertions are all equivalent;

i) $m \mid n$,
ii) $(p^m - 1)$ divides $(p^n - 1)$,
iii) $(x^{p^m} - x)$ divides $(x^{p^n} - x)$,
iv) $\prod_{\omega \in \mathbb{F}_{p^m}} (x - \omega)$ divides $\prod_{\omega \in \mathbb{F}_{p^n}} (x - \omega)$,
v) $\mathbb{F}_{p^m}$ is a subfield of $\mathbb{F}_{p^n}$.

□

Example B.9

It follows from Corollary B.38 that F_{2^4} contains F_{2^2} as a subfield, while it does not contain F_{2^3} as a subfield. From Table B.2 one can easily verify that the elements 0, 1, x^5 and x^{10} form a subfield of cardinality 2^2 in $(F_2[x]/(x^4 + x^3 + 1), +, \cdot)$.

B.4.6 Cyclotomic Polynomials

Consider a finite field F_q of characteristic p. So, $q = p^m$ for some $m > 0$. By Theorem B.5, every element in F_q has an order dividing $q - 1$. Let $n \mid (q - 1)$ and let ω be a primitive n-th root of unity in F_q. For instance, $\omega = \alpha^{(q-1)/n}$, where α is a primitive element in F_q. Let $d \mid n$ and put $\eta = \omega^{n/d}$. Then η is a primitive d-root of unity. Clearly, the d elements 1, η, ..., η^{d-1} are a zero of $x^d - 1$. By Theorem B.15, no other element in F_q is a zero of $x^d - 1$.

> **Definition B.19**
> Let $q = p^m$, p prime. For any $d \mid (q - 1)$ the p-ary cyclotomic polynomial $Q^{(d)}(x)$ is defined by
> $$Q^{(d)}(x) = \prod_{\xi \in GF(q) \text{ of order } q} (x - \xi).$$

If ξ had order d, $d \mid (q - 1)$, then by Lemma B.4 also ξ^p has order d. So, with ξ a zero of $Q^{(d)}(x)$ also its conjugates are zeros of $Q^{(d)}(x)$. It follows from Theorem B.32 that $Q^{(d)}(x)$ is the product of some minimal polynomials over F_p and thus that $Q^{(d)}(x)$ is a polynomial over F_p.

By Theorem B.21, $Q^{(d)}(x)$ has degree $\phi(d)$. Since ω is a primitive n-th root of unity, it follows that

$$x^n - 1 = \prod_{i=1}^{n-1}(x - \omega^i) = \prod_{\xi \in F_q, \xi \text{ has order } n}(x - \xi) =$$
$$= \prod_{d \mid n} \prod_{\xi \in F_q, \xi \text{ has order } d}(x - \xi) = \prod_{d \mid n} Q^{(d)}(x). \tag{B.8}$$

> **Theorem B.39**
> $$Q^{(n)}(x) = \prod_{d \mid n}(x^d - 1)^{\mu(n/d)}.$$

Proof: Apply the Multiplicative Möbius Inversion Formula (Corollary A.39) to (B.8).

□

Example B.10

$$Q^{(36)}(x) = \prod_{d \mid 36}(x^d - 1)^{\mu(36/d)} = \frac{(x^{36}-1)(x^6-1)}{(x^{18}-1)(x^{12}-1)} = \frac{x^{18}+1}{x^6+1} = x^{12} - x^6 + 1.$$

This can also be evaluated with Mathematica:

```
DivisorProduct[f_, n_] := Times @@ (f /@ Divisors[n])
```

```
n = 36; Clear[f, x];
f[d_] := (x^d - 1)^MoebiusMu[n/d]
DivisorProduct[f, n] // Simplify
```

$$1 - x^6 + x^{12}$$

or directly with the Mathematica function `Cyclotomic`:

```
Cyclotomic[36, x]
```

$$1 - x^6 + x^{12}$$

If $p = 2$, one can write $Q^{(36)}(x) = x^{12} + x^6 + 1$.

The expression for $Q^{(n)}(x)$ in Theorem B.39 seems to be independent of the finite field. This is not really true, because in the evaluation of that expression the characteristic does play a role.

All the irreducible factors of $Q^{(d)}(x)$ have the same degree, because all the zeros of $Q^{(d)}(x)$ have the same order d. Indeed, by Theorem B.32, each irreducible factor of $Q^{(d)}(x)$ has as degree the multiplicative order of p modulo d.

In particular we have the following theorem.

Theorem B.40
The number of primitive, p-ary, monic polynomials of degree m is
$$\frac{\phi(p^m - 1)}{m}.$$

Proof: A primitive, p-ary polynomial of degree m divides $Q^{(p^m-1)}(x)$ and this cyclotomic polynomial has only factors of this type. The degree of $Q^{(p^m-1)}(x)$ is $\phi(p^m - 1)$.

$\square$

Example B.11: $p = 2$

$$x^{16} - x = x(x^{15} - 1) = x\, Q^{(1)}(x) Q^{(3)}(x) Q^{(5)}(x) Q^{(15)}(x)$$

where

$$Q^{(1)}(x) = x + 1,$$
$$Q^{(3)}(x) = x^2 + x + 1,$$
$$Q^{(5)}(x) = x^4 + x^3 + x^2 + x + 1,$$
$$Q^{(15)}(x) = (x^4 + x + 1)(x^4 + x^3 + 1).$$

Indeed, there are $\phi(15)/4 = 2$ primitive polynomials of degree 4. See also Example B.6.

A way to find all primitive polynomials of degree m over $\mathbb{F}_p$ is to factor $Q^{(p^m-1)}(x)$.

Example B.12

```
p = 2; m = 6; n = p^m - 1;
Factor[Cyclotomic[n, x], Modulus → p]
```

$(1 + x + x^6) \ (1 + x + x^3 + x^4 + x^6) \ (1 + x^5 + x^6)$
$(1 + x + x^2 + x^5 + x^6) \ (1 + x^2 + x^3 + x^5 + x^6) \ (1 + x + x^4 + x^5 + x^6)$

Remark:

In this chapter we have viewed $\mathbb{F}_q$, $q = p^m$ and p prime, as an extension field of $\mathbb{F}_p$, however all the concepts defined in this chapter can also be generalized to $\mathbb{F}_q[x]$. So, one may want to count the number of irreducible polynomials of degree n in $\mathbb{F}_q[x]$ or discuss primitive polynomials over $\mathbb{F}_q$, etc. We leave it to the reader to verify that all the theorems in this appendix can indeed be generalized from $\mathbb{F}_p$ and $\mathbb{F}_{p^m}$ to $\mathbb{F}_q$ resp. $\mathbb{F}_{q^m}$ simply by replacing p by q and q by q^m.

Example B.13

The field $\mathbb{F}_{16}$ can be viewed as the residue class ring $\mathbb{F}_4[x]/(x^2 + x + \alpha)$, where α is an element in $\mathbb{F}_4$ satisfying $\alpha^2 + \alpha + 1$.

B.5 Problems

Problem B.1

Prove that $(\{x \in \mathbb{R} \mid x^2 \in \mathbb{Q},\ x \neq 0\},\ \cdot)$ is a group.

Problem B.2

Prove that the elements of a reduced residue class system modulo m form a multiplicative group.

Problem B.3

Let $(G, *)$ be a group and H a non-empty subset of G. Then $(H, *)$ is a subgroup of $(G, *)$ if and only if $h_1 * h_2^{-1} \in H$ for every $h_1, h_2 \in H$.

Problem B.4

Prove that there are essentially two different groups of order 4 (hint: each element has an order dividing 4).

Problem B.5

Find an element of order 12 in the group $(\mathbb{Z}_{13}^*, \times)$. Which powers of this element have order 12. Answer the same question for elements of order 6, 4, 3, 2 and 1.

Problem B.6

Let $(G, \cdot)$ denote a commutative group. Let a and b be two elements in G of order m resp. n.
a) Assume that $\gcd(m, n) = 1$. Show that $a \cdot b$ has order $m \times n$.
b) Assume no longer that $\gcd(m, n) = 1$. Determine integers s and t such that $s \mid m$, $t \mid n$, $\gcd(s, t) = 1$, and $\mathrm{lcm}[s, t] = \mathrm{lcm}[m, n]$.
c) Construct an element in G of order $\mathrm{lcm}[m, n]$.

Problem B.7^M

Find the multiplicative inverse of $1 + x^2 + x^3 \pmod{1 + x^2 + x^5}$ over GF(2) (hint1: Thm. B.13; hint2).

Problem B.8^M

How many binary, irreducible polynomials (hint1: Def.B.15; hint2: Thm. B.17) are there of degree 7 and 8?

Problem B.9

Make a log table of $GF(2)[x] /(1 + x^2 + x^5)$ (hint: x is a primitive element). Use this table to express $x^{10} + x^{20}$ as power of x.

Problem B.10

Let $\alpha \in GF(q)$ have order m, $m < q - 1$. What is the probability that a random non-zero element $\beta \in GF(q)$ has an order n dividing m? Give an upperbound on this probability.

Construct an element of order $\mathrm{lcm}[m, n]$ (hint: see Problem B.6).

(In fact, this method leads to an efficient to find a primitive element in a finite field. It is due to Gauss.)

Problem B.11

Which subfields are contained in GF(625)? Let α be a primitive element in GF(625). Which powers of α constitute the various subfields of GF(625)? (Hint: Cor. B.38.)

Problem B.12

Prove that over GF(2): $(x + y)^{2^k+1} = x^{2^k+1} + x^{2^k} \cdot y + x \cdot y^{2^k} + y^{2^k+1}$.
(Hint: use Cor. B.28.)

Problem B.13

How many binary, primitive polynomials are there of degree 10? (Hint: Thm. B.40.)

Problem B.14

Determine the binary, cyclotomic polynomial $Q^{(21)}(x)$ (hint: Thm. B.39). What is the degree of the binary factors of $Q^{(21)}(x)$.

Problem B.15

What is the degree of a binary, minimal polynomial of a primitive 17-th root of unity (hint: Thm. B.32)? How many such polynomials do exist? Prove that each is its own reciprocal. Determine these polynomials explicitly.

Problem B.16

The trace mapping Tr is defined on GF(p), p prime, by

$$\text{Tr}(x) = x + x^p + x^{p^2} + \ldots + x^{p^{m-1}}.$$

a) Prove that $\text{Tr}(x) \in \text{GF}(p)$, for every $x \in \text{GF}(p^m)$ (hint: Thm. B.29). So, Tr is a mapping from GF(p^m) to GF(p).

b) Prove that Tr is a linear mapping (hint: Cor. B.28).

c) Prove that Tr takes on every value in GF(p) equally often (hint: use Theorem B.15).

d) Replace p by q in this problem, where q is a prime power, and verify the same statements.

Appendix C   Relevant Famous Mathematicians

Euclid of Alexandria

Born: about 365 BC in Alexandria, Egypt

Died: about 300 BC

Euclid is the most prominent mathematician of antiquity best known for his treatise on geometry *The Elements*. The long lasting nature of *The Elements* must make Euclid the leading mathematics teacher of all time.

Little is known of Euclid's life except that he taught at Alexandria in Egypt. The picture of Euclid above is from the 18th Century and must be regarded as entirely fanciful.

Euclid's most famous work is his treatise on geometry *The Elements*. The book was a compilation of geometrical knowledge that became the centre of mathematical teaching for 2000 years. Probably no results in *The Elements* were first proved by Euclid but the organization of the material and its exposition are certainly due to him.

The Elements begins with definitions and axioms, including the famous fifth, or parallel, postulate that one and only one line can be drawn through a point parallel to a given line. Euclid's decision to make this an axiom led to Euclidean geometry. It was not until the 19th century that this axiom was dropped and non-euclidean geometries were studied.

Zeno of Sidon, about 250 years after Euclid wrote: „*The Elements*, seems to have been the first to show that Euclid's propositions were not deduced from the axioms alone, and Euclid does make other subtle assumptions."

The Elements is divided into 13 books. Books 1-6, plane geometry: books 7-9, number theory: book 10, 's theory of irrational numbers: books 11-13, solid geometry. The book ends with a discussion of the properties of the five regular polyhedra and a proof that there are precisely five. Euclid's *Elements* is remarkable for the clarity with which the theorems are stated and proved. The standard of rigour was to become a goal for the inventors of the calculus centuries later.

More than one thousand editions of *The Elements* have been published since it was first printed in 1482.

Euclid also wrote *Data* (with 94 propositions), *On Divisions*, Optics and *Phaenomena* which have survived. His other books *Surface Loci*, *Porisms*, *Conics*, *Book of Fallacies* and *Elements of Music* have all been lost.

Euclid may not have been a first class mathematician but the long lasting nature ofThe *Elements* must make him the leading mathematics teacher of antiquity.

The source of this information is the following webpage:

http://www-history.mcs.st-and.ac.uk/history/Mathematicians/Euclid.html

Leonhard Euler

Born: 15 April 1707 in Basel, Switzerland

Died: 18 Sept 1783 in St Petersburg, Russia

Euler made large bounds in modern analytic geometry and trigonometry. He made decisive and formative contributions to geometry, calculus and number theory.

Euler's father wanted his son to follow him into the church and sent him to the University of Basel to prepare for the ministry. However geometry soon became his favorite subject. Euler obtained his father's consent to change to mathematics after Johann Bernoulli had used his persuasion. Johann Bernoulli became his teacher.

He joined the St. Petersburg Academy of Science in 1727, two years after it was founded by Catherine I the wife of Peter the Great. Euler served as a medical lieutenant in the Russian navy from 1727 to 1730. In St Petersburg he lived with Daniel Bernoulli. He became professor of physics at the academy in 1730 and professor of mathematics in 1733. He married and left Johann Bernoulli's house in 1733. He had 13 children altogether of which 5 survived their infancy. He claimed that he made some of his greatest discoveries while holding a baby on his arm with other children playing round his feet.

The publication of many articles and his book *Mechanica* (1736-37), which extensively presented Newtonian dynamics in the form of mathematical analysis for the first time, started Euler on the way to major mathematical work.

In 1741, at the invitation of Frederick the Great, Euler joined the Berlin Academy of Science, where he remained for 25 years. Even while in Berlin he received part of his salary from Russia and never got on well with Frederick. During his time in Berlin, he wrote over 200 articles, three books on mathematical analysis, and a popular scientific publication *Letters to a Princess of Germany* (3 vols., 1768-72).

In 1766 Euler returned to Russia. He had been arguing with Frederick the Great over academic freedom and Frederick was greatly angered at his departure. Euler lost the sight of his right eye at the age of 31 and soon after his return to St Petersburg he became almost entirely blind after a cataract operation. Because of his remarkable memory was able to continue with his work on optics, algebra, and lunar motion. Amazingly after 1765 (when Euler was 58) he produced almost half his works despite being totally blind.

After his death in 1783 the St. Petersburg Academy continued to publish Euler's unpublished work for nearly 50 more years.

Euler made large bounds in modern analytic geometry and trigonometry. He made decisive and formative contributions to geometry, calculus and number theory. In number theory he did much work in correspondence with Goldbach. He integrated Leibniz's differential calculus and Newton's method of fluxions into mathematical analysis. In number theory he stated the prime number theorem and the law of biquadratic reciprocity.

He was the most prolific writer of mathematics of all time. His complete works contains 886 books and papers.

We owe to him the notations $f(x)$ (1734), e for the base of natural logs (1727), i for the square root of -1 (1777), π for pi, $\sum$ for summation (1755) etc. He also introduced beta and gamma functions, integrating factors for differential equations etc.

He studied continuum mechanics, lunar theory with Clairaut, the three body problem, elasticity, acoustics, the wave theory of light, hydraulics, music etc. He laid the foundation of analytical mechanics, especially in his *Theory of the Motions of Rigid Bodies* (1765).

The source of this information is the following webpage:

http://www-history.mcs.st-and.ac.uk/history/Mathematicians/Euler.html

Pierre de Fermat

Born: 17 Aug 1601 in Beaumont-de-Lomagne, France

Died: 12 Jan 1665 in Castres, France

Pierre Fermat's father was a wealthy leather merchant and second consul of Beaumont-de-Lomagne. Pierre had a brother and two sisters and was almost certainly brought up in the town of his birth. Although there is little evidence concerning his school education it must have been at the local Franciscan monastery.

He attended the University of Toulouse before moving to Bordeau in the second half of the 1620s. In Bordeau he began his first serious mathematical researches and in 1629 he gave a copy of his restoration of Apollonius's *Plane loci* to one of the mathematicians there. Certainly in Bordeau he was in contact with Beaugrand and during this time he produced important work on maxima and minima which he gave to Etienne d'Espagnet who clearly shared mathematical interests with Fermat.

From Bordeau Fermat went to Orléans where he studied law at the University. He received a degree in civil law and he purchased the offices of councillor at the parliament in Toulouse. So by 1631 Fermat was a lawyer and government official in Toulouse and because of the office he now held he became entitled to change his name from Pierre Fermat to **Pierre de Fermat**.

For the remainder of his life he lived in Toulouse but as well as working there he also worked in his home town of Beaumont-de-Lomagne and a nearby town of Castres. From his appointment on 14 May 1631 Fermat worked in the lower chamber of the parliament but on 16 January 1638 he was appointed to a higher chamber, then in 1652 he was promoted to the highest level at the criminal court. Still further promotions seem to indicate a fairly meteoric rise through the profession but promotion was done mostly on seniority and the plague struck the region in the early 1650s meaning that many of the older men died. Fermat himself was struck down by the plague and in 1653 his death was wrongly reported, then corrected:

I informed you earlier of the death of Fermat. He is alive, and we no longer fear for his health, even though we had counted him among the dead a short time ago.

The following report, made to Colbert the leading figure in France at the time, has a ring of truth:

Fermat, a man of great erudition, has contact with men of learning everywhere. But he is rather preoccupied, he does not report cases well and is confused.

Of course Fermat was preoccupied with mathematics. He kept his mathematical friendship with Beaugrand after he moved to Toulouse but there he gained a new mathematical friend in Carcavi. Fermat met Carcavi in a professional capacity since both were councillors in Toulouse but they both shared a love of mathematics and Fermat told Carcavi about his mathematical discoveries.

In 1636 Carcavi went to Paris as royal librarian and made contact with Mersenne and his group. Mersenne's interest was aroused by Carcavi's descriptions of Fermat's discoveries on falling bodies, and he wrote to Fermat. Fermat replied on 26 April 1636 and, in addition to telling Mersenne about errors which he believed that Galileo had made in his description of free fall, he also told Mersenne about his work on spirals and his restoration of Apollonius's *Plane loci*. His work on spirals had been motivated by considering the path of free falling bodies and he had used methods generalised from Archimedes' work On spirals to compute areas under the spirals. In addition Fermat wrote:

I have also found many sorts of analyses for diverse problems, numerical as well as geometrical, for the solution of which Viète's analysis could not have sufficed. I will share all of this with you whenever you wish and do so without any ambition, from which I am more exempt and more distant than any man in the world.

It is somewhat ironical that this initial contact with Fermat and the scientific community came through his study of free fall since Fermat had little interest in physical applications of mathematics. Even with his results on free fall he was much more interested in proving geometrical theorems than in their relation to the real world. This first letter did however contain two problems on maxima which Fermat asked Mersenne to pass on to the Paris mathematicians and this was to be the typical style of Fermat's letters, he would challenge others to find results which he had already obtained.

Roberval and Mersenne found that Fermat's problems in this first, and subsequent, letters were extremely difficult and usually not soluble using current techniques. They asked him to divulge his methods and Fermat sent *Method for determining Maxima and Minima and Tangents to Curved Lines*, his restored text of Apollonius's *Plane loci* and his algebraic approach to geometry *Introduction to Plane* and *Solid Loci* to the Paris mathematicians.

His reputation as one of the leading mathematicians in the world came quickly but attempts to get his work published failed mainly because Fermat never really wanted to put his work into a polished form. However some of his methods were published, for example Hérigone added a supplement containing Fermat's methods of maxima and minima to his major work *Cursus mathematicus*. The widening correspondence between Fermat and other mathematicians did not find universal praise. Frenicle de Bessy became annoyed at Fermat's problems which to him were impossible. He wrote angrily to Fermat but although Fermat gave more details in his reply, Frenicle de Bessy felt that Fermat was almost teasing him.

However Fermat soon became engaged in a controversy with a more major mathematician than Frenicle de Bessy. Having been sent a copy of Descartes' *La Dioptrique* by Beaugrand, Fermat paid it little attention since he was in the middle of a correspondence with Roberval and Etienne Pascal over methods of integration and using them to find centres of gravity. Mersenne asked him to give an opinion on La Dioptrique which Fermat did describing it as

groping about in the shadows.

He claimed that Descartes had not correctly deduced his law of refraction since it was inherent in his assumptions. To say that Descartes was not pleased is an understatement. Descartes soon found reason to feel even more angry since he viewed Fermat's work on maxima, minima and tangents as reducing the importance of his own work *La Géométrie* which Descartes was most proud of and which he sought to show that his *Discours de la méthod* alone could give.

Descartes attacked Fermat's method of maxima, minima and tangents. Roberval and Etienne Pascal became involved in the argument and eventually so did Desargues who Descartes asked to act as a referee. Fermat proved correct and eventually Descartes admitted this writing:-

… seeing the last method that you use for finding tangents to curved lines, I can reply to it in no other way than to say that it is very good and that, if you had explained it in this manner at the outset, I would have not contradicted it at all.

Did this end the matter and increase Fermat's standing? Not at all since Descartes tried to damage Fermat's reputation. For example, although he wrote to Fermat praising his work on determining the tangent to a cycloid (which is indeed correct), Descartes wrote to Mersenne claiming that it was incorrect and saying that Fermat was inadequate as a mathematician and a thinker. Descartes was important and respected and thus was able to severely damage Fermat's reputation.

The period from 1643 to 1654 was one when Fermat was out of touch with his scientific colleagues in Paris. There are a number of reasons for this. Firstly pressure of work kept him from devoting so much time to mathematics. Secondly the Fronde, a civil war in France, took place and from 1648 Toulouse was greatly affected. Finally there was the plague of 1651 which must have had great consequences both on life in Toulouse and of course its near fatal consequences on Fermat himself. However it was during this time that Fermat worked on number theory.

Fermat is best remembered for this work in number theory, in particular for Fermat's Last Theorem. This theorem states that $x^n + y^n = z^n$ has no non-zero integer solutions for x, y and z when $n > 2$. Fermat wrote, in the margin of Bachet's translation of Diophantus's *Arithmetica*

I have discovered a truly remarkable proof which this margin is too small to contain.

These marginal notes only became known after Fermat's son Samuel published an edition of Bachet's translation of Diophantus's *Arithmetica* with his father's notes in 1670.

It is now believed that Fermat's 'proof' was wrong although it is impossible to be completely certain. The truth of Fermat's assertion was proved in June 1993 by the British mathematician Andrew Wiles, but Wiles withdrew the claim to have a proof when problems emerged later in 1993. In November 1994 Wiles again claimed to have a correct proof which has now been accepted.

Unsuccessful attempts to prove the theorem over a 300 year period led to the discovery of commutative ring theory and a wealth of other mathematical discoveries.

Fermat's correspondence with the Paris mathematicians restarted in 1654 when Blaise Pascal, Etienne Pascal's son, wrote to him to ask for confirmation about his ideas on probability. Blaise Pascal knew of Fermat through his father, who had died three years before, and was well aware of Fermat's outstanding mathematical abilities. Their short correspondence set up the theory of probability and from this they are now regarded as joint founders of the subject. Fermat however, feeling his isolation and still wanting to adopt his old style of challenging mathematicians, tried to change the topic from probability to number theory. Pascal was not interested but Fermat, not realising this, wrote to Carcavi saying:

I am delighted to have had opinions conforming to those of M Pascal, for I have infinite esteem for his genius... the two of you may undertake that publication, of which I consent to your being the masters, you may clarify or supplement whatever seems too concise and relieve me of a burden that my duties prevent me from taking on.

However Pascal was certainly not going to edit Fermat's work and after this flash of desire to have his work published Fermat again gave up the idea. He went further than ever with his challenge problems however:

Two mathematical problems posed as insoluble to French, English, Dutch and all mathematicians of Europe by Monsieur de Fermat, Councillor of the King in the Parliament of Toulouse.

His problems did not prompt too much interest as most mathematicians seemed to think that number theory was not an important topic. The second of the two problems, namely to find all solutions of $N x^2 + 1 = y^2$ for N not a square, was however solved by Wallis and Brouncker and they developed continued fractions in their solution. Brouncker produced rational solutions which led to arguments. Frenicle de Bessy was perhaps the only mathematician at that time who was really interested in number theory but he did not have sufficient mathematical talents to allow him to make a significant contribution.

Fermat posed further problems, namely that the sum of two cubes cannot be a cube (a special case of Fermat's Last Theorem which may indicate that by this time Fermat realised that his proof of the general result was incorrect), that there are exactly two integer solutions of $x^2 + 4 = y$ and that the equation $x^2 + 2 = y^3$ has only one integer solution. He posed problems directly to the English. Everyone failed to see that Fermat had been hoping his specific problems would lead them to discover, as he had done, deeper theoretical results.

Around this time one of Descartes' students was collecting his correspondence for publication and he turned to Fermat for help with the Fermat - Descartes correspondence. This led Fermat to look again at the arguments he had used 20 years before and he looked again at his objections to Descartes' optics. In particular he had been unhappy with Descartes' description of refraction of light and he now settled on a principle which did in fact yield the sine law of refraction that Snell and Descartes had proposed. However Fermat had now deduced it from a fundamental property that he proposed, namely that light always follows the shortest possible path. Fermat's principle, now one of the most basic properties of optics, did not find favour with mathematicians at the time.

In 1656 Fermat had started a correspondence with Huygens. This grew out of Huygens interest in probability and the correspondence was soon manipulated by Fermat onto topics of number theory. This topic did not interest Huygens but Fermat tried hard and in *New Account of Discoveries in the Science of Numbers* sent to Huygens via Carcavi in 1659, he revealed more of his methods than he had done to others.

Fermat described his method of infinite descent and gave an example on how it could be used to prove that every number of the form $4k+1$ could be written as the sum of two squares. For suppose some number of the form $4k+1$ could not be written as the sum of two squares. Then there is a smaller number of the form $4k+1$ which cannot be written as the sum of two squares. Continuing the argument will lead to a contradiction. What Fermat failed to explain in this letter is how the smaller number is constructed from the larger. One assumes that Fermat did know how to make this step but again his failure to disclose the method made mathematicians lose interest. It was not until Euler took up these problems that the missing steps were filled in.

Fermat is described as

Secretive and taciturn, he did not like to talk about himself and was loath to reveal too much about his thinking. ... His thought, however original or novel, operated within a range of possibilities limited by that [1600-1650] time and that [France] place.

Carl B Boyer says:

Recognition of the significance of Fermat's work in analysis was tardy, in part because he adhered to the system of mathematical symbols devised by Francois Viète, notations that Descartes's Géométrie had rendered largely obsolete. The handicap imposed by the awkward notations operated less severely in Fermat's favourite field of study, the theory of numbers, but here, unfortunately, he found no correspondent to share his enthusiasm.

The source of this information is the following webpage:

http://www-history.mcs.st-and.ac.uk/history/Mathematicians/Fermat.html

Evariste Galois

Born: 25 Oct 1811 in Bourg La Reine (near Paris), France

Died: 31 May 1832 in Paris, France

Famous for his contributions to group theory, **Evariste Galois** produced a method of determining when a general equation could be solved by radicals.

Galois' father Nicholas Gabriel Galois and his mother Adelaide Marie Demante were both intelligent and well educated in philosophy, classical literature and religion. However there is no sign of any mathematical ability in any of Galois' family. His mother served as Galois' sole teacher until he was 12 years old. She taught him Greek, Latin and religion where she imparted her own scepticism to her son. Galois' father was an important man in the community and in 1815 he was elected mayor of Bourg-la-Reine.

The starting point of the historical events which were to play a major role in Galois' life is surely the storming of the Bastille on 14 July 1789. From this point the monarchy of Louis 16th was in major difficulties as the majority of Frenchmen composed their differences and united behind an attempt to destroy the privileged establishment of the church and the state.

Despite attempts at compromise Louis 16th was tried after attempting to flee the country. Following the execution of the King on 21 January 1793 there followed a reign of terror with many political trials. By the end of 1793 there were 4595 political prisoners held in Paris. However France began to have better times as their armies, under the command of Napoleon Bonaparte, won victory after victory.

Napoleon became 1st Consul in 1800 and then Emperor in 1804. The French armies continued a conquest of Europe while Napoleon's power became more and more secure. In 1811 Napoleon was at the height of his power. By 1815 Napoleon's rule was over. The failed Russian campaign of 1812 was followed by defeats, the Allies entering Paris on 31 March 1814. Napoleon abdicated on 6 April and Louis XVIII was installed as King by the Allies. The year 1815 saw the famous one hundred days. Napoleon entered Paris on March 20, was defeated at Waterloo on 18 June and abdicated for the second time on 22 June. Louis XVIII was reinstated as King but died in

September 1824, Charles X becoming the new King.

Galois was by this time at school. He had enrolled at the Lycée of Louis-le-Grand as a boarder in the 4 th class on 6 October 1823. Even during his first term there was a minor rebellion and 40 pupils were expelled from the school. Galois was not involved and during 1824-25 his school record is good and he received several prizes. However in 1826 Galois was asked to repeat the year because his work in rhetoric was not up to the required standard.

February 1827 was a turning point in Galois' life. He enrolled in his first mathematics class, the class of M. Vernier. He quickly became absorbed in mathematics and his director of studies wrote:

It is the passion for mathematics which dominates him, I think it would be best for him if his parents would allow him to study nothing but this, he is wasting his time here and does nothing but torment his teachers and overwhelm himself with punishments.

Galois' school reports began to describe him as singular, bizarre, original and closed . It is interesting that perhaps the most original mathematician who ever lived should be criticised for being original. M. Vernier reported however

Intelligence, marked progress but not enough method.

In 1828 Galois took the examination of the Ecole Polytechnique but failed. It was the leading University of Paris and Galois must have wished to enter it for academic reasons. However, he also wished to enter the this school because of the strong political movements that existed among its students, since Galois followed his parents example in being an ardent republican.

Back at Louis-le-Grand, Galois enrolled in the mathematics class of Louis Richard. However he worked more and more on his own researches and less and less on his schoolwork. He studied Legendre's *Géométrie* and the treatises of Lagrange. As Richard was to report

This student works only in the highest realms of mathematics.

In April 1829 Galois had his first mathematics paper published on continued fractions in the Annales de mathématiques . On 25 May and 1 June he submitted articles on the algebraic solution of equations to the Académie des Sciences. Cauchy was appointed as referee of Galois' paper.

Tragedy was to strike Galois for on 2 July 1829 his father committed suicide. The priest of Bourg-la-Reine forged Mayor Galois' name on malicious forged epigrams directed at Galois' own relatives. Galois' father was a good natured man and the scandal that ensued was more than he could stand. He hanged himself in his Paris apartment only a few steps from Louis-le-Grand where his son was studying. Galois was deeply affected by his father's death and it greatly influenced the direction his life was to take.

A few weeks after his father's death, Galois presented himself for examination for entry to the Ecole Polytechnique for the second time. For the second time he failed, perhaps partly because he took it under the worst possible circumstances so soon after his father's death, partly because he was never good at communicating his deep mathematical ideas. Galois therefore resigned himself to enter the Ecole Normale, which was an annex to Louis-le-Grand, and to do so he had to take his Baccalaureate examinations, something he could have avoided by entering the Ecole Polytechnique.

He passed, receiving his degree on 29 December 1829. His examiner in mathematics reported:

This pupil is sometimes obscure in expressing his ideas, but he is intelligent and shows a remarkable spirit of research.

His literature examiner reported:

This is the only student who has answered me poorly, he knows absolutely nothing. I was told that this student has an extraordinary capacity for mathematics. This astonishes me greatly, for, after his examination, I believed him to have but little intelligence.

Galois sent Cauchy further work on the theory of equations, but then learned from *Bulletin de Férussac* of a posthumous article by Abel which overlapped with a part of his work. Galois then took Cauchy's advice and submitted a new article *On the condition that an equation be soluble by radicals* in February 1830. The paper was sent to Fourier, the secretary of the Academy, to be considered for the Grand Prize in mathematics. Fourier died in April 1830 and Galois' paper was never subsequently found and so never considered for the prize.

Galois, after reading Abel and Jacobi's work, worked on the theory of elliptic functions and abelian integrals. With support from Jacques Sturm, he published three papers in *Bulletin de*

Férussac in April 1830. However, he learnt in June that the prize of the Academy would be awarded the Prize jointly to Abel (posthumously) and to Jacobi, his own work never having been considered.

July 1830 saw a revolution. Charles 10th fled France. There was rioting in the streets of Paris and the director of École Normale, M. Guigniault, locked the students in to avoid them taking part. Galois tried to scale the wall to join the rioting but failed. In December 1830 M. Guigniault wrote newspaper articles attacking the students and Galois wrote a reply in the *Gazette des Écoles* , attacking M. Guigniault for his actions in locking the students into the school. For this letter Galois was expelled and he joined the Artillery of the National Guard, a Republican branch of the militia. On 31 December 1830 the Artillery of the National Guard was abolished by Royal Decree since the new King Louis-Phillipe felt it was a threat to the throne.

Two minor publications, an abstract in *Annales de Gergonne* (December 1830) and a letter on the teaching of science in the *Gazette des Écoles* (2 January 1831) were the last publications during his life. In January 1831 Galois attempted to return to mathematics. He organised some mathematics classes in higher algebra which attracted 40 students to the first meeting but after that the numbers quickly fell off. Galois was invited by Poisson to submit a third version of his memoir on equation to the Academy and he did so on 17 January.

On 18 April Sophie Germain wrote a letter to her friend the mathematician Libri which describes Galois' situation.

... the death of M. Fourier, have been too much for this student Galois who, in spite of his impertinence, showed signs of a clever disposition. All this has done so much that he has been expelled form École Normale. He is without money... They say he will go completely mad. I fear this is true.

Late in 1830 19 officers from the Artillery of the National Guard were arrested and charged with conspiracy to overthrow the government. They were acquitted and on 9 May 1831 200 republicans gathered for a dinner to celebrate the acquittal. During the dinner Galois raised his glass and with an open dagger in his hand appeared to make threats against the King, Louis-Phillipe. After the dinner Galois was arrested and held in Sainte-Pélagie prison. At his trial on 15 June his defence lawyer claimed that Galois had said

To Louis-Phillipe, if he betrays

but the last words had been drowned by the noise. Galois, rather surprisingly since he essentially repeated the threat from the dock, was acquitted.

The 14th July was Bastille Day and Galois was arrested again. He was wearing the uniform of the Artillery of the National Guard, which was illegal. He was also carrying a loaded rifle, several pistols and a dagger. Galois was sent back to Sainte-Pélagie prison. While in prison he received a rejection of his memoir. Poisson had reported that:-

His argument is neither sufficiently clear nor sufficiently developed to allow us to judge its rigour.

He did, however, encourage Galois to publish a more complete account of his work. While in Sainte-Pélagie prison Galois attempted to commit suicide by stabbing himself with a dagger but the other prisoners prevented him. While drunk in prison he poured out his soul

Do you know what I lack my friend? I confide it only to you: it is someone whom I can love and love only in spirit. I have lost my father and no one has ever replaced him, do you hear me...?

In March 1832 a cholera epidemic swept Paris and prisoners, including Galois, were transferred to the pension Sieur Faultrier. There he apparently fell in love with Stephanie-Felice du Motel, the daughter of the resident physician. After he was released on 29 April Galois exchanged letters with Stephanie, and it is clear that she tried to distance herself from the affair.

The name Stephanie appears several times as a marginal note in one of Galois' manuscripts.

Galois fought a duel with Perscheux d'Herbinville on 30 May, the reason for the duel not being clear but certainly linked with Stephanie.

You can see a note in the margin of the manuscript that Galois wrote the night before the duel. It reads

There is something to complete in this demonstration. I do not have the time. (Author's note).

It is this which has led to the legend that he spent his last night writing out all he knew about group theory. This story appears to have been exaggerated.

Galois was wounded in the duel and was abandoned by d'Herbinville and his own seconds and found by a peasant. He died in Cochin hospital on 31 May and his funeral was held on 2 June. It was the focus for a Republican rally and riots followed which lasted for several days.

Galois' brother and his friend Chevalier copied his mathematical papers and sent them to Gauss, Jacobi and others. It had been Galois' wish that Jacobi and Gauss should give their opinions on his work. No record exists of any comment these men made. However the papers reached Liouville who, in September 1843, announced to the Academy that he had found in Galois' papers a concise solution

...as correct as it is deep of this lovely problem: Given an irreducible equation of prime degree, decide whether or not it is soluble by radicals.

Liouville published these papers of Galois in his Journal in 1846.

The source of this information is the following webpage:

http://www-history.mcs.st-and.ac.uk/history/Mathematicians/Galois.html

Johann Carl Friedrich Gauss

Born: 30 April 1777 in Brunswick, Duchy of Brunswick (now Germany)

Died: 23 Feb 1855 in Göttingen, Hanover (now Germany)

Carl Friedrich Gauss worked in a wide variety of fields in both mathematics and physics incuding number theory, analysis, differential geometry, geodesy, magnetism, astronomy and optics. His work has had an immense influence in many areas.

At the age of seven, Carl Friedrich started elementary school, and his potential was noticed almost immediately. His teacher, Büttner, and his assistant, Martin Bartels, were amazed when Gauss summed the integers from 1 to 100 instantly by spotting that the sum was 50 pairs of numbers each pair summing to 101.

In 1788 Gauss began his education at the Gymnasium with the help of Büttner and Bartels, where he learnt High German and Latin. After receiving a stipend from the Duke of Brunswick-Wolfenbüttel, Gauss entered Brunswick Collegium Carolinum in 1792. At the academy Gauss independently discovered Bode's law, the binomial theorem and the arithmetic- geometric mean, as well as the law of quadratic reciprocity and the prime number theorem.

In 1795 Gauss left Brunswick to study at Göttingen University. Gauss's teacher there was Kaestner, whom Gauss often ridiculed. His only known friend amongst the students was Farkas Bolyai. They met in 1799 and corresponded with each other for many years.

Gauss left Göttingen in 1798 without a diploma, but by this time he had made one of his most important discoveries - the construction of a regular 17-gon by ruler and compasses This was the most major advance in this field since the time of Greek mathematics and was published as Section VII of Gauss's famous work, *Disquisitiones Arithmeticae*.

Gauss returned to Brunswick where he received a degree in 1799. After the Duke of Brunswick had agreed to continue Gauss's stipend, he requested that Gauss submit a doctoral dissertation to the University of Helmstedt. He already knew Pfaff, who was chosen to be his advisor. Gauss's dissertation was a discussion of the fundamental theorem of algebra.

With his stipend to support him, Gauss did not need to find a job so devoted himself to research. He published the book *Disquisitiones Arithmeticae* in the summer of 1801. There were seven sections, all but the last section, referred to above, being devoted to number theory.

In June 1801, Zach, an astronomer whom Gauss had come to know two or three years previously, published the orbital positions of Ceres, a new 'small planet' which was discovered by G Piazzi, an Italian astronomer on 1 January, 1801. Unfortunately, Piazzi had only been able to observe 9 degrees of its orbit before it disappeared behind the Sun. Zach published several predictions of its position, including one by Gauss which differed greatly from the others. When Ceres was rediscovered by Zach on 7 December 1801 it was almost exactly where Gauss had predicted. Although he did not disclose his methods at the time, Gauss had used his least squares approximation method.

In June 1802 Gauss visited Olbers who had discovered Pallas in March of that year and Gauss investigated its orbit. Olbers requested that Gauss be made director of the proposed new observatory in Göttingen, but no action was taken. Gauss began corresponding with Bessel, whom he did not meet until 1825, and with Sophie Germain.

Gauss married Johanna Ostoff on 9 October, 1805. Despite having a happy personal life for the first time, his benefactor, the Duke of Brunswick, was killed fighting for the Prussian army. In 1807 Gauss left Brunswick to take up the position of director of the Göttingen observatory.

Gauss arrived in Göttingen in late 1807. In 1808 his father died, and a year later Gauss's wife Johanna died after giving birth to their second son, who was to die soon after her. Gauss was shattered and wrote to Olbers asking him give him a home for a few weeks,

to gather new strength in the arms of your friendship - strength for a life which is only valuable because it belongs to my three small children.

Gauss was married for a second time the next year, to Minna the best friend of Johanna, and although they had three children, this marriage seemed to be one of convenience for Gauss.

Gauss's work never seemed to suffer from his personal tragedy. He published his second book, *Theoria motus corporum coelestium in sectionibus conicis Solem ambientium*, in 1809, a major two volume treatise on the motion of celestial bodies. In the first volume he discussed differential equations, conic sections and elliptic orbits, while in the second volume, the main part of the work, he showed how to estimate and then to refine the estimation of a planet's orbit. Gauss's contributions to theoretical astronomy stopped after 1817, although he went on making observations until the age of 70.

Much of Gauss's time was spent on a new observatory, completed in 1816, but he still found the time to work on other subjects. His publications during this time include *Disquisitiones generales circa seriem infinitam*, a rigorous treatment of series and an introduction of the hypergeometric function, *Methodus nova integralium valores per approximationem inveniendi*, a practical essay on approximate integration, *Bestimmung der Genauigkeit der Beobachtungen*, a discussion of statistical estimators, and Theoria attractionis corporum sphaeroidicorum ellipticorum homogeneorum methodus nova tractata. The latter work was inspired by geodesic problems and was principally concerned with potential theory. In fact, Gauss found himself more and more interested in geodesy in the 1820's.

Gauss had been asked in 1818 to carry out a geodesic survey of the state of Hanover to link up with the existing Danish grid. Gauss was pleased to accept and took personal charge of the survey, making measurements during the day and reducing them at night, using his extraordinary mental capacity for calculations. He regularly wrote to Schumacher, Olbers and Bessel, reporting on his progress and discussing problems.

Because of the survey, Gauss invented the heliotrope which worked by reflecting the Sun's rays using a design of mirrors and a small telescope. However, inaccurate base lines were used for the survey and an unsatisfactory network of triangles. Gauss often wondered if he would have been better advised to have pursued some other occupation but he published over 70 papers between 1820 and 1830.

In 1822 Gauss won the Copenhagen University Prize with *Theoria attractionis...* together with the idea of mapping one surface onto another so that the two are similar in their smallest parts . This paper was published in 1825 and led to the much later publication of *Untersuchungen über Gegenstände der Höheren Geodäsie* (1843 and 1846). The paper *Theoria combinationis observationum erroribus minimis obnoxiae* (1823), with its supplement (1828), was devoted to mathematical statistics, in particular to the least squares method.

From the early 1800's Gauss had an interest in the question of the possible existence of a non-Euclidean geometry. He discussed this topic at length with Farkas Bolyai and in his correspondence with Gerling and Schumacher. In a book review in 1816 he discussed proofs which deduced the axiom of parallels from the other Euclidean axioms, suggesting that he believed in the existence of non-Euclidean geometry, although he was rather vague. Gauss confided in Schumacher, telling him that he believed his reputation would suffer if he admitted in public that he believed in the existence of such a geometry.

In 1831 Farkas Bolyai sent to Gauss his son János Bolyai's work on the subject. Gauss replied

to praise it would mean to praise myself.

Again, a decade later, when he was informed of Lobachevsky's work on the subject, he praised its "genuinely geometric" character, while in a letter to Schumacher in 1846, states that he

had the same convictions for 54 years

indicating that he had known of the existence of a non-Euclidean geometry since he was 15 years of age (this seems unlikely).

Gauss had a major interest in differential geometry, and published many papers on the subject. *Disquisitiones generales circa superficies curva* (1828) was his most renowned work in this field. In fact, this paper rose from his geodesic interests, but it contained such geometrical ideas as Gaussian curvature. The paper also includes Gauss's famous theorema egregrium:

If an area in E^3 can be developed (i.e. mapped isometrically) into another area of E^3, the values of the Gaussian curvatures are identical in corresponding points.

The period 1817-1832 was a particularly distressing time for Gauss. He took in his sick mother in 1817, who stayed until her death in 1839, while he was arguing with his wife and her family about whether they should go to Berlin. He had been offered a position at Berlin University and Minna and her family were keen to move there. Gauss, however, never liked change and decided to stay in Göttingen. In 1831 Gauss's second wife died after a long illness.

In 1831, Wilhelm Weber arrived in Göttingen as physics professor filling Tobias Mayer's chair. Gauss had known Weber since 1828 and supported his appointment. Gauss had worked on physics before 1831, publishing *Uber ein neues allgemeines Grundgesetz der Mechanik*, which contained the principle of least constraint, and *Principia generalia theoriae figurae fluidorum in statu aequilibrii* which discussed forces of attraction. These papers were based on Gauss's potential theory, which proved of great importance in his work on physics. He later came to believe his potential theory and his method of least squares provided vital links between science and nature.

In 1832, Gauss and Weber began investigating the theory of terrestrial magnetism after Alexander von Humboldt attempted to obtain Gauss's assistance in making a grid of magnetic observation points around the Earth. Gauss was excited by this prospect and by 1840 he had written three important papers on the subject: *Intensitas vis magneticae terrestris ad mensuram absolutam revocata* (1832), *Allgemeine Theorie des Erdmagnetismus* (1839) and *Allgemeine Lehrsätze in Beziehung auf die im verkehrten Verhältnisse des Quadrats der Entfernung wirkenden Anziehungs- und Abstossungskräfte* (1840). These papers all dealt with the current theories on terrestrial magnetism, including Poisson's ideas, absolute measure for magnetic force and an empirical definition of terrestrial magnetism. Dirichlet's principal was mentioned without proof.

Allgemeine Theorie... showed that there can only be two poles in the globe and went on to prove an important theorem, which concerned the determination of the intensity of the horizontal component of the magnetic force along with the angle of inclination. Gauss used the Laplace equation to aid him with his calculations, and ended up specifying a location for the magnetic South pole.

Humboldt had devised a calendar for observations of magnetic declination. However, once Gauss's new magnetic observatory (completed in 1833 - free of all magnetic metals) had been built, he proceeded to alter many of Humboldt's procedures, not pleasing Humboldt greatly. However, Gauss's changes obtained more accurate results with less effort.

Gauss and Weber achieved much in their six years together. They discovered Kirchhoff's laws, as well as building a primitive telegraph device which could send messages over a distance of 5000 ft. However, this was just an enjoyable pastime for Gauss. He was more interested in the task of establishing a world-wide net of magnetic observation points. This occupation produced many concrete results. The *Magnetischer Verein* and its journal were founded, and the atlas of geomagnetism was published, while Gauss and Weber's own journal in which their results were published ran from 1836 to 1841.

In 1837, Weber was forced to leave Göttingen when he became involved in a political dispute and, from this time, Gauss's activity gradually decreased. He still produced letters in response to fellow scientists' discoveries usually remarking that he had known the methods for years but had never felt the need to publish. Sometimes he seemed extremely pleased with advances made by other mathematicians, particularly that of Eisenstein and of Lobachevsky.

Gauss spent the years from 1845 to 1851 updating the Göttingen University widow's fund. This work gave him practical experience in financial matters, and he went on to make his fortune through shrewd investments in bonds issued by private companies.

Two of Gauss's last doctoral students were Moritz Cantor and Dedekind. Dedekind wrote a fine description of his supervisor

... usually he sat in a comfortable attitude, looking down, slightly stooped, with hands folded above his lap. He spoke quite freely, very clearly, simply and plainly: but when he wanted to emphasise a new viewpoint ... then he lifted his head, turned to one of those sitting next to him, and gazed at him with his beautiful, penetrating blue eyes during the emphatic speech. ... If he proceeded from an explanation of principles to the development of mathematical formulas, then he got up, and in a stately very upright posture he wrote on a blackboard beside him in his peculiarly beautiful handwriting: he always succeeded through economy and deliberate arrangement in making do with a rather small space. For numerical examples, on whose careful completion he placed special value, he brought along the requisite data on little slips of paper.

Gauss presented his golden jubilee lecture in 1849, fifty years after his diploma had been granted by Hemstedt University. It was appropriately a variation on his dissertation of 1799. From the mathematical community only Jacobi and Dirichlet were present, but Gauss received many messages and honours.

From 1850 onwards Gauss's work was again of nearly all of a practical nature although he did approve Riemann's doctoral thesis and heard his probationary lecture. His last known scientific exchange was with Gerling. He discussed a modified Foucalt pendulum in 1854. He was also able to attend the opening of the new railway link between Hanover and Göttingen, but this proved to be his last outing. His health deteriorated slowly, and Gauss died in his sleep early in the morning of 23 February, 1855.

The source of this information is the following webpage:

http://www-history.mcs.st-and.ac.uk/history/Mathematicians/Gauss.html

Karl Gustav Jacob Jacobi

Born: 10 Dec 1804 in Potsdam, Prussia (now Germany)

Died: 18 Feb 1851 in Berlin, Germany

Karl Jacobi founded the theory of elliptic functions.

Jacobi's father was a banker and his family were prosperous so he received a good education at the University of Berlin. He obtained his Ph.D. in 1825 and taught mathematics at the University of Königsberg from 1826 until his death, being appointed to a chair in 1832.

He founded the theory of elliptic functions based on four theta functions. His *Fundamenta nova theoria functionum ellipticarum* in 1829 and its later supplements made basic contributions to the theory of elliptic functions.

In 1834 Jacobi proved that if a single-valued function of one variable is doubly periodic then the ratio of the periods is imaginary. This result prompted much further work in this area, in particular by Liouville and Cauchy.

Jacobi carried out important research in partial differential equations of the first order and applied them to the differential equations of dynamics.

He also worked on determinants and studied the functional determinant now called the Jacobian. Jacobi was not the first to study the functional determinant which now bears his name, it appears first in a 1815 paper of Cauchy. However Jacobi wrote a long memoir *De determinantibus functionalibus* in 1841 devoted to the this determinant. He proves, among many other things, that if a set of n functions in n variables are functionally related then the Jacobian is identically zero, while if the functions are independent the Jacobian cannot be identically zero.

Jacobi's reputation as an excellent teacher attracted many students. He introduced the seminar method to teach students the latest advances in mathematics.

The source of this information is the following webpage:

http://www-history.mcs.st-and.ac.uk/history/Mathematicians/Jacobi.html

Adrien-Marie Legendre

Born: 18 Sept 1752 in Paris, France

Died: 10 Jan 1833 in Paris, France

Legendre's major work on elliptic integrals provided basic analytical tools for mathematical physics.

Legendre was educated at Collège Mazarin in Paris. From 1775 to 1780 he taught with Laplace at École Militaire where his appointment was made on the advice of d'Alembert. Legendre was appointed to the Académie des Sciences in 1783 and remained there until it closed in 1793.

In 1782 Legendre determined the attractive force for certain solids of revolution by introducing an infinite series of polynomials P_n which are now called Legendre polynomials.

His major work on elliptic functions in *Exercises du Calcul Intégral* (1811,1817,1819) and elliptic integrals in *Traité des Fonctions Elliptiques* (1825,1826,1830) provided basic analytical tools for mathematical physics.

In his famous textbook *Éléments de géométrie* (1794) he gave a simple proof that π is irrational as well as the first proof that π^2 is irrational and conjectured that is not the root of any algebraic equation of finite degree with rational coefficients i.e. is not algebraic.

His attempt to prove the parallel postulate extended over 40 years.

In 1824 Legendre refused to vote for the government's candidate for Institut National. Because of this his pension was stopped and he died in poverty. Abel wrote in October 1826

Legendre is an extremely amiable man, but unfortunately as old as the stones.

The source of this information is the following webpage:

http://www-history.mcs.st-and.ac.uk/history/Mathematicians/Legendre.html

August Ferdinand Möbius

Born: 17 Nov 1790 in Schulpforta, Saxony (now Germany)

Died: 26 Sept 1868 in Leipzig, Germany

August Möbius is best known for his work in topology, especially for his conception of the Möbius strip, a two dimensional surface with only one side.

August was the only child of Johann Heinrich Möbius, a dancing teacher, who died when August was three years old. His mother was a descendant of Martin Luther. Möbius was educated at home until he was 13 years old when, already showing an interest in mathematics, he went to the College in Schulpforta in 1803.

In 1809 Möbius graduated from his College and he became a student at the University of Leipzig. His family had wanted him study law and indeed he started to study this topic. However he soon discovered that it was not a subject that gave him satisfaction and in the middle of his first year of study he decided to follow him own preferences rather than those of his family. He therefore took up the study of mathematics, astronomy and physics.

The teacher who influenced Möbius most during his time at Leipzig was his astronomy teacher Karl Mollweide. Although an astronomer, Mollweide is well known for a number of mathematical discoveries in particular the Mollweide trigonometric relations he discovered in 1807-09 and the Mollweide map projection which preserves angles and so is a conformal projection.

In 1813 Möbius travelled to Göttingen where he studied astronomy under Gauss. Now Gauss was the director of the Observatory in Göttingen but of course the greatest mathematician of his day, so again Möbius studied under an astronomer whose interests were mathematical. From Göttingen Möbius went to Halle where he studied under Johann Pfaff, Gauss's teacher. Under Pfaff he studied mathematics rather than astronomy so by this stage Möbius was very firmly working in both fields.

In 1815 Möbius wrote his doctoral thesis on *The occultation of fixed stars* and began work on his Habilitation thesis. In fact while he was writing this thesis there was an attempt to draft him into the Prussian army. Möbius wrote

This is the most horrible idea I have heard of, and anyone who shall venture, dare, hazard, make bold and have the audacity to propose it will not be safe from my dagger.

He avoided the army and completed his Habilitation thesis on *Trigonometrical equations*. Mollweide's interest in mathematics was such that he had moved from astronomy to the chair of mathematics at Leipzig so Möbius had high hopes that he might be appointed to a professorship in astronomy at Leipzig. Indeed he was appointed to the chair of astronomy and higher mechanics at the University of Leipzig in 1816. His initial appointment was as Extraordinary Professor and it was an appointment which came early in his career.

However Möbius did not receive quick promotion to full professor. It would appear that he was not a particularly good lecturer and this made his life difficult since he did not attract fee paying students to his lectures. He was forced to advertise his lecture courses as being free of charge before students thought his courses worth taking.

He was offered a post as an astronomer in Greifswald in 1916 and then a post as a mathematician at Dorpat in 1819. He refused both, partly through his belief in the high quality of Leipzig

University, partly through his loyalty to Saxony. In 1825 Mollweide died and Möbius hoped to transfer to his chair of mathematics taking the route Mollweide had taken earlier. However it was not to be and another mathematician was preferred for the post.

By 1844 Möbius's reputation as a researcher led to an invitation from the University of Jena and at this stage the University of Leipzig gave him the Full Professorship in astronomy which he clearly deserved.

From the time of his first appointment at Leipzig Möbius had also held the post of Observer at the Observatory at Leipzig. He was involved the rebuilding of the Observatory and, from 1818 until 1821, he supervised the project. He visited several other observatories in Germany before making his recommendations for the new Observatory. In 1820 he married and he was to have one daughter and two sons. In 1848 he became director of the Observatory.

In 1844 Grassmann visited Möbius. He asked Möbius to review his major work *Die lineale Ausdehnundslehre, ein neuer Zweig der Mathematik* (1844) which contained many results similar to Möbius's work. However Möbius did not understand the significance of Grassmann's work and did not review it. He did however persuade Grassmann to submit work for a prize and, after Grassmann won the prize, Möbius did write a review of his winning entry in 1847.

Although his most famous work is in mathematics, Möbius did publish important work on astronomy. He wrote *De Computandis Occultationibus Fixarum per Planetas* (1815) concerning occultations of the planets. He also wrote on the principles of astronomy, *Die Hauptsätze der Astronomie* (1836) and on celestial mechanics *Die Elemente der Mechanik des Himmels* (1843).

Möbius's mathematical publications, although not always original, were effective and clear presentations. His contributions to mathematics are described by his biographer Richard Baltzer in as follows:

The inspirations for his research he found mostly in the rich well of his own original mind. His intuition, the problems he set himself, and the solutions that he found, all exhibit something extraordinarily ingenious, something original in an uncontrived way. He worked without hurrying, quietly on his own. His work remained almost locked away until everything had been put into its proper place. Without rushing, without pomposity and without arrogance, he waited until the fruits of his mind matured. Only after such a wait did he publish his perfected works...

Almost all Möbius's work was published in *Crelle's Journal*, the first journal devoted exclusively to publishing mathematics. Möbius's 1827 work *Der barycentrische Calkul*, on analytical geometry, became a classic and includes many of his results on projective and affine geometry. In it he introduced homogeneous coordinates and also discussed geometric transformations, in particular projective transformations. He introduced a configuration now called a Möbius net, which was to play an important role in the development of projective geometry.

Möbius's name is attached to many important mathematical objects such as the Möbius function which he introduced in the 1831 paper *Uber eine besondere Art von Umkehrung der Reihen* and the Möbius inversion formula.

In 1837 he published *Lehrbuch der Statik* which gives a geometric treatment of statics. It led to the study of systems of lines in space.

Before the question on the four colouring of maps had been asked by Francis Guthrie, Möbius had posed the following, rather easy, problem in 1840.

There was once a king with five sons. In his will he stated that on his death his kingdom should be divided by his sons into five regions in such a way that each region should have a common boundary with the other four. Can the terms of the will be satisfied?

The answer, of course, is negative and easy to show. However it does illustrate Möbius's interest in topological ideas, an area in which he most remembered as a pioneer. In a memoir, presented to the Académie des Sciences and only discovered after his death, he discussed the properties of one-sided surfaces including the Möbius strip which he had discovered in 1858. This discovery was made as Möbius worked on a question on the geometric theory of polyhedra posed by the Paris Academy.

Although we know this as a Möbius strip today it was not Möbius who first described this object, rather by any criterion, either publication date or date of first discovery, precedence goes to Listing.

A Möbius strip is a two-dimensional surface with only one side. It can be constructed in three dimensions as follows. Take a rectangular strip of paper and join the two ends of the strip together so that it has a 180 degree twist. It is now possible to start at a point A on the surface and trace out a path that passes through the point which is apparently on the other side of the surface from A.

The source of this information is the following webpage:

http://www-history.mcs.st-and.ac.uk/history/Mathematicians/Mobius.html

Joseph Henry Maclagen Wedderburn

Born: 2 Feb 1882 in Forfar, Angus, Scotland

Died: 9 Oct 1948 in Princeton, New Jersey, USA

Joseph Wedderburn made important advances in the theory of rings, algebras and matrix theory.

He entered Edinburgh University in 1898, obtaining a degree in mathematics in 1903. Wedderburn then pursued postgraduate studies in Germany spending 1903-1904 at the University of Leipzig and then a semester at the University of Berlin.

He was awarded a Carnegie Scholarship to study in the USA and he spent 1904-1905 at the University of Chicago where he did joint work with Veblen. Returning to Scotland he worked for 4 years at Edinburgh as assistant to George Chrystal. From 1906 to 1908 he served as editor of the Proceedings of the Edinburgh Mathematical Society.

In 1909 Wedderburn was appointed a Preceptor in Mathematics at Princeton where he joined Veblen. However World War I saw Wedderburn volunteer for the British Army and he served, partly in France, until the end of the war.

On his return to Princeton he was soon promoted obtaining permanent tenure in 1921. He served as Editor of the Annals of Mathematics from 1912 to 1928. From about the end of this period Wedderburn seemed to suffer a mild nervous breakdown and became an increasingly solitary figure. By 1945 the Priceton gave him early retirement in his own best interests.

Wedderburn's best mathematical work was done before his war service. In 1905 he showed that a non-commutatiove finite field could not exist. This had as a corollary the complete structure of all finite projective geometries, showing that in all these geometries Pascal's theorem is a consequence of Desargues' theorem.

In 1907 he published what is perhaps his most famous paper on the classification of semisimple algebras. He showed that every semisimple algebra is a direct sum of simple algebras and that a simple algebra was a matrix algebra over a division ring.

In total he published around 40 works mostly on rings and matrices. His most famous book is *Lectures on Matrices* (1934).

The source of this information is the following webpage:

http://www-history.mcs.st-and.ac.uk/history/Mathematicians/Wedderburn.html

Appendix D   New Functions

□ **AddTwoLetters**

AddTwoLetters adds two letters modulo 26, where $a = 0$, $b = 1$, ..., $z = 25$.

```
AddTwoLetters[a_, b_] := FromCharacterCode[
  Mod[(ToCharacterCode[a] - 97) +
    (ToCharacterCode[b] - 97), 26] + 97]
```

Example:

```
AddTwoLetters["b", "c"]
```

```
d
```

□ **CaesarCipher**

Applies the Caesar cipher with a given key to a given plaintext of small letters.

```
CaesarCipher[plaintext_, key_] :=    FromCharacterCode[
  Mod[ToCharacterCode[plaintext] - 97 + key, 26] + 97]
```

Example:

```
plaintext = "typehereyourplaintextinsmallletters";
key = 24;
CaesarCipher[plaintext, key]
```

```
rwncfcpcwmspnjyglrcvrglqkyjjjcrrcpq
```

□ **ColumnSwap**

ColumnSwap interchanges columns i and j in matrix B.

```
ColumnSwap[B_, i_, j_] := Module[{U, V}, U = Transpose[B];
  V = U[[i]]; U[[i]] = U[[j]]; U[[j]] = V; Transpose[U]]
```

Example:

```
Clear[a, b, c, d, e, f, g, h, i, j, k, l]; A = ( a  b  c  d )
                                              ( e  f  g  h ) ;
                                              ( i  j  k  l )

AA = ColumnSwap[A, 1, 4]; MatrixForm[AA]
```

$$\begin{pmatrix} d & b & c & a \\ h & f & g & e \\ l & j & k & i \end{pmatrix}$$

□ **CoPrimeQ**

CoPrime test if two integers are coprime, i.e. have gcd 1.

```
CoPrimeQ[n_Integer, m_Integer] := GCD[n, m] == 1
```

Example:

```
CoPrimeQ[35, 91]
CoPrimeQ[36, 91]
```

```
False
```

```
True
```

□ **CoPrimes**

CoPrimes generates a list of all integers in between 1 and n that are coprime with n. In other words, it generates a reduced residue system modulo n.

Coprimes makes use of the function CoPrimeQ defined earlier.

```
CoPrimes[n_Integer?Positive] :=
  Select[Range[n], CoPrimeQ[n, #] &]
```

Example:

```
CoPrimes[15]
```

```
{1, 2, 4, 7, 8, 11, 13, 14}
```

□ **DivisorProduct**

DivisorProduct calculates $\prod_{d|n} f[d]$.

```
DivisorProduct[f_, n_] := Times @@ (f /@ Divisors[n])
```

Example:

```
f[n_] := n
DivisorProduct[f, 25]
```

```
125
```

□ **DivisorSum**

DivisorSum calculates $\sum_{d|n} f[d]$.

```
DivisorSum[f_, n_] := Plus @@ (f /@ Divisors[n])
```

Example:

```
f[n_] := n
DivisorSum[f, 15]
```

```
24
```

□ **EllipticAdd**

EllipticAdd evaluates the sum of the points P and Q on an elliptic curve over $\mathbf{Z}_p$ given by the equation $y^2 = x^3 + a.x^2 + b.x + c$. Here p is prime, $p > 2$.

```
EllipticAdd[p_, a_, b_, c_, P_List, Q_List] :=
 Module[{lam, x3, y3, P3},
  If[P == {O}, R = Q, If[Q == {O}, R = P,      If[P[[1]] !=
     Q[[1]],                                   {lam = Mod[
       (Q[[2]] - P[[2]]) * PowerMod[Q[[1]] - P[[1]], p - 2, p],
       p];                                     x3 = Mod[lam^2 - a -
       P[[1]] - Q[[1]], p];
     y3 = Mod[-(lam (x3 - P[[1]]) + P[[2]]),
       p];                                     R = {x3, y3}},
    If[(P == Q) ∧ (P != {O}),
      {lam = Mod[(3 * P[[1]]^2 + 2 a * P[[1]] + b) * PowerMod[
          2 P[[2]], p - 2, p], p];
      x3 = Mod[lam^2 - a - P[[1]] - Q[[1]],
        p];
      y3 = Mod[-(lam (x3 - P[[1]]) + P[[2]]),
        p];                                    R = {x3, y3}}, If[
      (P[[1]] == Q[[1]]) ∧ (P[[2]] != Q[[2]]), R = {O}]]]]]; R]
```

Example:

```
p = 11; a = 0; b = 6; c = 3;
EllipticAdd[p, a, b, c, {4, 6}, {9, 4}]
```

```
{3, 9}
```

□ Entropy

Computes the entropy $-p.\log_2 p - (1 - p).\log_2(1 - p)$ function.

```
Entropy[p_] = -p * Log[2, p] - (1 - p) Log[2, 1 - p];
```

Example:

```
Entropy[1 / 2]
```

```
1
```

□ ListQuadRes

ListQuadRes gives a listing of all the quadratic residues modulo p.

```
ListQuadRes[p_] :=
  Select[Range[p], JacobiSymbol[#1, p] == 1 &]
```

Example:

```
p = 17;
ListQuadRes[p]
```

```
{1, 2, 4, 8, 9, 13, 15, 16}
```

□ MultiEntropy

MultiEntropy evaluates $-\sum_{i=1}^{n} p_i \log_2 p_i$ for a list $\{p_1, p_2, ..., p_n\}$.

```
                      Length[p]
MultiEntropy[p_List] := -   ∑   p[[i]] * Log[2, p[[i]]]
                         i=1
```

Example:

```
p = {1 / 4, 1 / 4, 1 / 4, 1 / 4};
MultiEntropy[p]
```

```
2
```

□ MultiplicativeOrder

MultiplicativeOrder computes the multiplicative order of an integer a modulo n, assuming that they are coprime. So, it outputs the smallest positive integer m such that $a^m \equiv 1 \pmod{n}$.

```
MultiplicativeOrder[a_, n_] :=
   If[GCD[a, n] == 1,              Divisors[ EulerPhi[n] ] //.
      {x_, y___} -> If[PowerMod[a, x, n] == 1, x, {y}] ];
```

Example:

```
MultiplicativeOrder[2, 123456789123]
```

```
1285901112
```

□ KnapsackForSuperIncreasingSequence

KnapsackForSuperIncreasingSequence finds the {0, 1}-solution of the knapsack problem $\sum_{i=1}^{n} x_i.a_i = S$, where $\{a_i\}_{i=1}^{n}$ is a superincreasing sequence.

```
KnapsackForSuperIncreasingSequence[a_List, S_] :=
   Module[{n, x, X, T},           n = Length[a]; X = {};
     T = S;             While[n ≥ 1,
      If[T ≥ a[[n]], x = 1, x = 0];
      T = T - x * a[[n]];                     X = Join[{x}, X];
      n = n - 1];             If[T != 0, Print["No solution"], X]]
```

Example:

```
a = {22, 89, 345, 987, 4567, 45678}; S = 5665;
X = KnapsackForSuperIncreasingSequence[a, S]
```

```
{1, 1, 0, 1, 1, 0}
```

□ RowSwap

RowSwaps interchanges rows i and j in matrix B.

```
RowSwap[B_, i_, j_] :=
 Module[{U, V}, U = B; V = U[[i]]; U[[i]] = U[[j]]; U[[j]] = V; U]
```

Example:

```
Clear[a, b, c, d, e, f, g, h, i, j, k, l]; A = ( a b c
                                                 d e f
                                                 g h i
                                                 j k l );

AA = RowSwap[A, 1, 4]; MatrixForm[AA]
```

$$
\begin{pmatrix}
j & k & l \\
d & e & f \\
g & h & i \\
a & b & c
\end{pmatrix}
$$

References

[Adle79] Adleman, L.M., *A subexponential algorithm for the discrete logarithm problem with applications to cryptography*, in Proc. IEEE 20-th Annual Symp. on Found. of Comp. Science, pp. 55-60, 1979.

[Adle83] Adleman, L.M., *On breaking the iterated Merkle-Hellman public key cryptosystem*, in Proc. 15-th Annual ACM Symp. Theory of Computing, pp. 402-412, 1983.

[Adle94] Adleman, L.M., *The function field sieve*, Lecture Notes in Computer Science 877, Springer Verlag, Berlin, etc., pp. 108-121, 1995.

[AdDM93] Adleman, L.M. and J. DeMarrais, A subexponential algorithm for discrete logarithms over all finite fields, Mathematics of Computation, 61, pp. 1-15, 1993.

[AdPR83] Adleman, L.M., C. Pomerance, and R. Rumely, *On distinguishing prime numbers from composite numbers*, Annals of Math. 17, pp. 173-206, 1983.

[Aign79] Aigner, M., *Combinatorial Theory*, Springer Verlag, Berlin, etc., 1979.

[BaKT99] Barg, A., E. Korzhik and H.C.A. van Tilborg, *On the complexity of minimum distance decoding of long linear codes*, to appear in the IEEE Transactions on Information Theory.

[Baue97] Bauer, F.L., *Decrypted Secrets; Methods and Maxims of Cryptology*, Springer Verlag, Berlin, etc., 1997.

[BekP82] Beker, H. and F. Piper, *Cipher Systems, the Protection of Communications*, Northwood Books, London, 1982.

[Berl68] Berlekamp, E.R., *Algebraic Coding Theory*, McGraw-Hill Book Company, New York, etc., 1968

[BeMT78] Berlekamp, E.R., R.J. McEliece and H.C.A. van Tilborg, *On the inherent intractability of certain coding problems*, IEEE Transactions on Information Theory, IT-24, pp. 384-386, May 1978.

[BeJL86] Beth, T., D. Jungnickel, and H. Lenz, *Design Theory*, Cambridge University Press, Cambridge, etc., 1986.

[BihS93], Biham E. and A. Shamir, *Differential Cryptanalysis of the Data Encryption Standard*, Spinger Verlag, New York etc., 1993.

[BoDML97] Boneh, D., R.A. DeMillo, and R.J. Lipton, *On the importance of checking cryptographic protocols for faults*, Advances in Cryptology: Proc. of Eurocrypt'97, W. Fumy, Ed., Lecture Notes in Computer Science 1233, Springer Verlag, Berlin, etc., pp. 37-51, 1997.

[Bos92] Bos, J.N.E., *Practical privacy*, Ph.D. Thesis, Eindhoven University of Technology, the Netherlands, 1992.

[Bric85] Brickell, E.F., *Breaking iterated knapsacks*, in Advances in Cryptography: Proc. of Crypto '84, G.R. Blakley and D. Chaum, Eds., Lecture Notes in Computer Science 196, Springer Verlag, Berlin etc., pp. 342-358, 1985.

[Bric89] Brickell, E.F., *Some ideal secret sharing schemes*, The Journal of Combinatorial Mathematics and Combinatorial Computing, Vol. 6, pp. 105-113, 1989.

[CanS98] Canteaut, A. and N. Sendrier, *Cryptanalysis of the original McEliecese cryptosystem*, Advances in Cryptology: Proc. AsiaCrypt'98, K. Ohta and D. Pei, Eds., Lecture Notes in Computer Science 1514, Springer, Berlin etc., pp. 187-199, 1998.

[ChoR85] Chor, B. and R.L. Rivest, *A knapsack type public key cryptosystem based on arithmetic in finite fields*, in Advances in Cryptography: Proc. of Crypto '84, G.R. Blakley and D. Chaum, Eds., Lecture Notes in Computer Science 196, Springer Verlag, Berlin etc., pp. 54-65, 1985.

[CohL82] Cohen, H. and H.W. Lenstra Jr., *Primality testing and Jacobi sums*, Report 82-18, Math. Inst., Univ. of Amsterdam, Oct. 1982.

[Cohn77] Cohn, P.M., *Algebra Vol.2*, John Wiley & Sons, London, etc., 1977.

[Copp84] Coppersmith, D., *Fast evaluation of logarithms in fields of characteristic two*, IEEE Transactions on Infprmation Theory, IT-30, pp. 587-594, July 1984.

[CopFPR96] Coppersmith, D., M. Franklin, J. Patarin, and M. Reiter, *Low-exponent RSA with Related Messages*, Advances in Cryptology: Proc. of Eurocrypt'96, U. Maurer, Ed., Lecture Notes in Computer Science 1070, Springer Verlag, Berlin, etc., pp. 1-9, 1996.

[CovM67] Coveyou, R.R. and R.D. McPherson, *Fourier analysis of uniform random number generators*, J. Assoc. Comput. Mach., 14, pp. 100-119, 1967.

[Demy94] Demytko, N., *A new elliptic curve based analogue of RSA*, Advances in Cryptology: Proc. of Eurocrypt'93, T. Helleseth, Ed., Lecture Notes in Computer Science 765, Springer Verlag, Berlin, etc., pp. 40-49, 1994.

[Denn82] Denning, D.E.R., *Cryptography and Data Security*, Addison-Wesley publ. Comp., Reading Ma, etc., 1982.

[DifH76] Diffie, W. and M.E. Hellman, *New directions in cryptography*, IEEE Transactions on Information Theory, IT-22, pp. 644-654, Nov. 1976.

[Dijk97] Dijk, M. van, *Secret Key Sharing and Secret Key Generation*, Ph.D. Thesis, Eindhoven University of Technology, the Netherlands, 1997.

[ElGa85] ElGamal, T., *A public-key cryptosystem and a signature scheme based on discrete logarithms*, Advances in Cryptology: Proc. of Crypto'84, G.R. Blakley and D. Chaum, Eds., Lecture Notes in Computer Science 196, Springer Verlag, Berlin, etc., pp. 10-18, 1985.

[FiaS87] Fiat, A. and A. Shamir, *How to prove yourself: Practical solutions to identification and signature problems*, Advances in Cryptology: Proc. of Crypto'86, A.M. Odlyzko, Ed., Lecture Notes in Computer Science 263, Springer Verlag, Berlin, etc., pp. 186-194, 1987.

[FIPS94] FIPS 186, *Digital Signature Standard*, Federal Information Processing Standards Publication 186, U.S. Department of Commerce/N.I.S.T., National Technical Information Service, Springfield, Virginia, 1994.

[Frie73] Friedman, W.F., *Cryptology*, in Encyclopedia Brittanica, p. 848, 1973.

[GarJ79] Garey, M.R. and D.S. Johnson, *Computers and Intractability: A Guide to the Theory of NP-Completeness*, W.H. Freeman and Co., San Fransisco, 1979.

[GilMS74] Gilbert, E.N., F.J. MacWilliams, and N.J.A. Sloane, *Codes which detect deception*, Bell System Technical Journal, Vol. 53, pp. 405-424, 1974.

[Golo67] Golomb, S.W., *Shift Register Sequences*, Holden-Day, San Fransisco, 1967.

[Hall67] Hall, Jr., M., *Combinatorial Theory*, Blaisdell Publishing Company, Waltham, Ma., 1967

[HarW45] Hardy, G.H. and E.M. Wright, *An Introduction to the Theory of Numbers*, Clarendon Press, Oxford, 1945.

[Håst88] Håstad, J., *Solving simultaneous modular equations of low degree*, SIAM Journal on Computing, 17, pp. 336-341, 1988.

[HelR83] Hellman, M.E. and J.M. Reyneri, *Fast computation of discrete logarithms over GF(q)*, in Advances in Cryptography: Proc. of Crypto '82, D. Chaum, R. Rivest and A. Sherman, Eds., Plenum Publ. Comp., New York, pp. 3-13, 1983.

[Huff52] Huffman, D.A., *A method for the construction of minimum-redundancy codes*, Proc. IRE, 14, pp. 1098-1101, 1952.

[Joha94a] Johansson, T., *A shift register of unconditionally secure authentication codes*, Designs, Codes and Cryptography, 4, pp. 69-81, 1994.

[Joha94b] Johansson, T., *Contributions to Unconditionally Secure Authentication*, KF Sigma, Lund, 1994.

[JohKS93] Johansson, T., G. Kabatianskii, and B. Smeets, *On the relation between A-codes and codes correcting independent errors*, Advances in Cryptography: Proc. of Eurocrypt '93, T. Helleseth, Edt., Lecture Notes in Computer Science 765, Springer Verlag, Berlin etc., pp. 1-10, 1993.

[Kahn67] Kahn, D., *The Codebreakers, the Story of Secret Writing*, Macmillan Company, New York, 1967.

[Khin57] Khinchin. A.I., *Mathematical Foundations of Information Theory*, Dover Publications, New York, 1957.

[Knud94] Knudsen, L.R., *Block Ciphers—Analysis, Designs and Applications*, PhD Thesis, Computer Science Department, Aarhus University, Denmark, 1994.

[Knut69] Knuth, D.E., *The Art of Computer Programming, Vol.2, Semi-numerical Algorithms*, Addison-Wesley, Reading, MA., 1969.

[Knut73] Knuth, D.E., *The Art of Computer Programming, Vol.3, Sorting and searching*, Addison-Wesley, Reading, M.A., 1973.

[Knut81] Knuth, D.E., *The Art of Computer Programming, Vol.2, Semi-Numerical Algorithms*, Second Edition, Addison-Wesley, Reading, MA., 1981.

[Koch96] Kocher, P.C., *Timing attacks on implementations of Diffie-Hellman, RSA, DSS, and Other Systems*, Advances in Cryptology: Proc. of Crypto'96, N. Koblitz, Ed., Lecture Notes in Computer Science 1109, Springer Verlag, Berlin etc., pp. 104-113 , 1996.

[Konh81] Konheim, A.G., *Cryptography, a Primer*, John Wiley & Sons, New York, etc., 1981.

[Kraf49] Kraft, L.G., *A Device for Quantizing, Grouping and Coding Amplitude Modulated Pulses*, MS Thesis, Dept. of EE, MIT, Cambridge, Mass., 1949.

[LagO83] Lagarias, J.C. and A.M. Odlyzko, *Solving low-density subset problems*, Proc. 24th Annual IEEE Symp. on Found. of Comp. Science, pp. 1-10, 1983.

[Lai92] Lai, X., *On the design and security of block ciphers*, ETH Series in Information Processing, J.J. Massey, Ed., vol. 1, Hartung-Gorre Verlag, Konstantz, 1992)

[LeeB88] Lee, P.J. and E.F. Brickell, *An observation on the security of McEliece's public-key cryptosystem*, in Advances in Cryptography: Proc. of Eurocrypt'88, C.G. Günther, Ed., Lecture Notes in Computer Science 330, Springer Verlag, Berlin etc., pp. 275-280, 1988.

[Lehm76] Lehmer, D.H., *Strong Carmichael numbers*, J. Austral. Math. Soc., Ser. A 21, pp. 508-510, 1976.

[LensA96] Lenstra, A.K., *Memo on RSA signature generation in the presence of faults*, Sept. 1996.

[LenLL82] Lenstra, A.K., H.W. Lenstra, Jr., and L. Lovász, *Factoring polynomials with rational coefficients*, Math. Annalen, 261, pp. 515-534, 1982.

[LensH83] Lenstra, H.W. Jr., *Fast prime number tests*, Nieuw Archief voor Wiskunde (4) 1, pp. 133-144, 1983.

[LensH86] Lenstra, H.W. Jr., *Factoring integers with elliptic curves*, Report 86-16, Dept. of Mathematics, University of Amsterdam, Amsterdam, the Netherlands.

[Liu68] Liu, C.L., *Introduction to combinatorial mathematics*, McGraw-Hill, New York, 1968.

[Lüne87] Lüneberg H., *On the Rational Normal Form of Endomorphisms; a Primer to Constructive Algebra*, BI Wissenschaftsverlag, Mannheim etc., 1987.

[MacWS77] MacWilliams, F.J. and N.J.A. Sloane, *The Theory of Error-Correcting Codes*, North-Holland Publ. Comp., Amsterdam, etc., 1977.

[Mass69] Massey, J.L., *Shift-register synthesis and BCH decoding*, IEEE Transactions on Information Theory, IT-15, pp. 122-127, Jan. 1969.

[MatY93] Matsui, M. and A. Yamagishi, *A new method for known plaintext attack of FEAL cipher*, Advances in Cryptology: Proc. Eurocrypt'92, R.A. Rueppel, Ed., Lecture Notes in Computer Science 658, Springer, Berlin etc., pp. 81-91, 1993.

[Maur92] Maurer, U., *A universal statistical test for random bit generators*, Journal of Cryptology, 5, pp. 89-105, 1992.

[McEl78] McEliece, R.J., *A public-key cryptosystem based on algebraic coding theory*, JPL DSN Progress Report 42-44, pp. 114-116, Jan-Febr. 1978.

[McEl81] McEliece, R.J. and D.V. Sarwate, *On sharing secrets and Reed-Solomon codes*, Comm. ACM, vol. 24, pp. 583-584, Sept. 1981.

[McMi56] McMillan, B., *Two inequalities implied by unique decipherability*, IEEE Trans. Inf. Theory, IT-56, pp. 115-116, Dec. 1956.

[Mene93] Menezes, A.J., *Elliptic Curve Public Key Cryptosystems*, Kluwer Academic Publishers, Boston etc., MA, 1993.

[MeOkV93] Menezes, A.J., T. Okamoto, and S.A. Vanstone, *Reducing elliptic curve logarithms to logarithms in a finite filed*, IEEE Transactions on Information Theory, IT-39, 1639-1646, 1993.

[MeOoV97] Menezes, A.J., P.C. van Oorschot, and S.A. Vanstone, *Handbook of Applied Cryptography*, CRC Press, Boca Raton, etc. 1997.

[MerH78] Merkle, R.C. and M.E. Hellman, *Hiding information and signatures in trapdoor knapsacks*, IEEE Transactions on Information Theory, IT-24, pp. 525-530, Sept. 1978.

[MeyM82] Meyer, C.H. and S.M. Matyas, *Cryptography: a New Dimension in Computer Data Security*, John Wiley & Sons, New York, etc., 1982

[Mill76] Miller, G.L., *Riemann's hypothesis and tests for primality*, Journal of Computer and System Sciences, 13, pp. 300-317, 1976.

[Mill86] Miller, G.L., *Use of elliptic curves in cryptography*, Advances in Cryptology: Proc. Crypto'85, H.C. Williams, Ed., Lecture Notes in Computer Science 218, Springer, Berlin etc., pp. 417-426, 1986.

[Moni80] Monier, L., *Evaluation and comparison of two efficient probabilistic primality testing algorithms*, Theoretical Computer Science, 12, pp. 97-108, 1980

[MorB75] Morrison, M.A. and J. Brillhart, *A method of factoring and the factorization of F_7*, Math. Comp. 29, pp. 183-205, 1975.

[Nied86] Niederreiter, H., *Knapsack type cryptosystems and algebraic coding theory*, Problems of Control and Information Theory, 15, pp. 159-166, 1986.

[NybR93] Nyberg, K. and R.A. Rueppel, *A new signature scheme based on the DSA giving message recovery*, 1st ACM Conference on Computer and Communications Security, ACM Press, 1993, pp. 58-61.

[Odly85] Odlyzko, A.M., *Discrete logarithms in finite fields and their cryptographic significance*, Advances in Cryptology: Proc. Eurocrypt '84, T. Beth, N. Cot and I. Ingemarsson, Eds., Lecture Notes in Computer Science 209, Springer, Berlin etc., pp. 224-314, 1985.

[Patt75] Patterson N.J., *The algebraic decoding of Goppa codes*, IEEE Transactions on Information Theory, IT-21, pp. 203-207, Mar. 1975.

[Pera86] Peralta, R., *A simple and fast probablistic algorithm for computing square roots modulo a prime number*, presented at Eurocrypt'86, J.L. Massey, Ed., no proceedings published.

[PohH78] Pohlig, S.C. and M.E. Hellman, *An improved algorithm for computing logarithms over $GF(p)$ and its cryptographic significance*, IEEE Transactions on Information Theory, IT-24, pp. 106-110, Jan. 1978.

[Poll75] Pollard, J.M., *A Monte Carlo method for factoring*, BIT-15, pp. 331-334, 1975.

[Poll78] Pollard, J.M., *Monte Carlo methods for index computations (mod p)*, Mathematics of Computations 32, pp. 918-924, 1978.

[Rabi79] Rabin, M.O., *Digitalized signatures and public-key functions as intractable as factorization*, MIT/LCS/TR-212, MIT Lab. for Comp. Science, Cambridge, Mass., Jan. 1979.

[Rabi80a] Rabin, M.O., *Probabilistic algorithms for testing primality*, Journal of Number Theory, 12, pp. 128-138, 1980.

[Rabi80b] Rabin, M.O., *Probabilistic algorithms in finite fields*, SIAM J. Comput. 80, pp. 273-280, 1980.

[RisL79] Rissanen, J. and G. Langdon, *Arithmetic coding*, IBM Journal of Research and Development, 23, pp. 149-162, 1979.

[RivSA78] Rivest, R.L., A. Shamir and L. Adleman, *A method for obtaining digital signatures and public-key cryptosystems*, Comm. ACM, Vol. 21, pp. 120-126, Febr. 1978.

[Rose84] Rosen, K.H., *Elementary Number Theory*, Addison-Wesley Publ. Comp., Reading, Mass, 1984.

[Ruep86] Rueppel, R.A., *Analysis and Design of Streamciphers*, Springer-Verlag, Berlin etc., 1986.

[SatA98] T. Satoh and K. Araki, *Fermat quotients and the polynomial time discrete log algorithm for anomalous elliptic curves*, Commentarii Mathematici Universitatis Sancti Pauli 47, pp. 81-92, 1998.

[Schne96] Schneier, B., *Applied Cryptography, 2nd Edition*, John Wiley & Sons, New York, etc., 1996.

[Schno90] Schnorr, C.P., *Efficient identification and signatures for smart cards*, In: Advances in Cryptology-Crypto'89, Ed. G. Brassard, Lecture Notes in Computer Science 435, Springer Verlag, Berlin, etc., pp.239-252, 1990.

[Schno91] Schnorr, C.P., *Efficient signature generation by smart cards*, Journal of Cryptology 4, pp. 161-174, 1991.

[Scho95] Schoof, R., *Counting points on elliptic curves over finite fields*, Journal de Théorie des Nombres de Bordeaux, 7, pp. 219-254, 1995.

[Sham79] Shamir, A., *How to share a secret*, Communications of the A.C.M., Vol. 22, pp. 612-613, Nov. 1979.

[Sham82] Shamir, A., *A polynomial time algorithm for breaking the basic Merkle-Hellman cryptosystem*, in Proc. 23-rd IEEE Symp. Found. Computer Sci., pp. 145-152, 1982.

[Sham49] Shannon, C.E., Communication Theory and Secrecy Systems, B.S.T.J. 28, pp. 656-715, Oct. 1949.

[Shap83] Shapiro, H.N., *Introduction to the Theory of Numbers*, John Wiley & Sons, New York, etc., 1983.

[Silv86] Silverman, J.H., *The Arithmetic of Elliptic Curves*, Springer Verlag, Berlin, etc., 1986.

[Silv98] Silverman, J.H., *The XEDNI calculus and the elliptic curve discrete logarithm problem*, preprint.

[SilT92] Silverman J.H. and J. Tate, *Rational Points on Elliptic Curves*, Undergraduate Texts in Mathematics, Springer-Verlag New York Inc.,1992.

[Simm92] Simmons, G.J., *A survey of information authentication*, in Contemporary Cryptology: the Science of Information Integrity, G.J. Simmons, Ed., IEEE Press, New York, pp. 379-419, 1992.

[Smar98] N. Smart, *The discrete logarithm problem on elliptic curves of trace one*, Journal of Cryptology, to appear.

[SolS77] Solovay, R. and V. Strassen, *A fast Monte-Carlo test for primality*, SIAM J. Comput 6, pp. 84-85, March 1977.

[Stin95] Stinson, D.R., *Cryptography: Theory and Practice*, CRC Press, Inc., Boca Raton, 1995.

[SugK76] Sugiyama, Y., M. Kashara, S. Hirasawa and T. Namekawa, *An erasures-and-errors decoding algorithm for Goppa codes*, IEEE Transactions on Information Theory, IT-22, pp. 238-241, Mar. 1976.

[vTbu88] van Tilburg, H., *On the McEliece public-key cryptosystem*, Advances in Cryptography: Proc. of Crypto '88, S. Goldwasser, Ed., Lecture Notes in Computer Science 403, Springer Verlag, Berlin etc., pp. 119-131, 1989.

[Vaud98] Vaudenay, S., *Cryptanalysis of the Chor-Rivest cryptosystem*, Advances in Cryptography: Proc. of Crypto '98, H. Krawczyk, Ed., Lecture Notes in Computer Science 1462, Springer Verlag, Berlin etc., pp. 243-256, 1998.

[VerT97] Verheul, E.R. and H.C.A. van Tilborg, *Constructions and properties of k out of n visual secret sharing shemes*, Designs, Codes and Cryptography, Vol. 11, No. 2, pp.179-196, May 1997.

[Well99] Wells, R.B., *Applied Coding and Information Theory*, Prentice Hall, Upper Saddle River NJ, 1999.

[Wien90] Wiener, M.J., *Cryptanalysis of Short RSA Secret Exponents*, IEEE Transactions on Information Theory, IT-36, pp. 553-558, May 1990.

[ZivL77] Ziv, J. and A. Lempel, *A universal algorithm for sequential data compression*, IEEE Transactions on Information Theory, IT-23, pp. 337-343, 1977.

[ZivL78] Ziv, J. and A. Lempel, *Compression of individual sequences by variable rate coding*, IEEE Transactions on Information Theory, IT-24, pp. 530-536, 1978.

Symbols and Notations

(a, b) greatest common divisor, 344, 345

$[a, b]$ least common multiple, 345

$\left(\frac{u}{m}\right)$ Jacobi symbol, 364

R/S residue class ring, 388

$(s(x))$ ideal generated by $s(x)$, 398

$\equiv$ ·   congruent, 352

$\|v\|$ length of vector, 393

$U^{\perp}$ orthogonal complement, 394

$\Gamma(p_U(x), GF(2^m))$ Goppa code, 236

μ Möbius function, 378

$\pi(x)$ number of primes $\leq x$, 344

φ Euler totient function, 354

χ Legendre symbol, 364

$\Omega(f)$ output space of LFSR, 35

$AC(k)$ auto-correlation, 28

D_n redundancy, 79

$d(u)$ density of a knapsack, 269

$\mathcal{E}$ elliptic curve, 213

gcd   greatest common divisor, 344, 345

f^* minimal characteristic polynomial, 36

$f^{(k)}$ linear complexity, 52

$F[x]$ ring of polynomials over F, 395

$\mathbb{F}_q$ finite field of q elements, 387

GF   Galois field, 387

$h(p)$ entropy, 76

H(X)   entropy, 76

$H(X \mid Y)$ conditional entropy, 81

$I_q(n)$ number of irreducible polynomials of degree n over $\mathbb{F}_q$, 401

$I(n)$ number of binary, irreducible polynomials of degree n, 401

I(X,Y) mutual information, 82

lcm   least common multiple , 345, 344

L_k linear complexity, 52

$\mathcal{N}$ non-privileged set (of an access system), 322

NQR   quadratic non-residue , 364

P_D probability of a successful deception, 293

P_I probability of a successful impersonation attack, 293

P_S probability of a successful substitution attack, 293

$\mathcal{P}$ privileged set (of an access system), 322

$Q^{(d)}$ cyclotomic polynomial, 420

QR   quadratic residue , 364

Tr   trace function, 424

V(n,q) n-dimensional vectorspace over GF(q), 309

$w(x)$ weight of a vector, 242

$\mathbf{Z}_p$ integers modulo p, 395

Index

A

Abelian group, 385

access structure, 322

    complete, 322

    perfect , 322

A-code (for message authentication), 292

    Johansson's construction of A-code from EC-code, 309

    from orthogonal array, 305

active cryptanalist, 3

addition of points on an elliptic curve, 225

addition chain, 113

additive group, 385

address, 98

alphabet, 2

algorithm

    addition of points on an elliptic curve , 225

    Baby-step Giant-step (for taking discrete logarithms), 130

    Berlekamp-Massey, 56

    bit swapping, 255

    Cohen and Lenstra (deterministic primality test 1), 194

    continued fraction, 371

    conversion from integer to binary weight k vector, 283

    decryption of Chor-Rivest, 284

    Euclid (simple version), 348

        (extended version), 349

    factoring algorithms

        Pollard p-1, 159

        Pollard-ϱ, 161

        quadratic sieve, 167

        random squares method, 163

        Gauss (to find a primitive element), 423

    Gram-Schmidt (for orthogonalization process), 272

    Huffman (for data compression), 93

    index-calculus (for taking discrete logarithms), 135

    Floyd's cycle-finding algorithm, 133

    knapsack problem for superincreasing sequences, 264

C

D

E

I

N

O

Q

R

S

DISCLAIMER